The CRC
Materials Science and Engineering
Handbook

Editor

James F. Shackelford

Professor of Materials Science and Engineering

Division of Materials Science and Engineering

and

Associate Dean of the College of Engineering

University of California, Davis

Associate Editor

William Alexander

Research Engineer

Division of Materials Science and Engineering

University of California, Davis

CRC Press
Boca Raton Ann Arbor London

Library of Congress Cataloging-in-Publication Data

Catalog record is available from the Library of Congress

ISBN 0-8493-4276-7

Direct all inquiries to CRC Press, Inc., 2000 Corporate Blvd., N. W., Boca Raton, Florida, 33431.

© 1992 by CRC Press, Inc.

International Standard Book Number 0-8493-4276-7

Printed in the United States 0 1 2 3 4 5 6 7 8 9

TABLE OF CONTENTS

Table of Contents (Continued)

Table of Contents (Continued)

Table of Contents (Continued)

Table of Contents (Continued)

PREFACE

The CRC Materials Science and Engineering Handbook began as a project of relatively limited scope. Originally, we agreed to review a number of existing CRC Press handbooks in the areas of materials science and engineering and to produce an up-to-date single volume. However, the publication of this book now represents the first step in a planned and greatly expanded publishing program.

This handbook fulfills our initial goals of providing; *i)* a comprehensive source for data on engineering materials, *ii)* an easy to follow organization based on materials properties, *iii)* a significant number of graphical representations to supplement and enhance the tabular data, and *iv)* a solid foundation from which to build an electronic CRC Press materials data base.

In the early stages of the preparation of this handbook, it became apparent that the core data base would require revision and expansion on a regular basis. The primary sources for new data and for verification of existing data are the major professional societies in the materials field, such as ASM International and the American Ceramic Society. From the outset, data storage on magnetic media has been utilized. Relevant tables from existing CRC Press handbooks have been converted to this format and all new data from other sources have been similarly stored. This electronic data base allows for regular updating and has led to the initiation of an ongoing project to produce a series of related CRC Press handbooks. This book will be followed by *The CRC Practical Handbook of Materials Science and Engineering.*

The *Practical Handbook* will have a more condensed treatment of fundamental tabular material and a greater emphasis on property data that is relevant to the selection of engineering materials. The *Practical Handbook* will also contain up-to-date information such as annual materials cost surveys, annual materials engineering hiring surveys and annual materials engineers' salary surveys. Regular up-dates and expansion of the materials data base will be enhanced by the development of numerous more specialized CRC Press handbooks dealing with engineering materials, e.g. advanced ceramics and composites.

The development of an extended series of CRC Press handbooks dealing with increasingly specialized components of the already specialized field of materials science parallels an exciting new project, *The Engineering Handbook.* We are serving as Editor-in-Chief and Managing Editor, respectively, of this new handbook. It will provide a useful one-volume reference for engineers. In addition, 1992 will see the publication of *The Electrical Engineering Handbook,* edited by Professor Richard Dorf. Also, major single source references are being developed for Mechanical Engineering, Civil Engineering, Chemical Engineering and Industrial Engineering. These books will be the core from which *The Engineering Handbook* will be drawn.

Every effort has been made to provide a presentation of the data in *The CRC Materials Science and Engineering Handbook* in an easy-to-follow organization based on materials properties. Data storage technology has permitted a visual presentation of the data in a consistent and readable format. We have attempted to present data as accurately as possible. This is, however, an ongoing effort and we would greatly appreciate your input. Suggestions and corrections can be sent to:

Dr. James F. Shackelford
College of Engineering
University of California
Davis, CA 95616-5294.

If you are interested in contributing to the upcoming series of CRC materials handbooks related to this volume, you may contact the above address or:

Mr. Joel Claypool
Publisher
CRC Press, Inc.
2000 Corporate Blvd., N.W.
Boca Raton, FL 33431.

Finally, we would like to emphasize that, although this book represents the first of a series of new reference volumes, it is our intent that it also serve by itself as a useful source of scientific and engineering information for many years to come. "Materials science and engineering" has come to be recognized as one of the most exciting and dynamic fields in the practice of engineering. We sincerely hope that this volume will be a useful reference for those fortunate enough to be involved in that field.

James F. Shackelford

William Alexander

ACKNOWLEDGMENTS

We wish to acknowledge those who have contributed to the production of this Handbook, especially Dr. Jun S. Park, who has served as a research assistant in the production of this first volume of this new series of CRC Handbooks on materials in particular and engineering in general. We have also had an effective working relationship with CRC Editors Russ Hall and Joel Claypool. We greatly appreciate their enthusiasm and creativity in this effort. Finally, we especially thank our families whose love and support ultimately made this project a reality.

DEDICATION

To Penelope and Scott

and Li-Li, Cassie, and Jie-Ying

The Elements

ELEMENTS FOR ENGINEERING MATERIALS

Although 107 elements are known to exist, only a limited number are used in engineering materials. Of the elements occurring in nature, hydrogen accounts for more than 90% of the atoms or about 75% of the mass of the universe. The remaining atoms are mainly Helium, with all the other elements contributing less than 1% to the total mass of the universe.

Distribution of Atomic Mass in the Universe

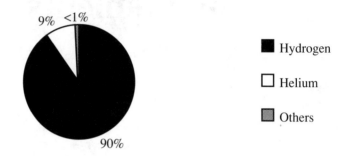

Within the earth's crust, the more common elements are those found in common engineering materials and applications. Oxygen accounts for about 47% of the crust's weight, while silicon contributes about 28% and aluminum about 8%. These elements, in addition to iron, calcium, sodium, potassium, and magnesium, account for about 99% of the weight composition of the crust. Other elements such as tin, copper, zinc, lead, mercury, silver, platinum, antimony, arsenic, and gold, which are vital for our current technology and social systems, are among some of the rarest elements in the earth's crust.

All atomic mass numbers from 1 to 238 are found naturally on earth, except for mass numbers 5 and 8. About 347 isotopes occur on earth, 280 are stable and 67 are naturally radioactive. Additionally, the transuranic elements up to atomic number 107 have been produced artificially. It is still believed that the production of elements beyond Element 109 is possible. Calculations indicate that Element 110, a homolog of platinum, may have a half-life of as long as 100 million years. The artificially produced neutron, technetium, and promethium may also be considered elements as is the electron. Several compounds known as "electrides" have recently been made of alkaline metal elements and electrons.

Each element from atomic number 1 to 107 is known to have at least one radioactive isotope. About 1700 different nuclides (the name given to different kinds of nuclei,whether they are of the same or different elements) are now recognized. Many stable and radioactive isotopes are now produced and distributed by the Oak Ridge National Laboratory, Oak Ridge. Tenn., U.S.A., to customers licensed by the U.S. Department of Energy.

One of the foundations of materials science and engineering is that the properties of an elemental material may change drastically by the presence of small amounts of impurities. New methods of purification can produce elements with 99.9999% purity, and consequently the values of elemental properties may change as purification methods improve. In general, the values of physical properties reported here are for the currently available level of elemental purity.

In the human body, the four most abundant elements are hydrogen, oxygen, carbon, and nitrogen. The seven next most common, in order of abundance, are calcium, phosphorus, chlorine, potassium, sulfur, sodium, and magnesium. Iron, copper, zinc, silicon, iodine, cobalt, manganese, molybdenum, fluorine, tin, chromium, selenium, and vanadium are needed and play a role in living matter. Boron is also thought essential for some plants, and it is possible that aluminum, nickel, and germanium may turn out to be necessary.

Many of the chemical elements and their compounds are either toxic or are not biologically compatible. These materials should be handled with due respect and care as there has been a greatly increased knowledge and awareness of the health hazards associated with chemicals, radioactive materials, and other agents. Anyone working with the elements and certain engineering materials should become thoroughly familiar with the proper safeguards to be taken.

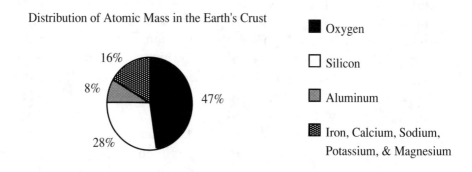

Distribution of Atomic Mass in the Earth's Crust

16%
8%
47%
28%

- Oxygen
- Silicon
- Aluminum
- Iron, Calcium, Sodium, Potassium, & Magnesium

Condensed from: C. R. Hammond, "The Elements", in Lide, David R., *CRC Handbook of Chemistry and Physics,* CRC Press, Boca Raton, (1990).

ELEMENTS IN THE EARTH'S CRUST
A. Demay

Element	No.	Concentration (mg/kg)	Element	No.	Concentration (mg/kg)
Oxygen	8	4.64×10^5	Lead	82	1.25×10^1
Silicon	14	2.82×10^5	Boron	5	1.00×10^1
Aluminum	13	8.32×10^4	Thorium	90	9.60×10^0
Iron	26	5.63×10^4	Praseodymium	59	8.20×10^0
Calcium	20	4.15×10^4	Samarium	62	6.00×10^0
Sodium	11	2.36×10^4	Gadolinium	64	5.40×10^0
Magnesium	12	2.33×10^4	Germanium	32	5.40×10^0
Potassium	19	2.09×10^4	Argon	18	3.50×10^0
Titanium	22	5.70×10^3	Dysprosium	66	3.00×10^0
Hydrogen	1	1.40×10^3	Ytterbium	70	3.00×10^0
Phosphorus	15	1.05×10^3	Beryllium	4	2.80×10^0
Manganese	25	9.50×10^2	Erbium	68	2.80×10^0
Fluorine	9	6.25×10^2	Uranium	92	2.70×10^0
Barium	56	4.25×10^2	Bromine	35	2.50×10^0
Strontium	38	3.75×10^2	Tantalum	73	2.00×10^0
Sulfur	16	2.60×10^2	Tin	50	2.00×10^0
Carbon	6	2.00×10^2	Arsenic	33	1.80×10^0
Zirconium	40	1.65×10^2	Molybdenum	42	1.50×10^0
Vanadium	23	1.35×10^2	Tungsten	74	1.50×10^0
Chlorine	17	1.30×10^2	(Wolfram)		
Chromium	24	1.00×10^2	Europium	63	1.20×10^0
Rubidium	37	9.00×10^1	Holmium	67	1.20×10^0
Nickel	28	7.50×10^1	Cesium	55	1.00×10^0
Zinc	30	7.00×10^1	Terbium	65	9.00×10^{-1}
Cerium	58	6.00×10^1	Iodine	53	5.00×10^{-1}
Copper	29	5.50×10^1	Lutetium	71	5.00×10^{-1}
Yttrium	39	3.30×10^1	Thulium	69	4.80×10^{-1}
Lanthanum	57	3.00×10^1	Thallium	81	4.50×10^{-1}
Neodymium	60	2.80×10^1	Antimony	51	2.00×10^{-1}
Cobalt	27	2.50×10^1	Cadmium	48	2.00×10^{-1}
Scandium	21	2.20×10^1	Bismuth	83	1.70×10^{-1}
Lithium	3	2.00×10^1	Indium	49	1.00×10^{-1}
Niobium	41	2.00×10^1	Mercury	80	8.00×10^{-2}
(Columbium)			Silver	47	7.00×10^{-2}
Nitrogen	7	2.00×10^1	Selenium	34	5.00×10^{-2}
Gallium	31	1.50×10^1	Palladium	46	1.00×10^{-2}

Elements in the Earth's Crust (Continued)

Element	No.	Concentration (mg/kg)	Element	No.	Concentration (mg/kg)
Helium	2	8.00×10^{-3}			
Neon	10	5.00×10^{-3}	Tellurium	52	1.00×10^{-3}
Platinum	78	5.00×10^{-3}	Krypton	36	1.00×10^{-4}
Rhenium	75	5.00×10^{-3}	Xenon	54	3.00×10^{-5}
Gold	79	4.00×10^{-3}	Protactinium	91	1.40×10^{-6}
Osmium	76	1.50×10^{-3}	Radium	88	9.00×10^{-7}
Iridium	77	1.00×10^{-3}	Actinium	89	5.50×10^{-10}
Rhodium	45	1.00×10^{-3}	Polonium	84	2.00×10^{-10}
Ruthenium	44	1.00×10^{-3}	Radon	86	4.00×10^{-13}

Weast, *Handbook of Chemistry and Physics.*

Concentration of Elements in the Earth's Crust

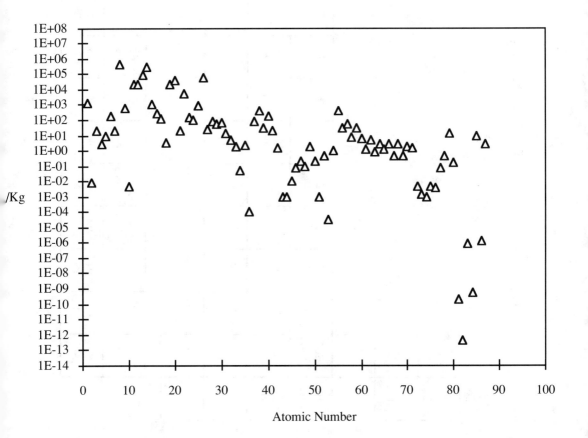

THE PERIODIC TABLE OF THE ELEMENTS

The Periodic Table of The Elements

1 IA	2 IIA	3 IIIB	4 IVB	5 VB	6 VIB	7 VIIB	8 ------	9 VIII	10 ------	11 IB	12 IIB	13 IIIA	14 IVA	15 VA	16 VIA	17 VIIA	18 VIIA
1 H																	2 He
3 Li	4 Be											5 B	6 C	7 N	8 O	9 F	10 Ne
11 Na	12 Mg											13 Al	14 Si	15 P	16 S	17 Cl	18 Ar
19 K	20 Ca	21 Sc	22 Ti	23 V	24 Cr	25 Mn	26 Fe	27 Co	28 Ni	29 Cu	30 Zn	31 Ga	32 Ge	33 As	34 Se	35 Br	36 Kr
37 Rb	38 Sr	39 Y	40 Zr	41 Nb	42 Mo	43 Tc	44 Ru	45 Rh	46 Pd	47 Ag	48 Cd	49 In	50 Sn	51 Sb	52 Te	53 I	54 Xe
55 Cs	56 Ba		72 Hf	73 Ta	74 W	75 Re	76 Os	77 Ir	78 Pt	79 Au	80 Hg	81 Tl	82 Pb	83 Bi	84 Po	85 At	86 Rn
87 Fr	88 Ra																

57 La	58 Ce	59 Pr	60 Nd	61 Pm	62 Sm	63 Eu	64 Gd	65 Tb	66 Dy	67 Ho	68 Er	69 Tm	70 Yb	71 Lu
89 Ac	90 Th	91 Pa	92 U	93 Np	94 Pu	95 Am	96 Cm	97 Bk	98 Cf	99 Es	100 Fm	101 Md	102 No	103 Lw

The Metallic Elements

1 IA	2 IIA	3 IIIB	4 IVB	5 VB	6 VIB	7 VIIB	8 ------	9 VIII	10 ------	11 IB	12 IIB	13 IIIA	14 IVA	15 VA	16 VIA	17 VIIA	18 VIIA
3 Li	4 Be											5 B					
11 Na	12 Mg											13 Al					
19 K	20 Ca	21 Sc	22 Ti	23 V	24 Cr	25 Mn	26 Fe	27 Co	28 Ni	29 Cu	30 Zn	31 Ga					
37 Rb	38 Sr	39 Y	40 Zr	41 Nb	42 Mo	43 Tc	44 Ru	45 Rh	46 Pd	47 Ag	48 Cd	49 In	50 Sn	51 Sb			
55 Cs	56 Ba		72 Hf	73 Ta	74 W	75 Re	76 Os	77 Ir	78 Pt	79 Au	80 Hg	81 Tl	82 Pb	83 Bi			
87 Fr	88 Ra																

57 La	58 Ce	59 Pr	60 Nd	61 Pm	62 Sm	63 Eu	64 Gd	65 Tb	66 Dy	67 Ho	68 Er	69 Tm	70 Yb	71 Lu
89 Ac	90 Th	91 Pa	92 U	93 Np	94 Pu	95 Am	96 Cm	97 Bk	98 Cf	99 Es	100 Fm	101 Md	102 No	103 Lw

The Metallic Elements

The Elements in Ceramic Materials

1 IA	2 IIA	3 IIIB	4 IVB	5 VB	6 VIB	7 VIIB	8 VIII	9 VIII	10 VIII	11 IB	12 IIB	13 IIIA	14 IVA	15 VA	16 VIA	17 VIIA	18 VIIA
	4 Be	12 Mg										5 B	6 C	7 N	8 O		
3 Li												13 Al	14 Si	14 Si	15 P		
11 Na		21 Sc	22 Ti	23 V	24 Cr	25 Mn	26 Fe	27 Co	28 Ni	29 Cu	30 Zn	31 Ga	32 Ge				
19 K	20 Ca	39 Y	40 Zr	41 Nb	42 Mo	43 Tc	44 Ru	45 Rh	46 Pd	47 Ag	48 Cd	49 In	50 Sn	51 Sb			
37 Rb	38 Sr		72 Hf	73 Ta	74 W	75 Re	76 Os	77 Ir	78 Pt	79 Au	80 Hg	81 Tl	82 Pb	83 Bi			
55 Cs	56 Ba																
87 Fr	88 Ra																

57 La	58 Ce	59 Pr	60 Nd	61 Pm	62 Sm	63 Eu	64 Gd	65 Tb	66 Dy	67 Ho	68 Er	69 Tm	70 Yb	71 Lu
89 Ac	90 Th	91 Pa	92 U	93 Np	94 Pu	95 Am	96 Cm	97 Bk	98 Cf	99 Es	100 Fm	101 Md	102 No	103 Lw

The Elements in Ceramic Materials

The Elements in Polymeric Materials

The Elements in Polymeric Materials

The Elements in Semiconducting Materials

The Elements in Semiconducting Materials

	VIA	VA	IVA	IIIA
	8 O			
	16 S	15 P	14 Si	13 Al
	34 Se	33 As	32 Ge	31 Ga
	52 Te	51 Sb	50 Sn	49 In

IIB
30 Zn
48 Cd
80 Hg

The Superconducting Elements

1 IA	2 IIA	3 IIIB	4 IVB	5 VB	6 VIB	7 VIIB	8 ----	9 VIII	10 ----	11 IB	12 IIB	13 IIIA	14 IVA	15 VA	16 VIA	17 VIIA	18 VIIIA
	4 Be																
												13 Al					
			22 Ti	23 V							30 Zn	31 Ga					
			40 Zr	41 Nb	42 Mo	43 Tc	44 Ru				48 Cd	49 In	50 Sn	51 Sb			
				73 Ta	74 W	75 Re	76 Os	77 Ir			80 Hg		82 Pb				

57 La	90 Th	91 Pa

AVAILABLE STABLE ISOTOPES OF THE ELEMENTS

Element	Mass No.	Natural Abundance (%)	Element	Mass No.	Natural Abundance (%)
Hydrogen	1	99.985	Aluminum	27	100.0
	2	0.015			
			Silicon	28	92.21
Helium	3	0.00013		29	4.70
	4	≈100.0		30	3.09
Lithium	6	7.42	Phosphorus	31	100.0
	7	92.58			
			Sulfur	32	95.0
Beryllium	9	100.0		33	0.76
				34	4.22
Boron	10	19.78		36	0.014
	11	80.22			
			Chlorine	35	75.53
Carbon	12	98.89		37	24.47
	13	1.11			
			Argon	36	0.34
Nitrogen	14	99.63		38	0.06
	15	0.37		40	99.60
Oxygen	16	99.76	Potassium	39	93.1
	17	0.04		40[a]	0.01
	18	0.20		41	6.9
Fluorine	19	100.0	Calcium	40	96.97
				42	0.64
Neon	20	90.92		43	0.14
	21	0.26		44	2.06
	22	8.82		46	0.003
				48	0.18
Sodium	23	100.0			
			Scandium	45	100.0
Magnesium	24	78.70			
	25	10.13			
	26	11.17			

Available Stable Isotopes of the Elements (Continued)

Element	Mass No.	Natural Abundance (%)	Element	Mass No.	Natural Abundance (%)
			Zinc	64	48.89
Titanium	46	7.93		66	27.81
	47	7.28		67	4.11
	48	73.94		68	18.57
	49	5.51		70	0.62
	50	5.34			
			Gallium	69	60.4
Vanadium	50	0.24		71	39.6
	51	99.76			
			Germanium	70	20.52
Chromium	50	4.31		72	27.43
	52	83.76		73	7.76
	53	9.55		74	36.54
	54	2.38		76	7.76
Manganese	55	100.0	Arsenic	75	100.0
Iron	54	5.82	Selenium	74	0.87
	56	91.66		76	9.02
	57	2.19		77	7.58
	58	0.33		78	23.52
				80	49.82
Cobalt	59	100.0		82	9.19
Nickel	58	67.84	Bromine	79	50.54
	60	26.23		81	49.46
	61	1.19			
	62	3.66	Krypton	78	0.35
	64	1.08		80	2.27
				82	11.56
Copper	63	69.09		83	11.55
	65	30.91		84	56.90
				86	17.37

Available Stable Isotopes of the Elements (Continued)

Element	Mass No.	Natural Abundance (%)	Element	Mass No.	Natural Abundance (%)
Rubidium	85	72.15	Palladium	102	0.96
	87	27.85		104	10.97
				105	22.23
Strontium	84	0.56		106	27.33
	86	9.86		108	26.71
	87	7.02		110	11.81
	88	82.56			
			Silver	107	51.82
Yttrium	89	100.0		109	48.18
Zirconium	90	51.46	Cadmium	106	1.22
	91	11.23		108	0.88
	92	17.11		110	12.39
	94	17.40		111	12.75
	96	2.80		112	24.07
				113	12.26
Niobium	93	100.0		114	28.86
				116	7.58
Molybdenum	92	15.84			
	94	9.04	Indium	113	4.28
	95	15.72		115	95.72
	96	16.53			
	97	9.46	Tin	112	0.96
	98	23.78		114	0.66
	100	9.63		115	0.35
				116	14.30
Ruthenium	96	5.51		117	7.61
	98	1.87		118	24.03
	99	12.72		119	8.58
	100	12.62		120	32.85
	101	17.07		122	4.72
	102	31.61		124	5.94
	104	18.60			
			Antimony	121	57.25
Rhodium	103	100.0		123	42.75
				145	8.30

Available Stable Isotopes of the Elements (Continued)

Element	Mass No.	Natural Abundance (%)	Element	Mass No.	Natural Abundance (%)
Tellurium	120	0.09	Cerium	136	0.193
	122	2.46		138	0.250
	123	0.87		140	88.48
	124	4.61		142[d]	11.07
	125	6.99			
	126	18.71	Praseodymium		
	128	31.79		141	100.0
	130	34.48			
			Neodymium	142	27.11
Iodine	127	100.0		143	12.17
				144	23.85
Xenon	124	0.096		146	17.22
	126	0.090		148	5.73
	128	1.92		150	5.62
	129	26.44			
	130	4.08	Samarium	144	3.09
	131	21.18		147[e]	14.97
	132	26.89		148[f]	11.24
	134	10.44		149[g]	13.83
	136	8.87		150	7.44
				152	26.72
Cesium	133	100.0		154	22.71
Barium	130	0.101	Europium	151	47.82
	132	0.097		153	52.18
	134	2.42			
	135	6.59	Gadolinium	152[h]	0.20
	136	7.81		154	2.15
	137	11.30		155	14.73
	138	71.66		156	20.47
				157	15.68
Lanthanum	138	0.09		158	24.87
	139	99.91		160	21.90
			Terbium	159	100.0

Available Stable Isotopes of the Elements (Continued)

Element	Mass No.	Natural Abundance (%)	Element	Mass No.	Natural Abundance (%)
Dysprosium	156[i]	0.052	Haffiium	174[k]	0.18
	158	0.090		176	5.20
	160	2.29		177	18.50
	161	18.88		178	27.14
	162	25.53		179	13.75
	163	24.97		180	35.24
	164	28.18			
			Tantalum	180	0.012
Holmium	165	100.0		181	99.988
	186	28.41			
			Tungsten	180	0.14
Erbium	162	0.136		182	26.41
	164	1.56		183	14.40
	166	33.41		184	30.64
	167	22.94			
	168	27.07	Rhenium	185	37.07
	170	14.88		187	62.93
	186	1.59			
			Osmium	184	0.018
Thulium	169	100.0		187	1.64
	189	16.1		188	13.3
				190	26.4
Ytterbium	168	0.135		192	41.0
	170	3.03			
	171	14.31	Iridium	191	37.3
	172	21.82		193	62.7
	173	16.13			
	174	31.84	Platinum	190[m]	0.013
	176	12.73		192	0.78
				194	32.9
Lutetium	175	97.40		195	33.8
	176[j]	2.60		196	25.3
				198	7.2

Available Stable Isotopes of the Elements (Continued)

Element	Mass No.	Natural Abundance (%)
Gold	197	100.0
Mercury	196	0.146
	198	10.02
	199	16.84
	200	23.13
	201	13.22
	202	29.80
	204	6.85
Thallium	203	29.50
	205	70.50
Lead	204	1.48
	206	23.6
	207	22.6
	208	52.3
Bismuth	209	100.0
Thorium	232[n][†]	100.0
Uranium	234[o][†]	0.0006
	235[p][†]	0.72
	238[q][†]	99.27

a   half-life = 1.3×10^9 y.

b   half-life $> 10^{15}$ y

c   half-life = 5×10^{14} y

d   half-life = 5×10^{14} y

e   half-life = 1.06×10^{11} y

f   half-life = 1.2×10^{13} y

g   half-life = 1.2×10^{14} y

h   half-life = 1.1×10^{14} y

i   half-life = 2×10^{14} y

j   half-life = 2.2×10^{10} y

k   half-life = 4.3×10^{15} y

l   half-life = 4×10^{10} y

m   half-life = 6×10^{11} y

n   half-life = 1.4×10^{10} y

o   half-life = 2.5×10^5 y

p   half-life = 7.1×10^8 y

q   half-life = 4.5×10^9 y

[†]naturally occurring.

From Wang, Y., Ed., *Handbook of Radioactive Nuclides,* The Chemical Rubber Co., Cleveland, 1969, 25.

ELECTRONIC STRUCTURE OF SELECTED ELEMENTS

At. No.	Element	Sym	1s	2s	2p	3s	3p	3d	4s	4p	4d	4f	5s	5p	5d	5f	6s	6p	6d	7s
1	Hydrogen	H	1																	
2	Helium	He	2																	
3	Lithium	Li	.	1																
4	Beryllium	Be	.	2																
5	Boron	B	.	2	1															
6	Carbon	C	.	2	2															
7	Nitrogen	N	.	2	3															
8	Oxygen	O	.	2	4															
9	Fluorine	F	.	2	5															
10	Neon	N	.	2	6															
11	Sodium	Na	.	.	.	1														
12	Magnesium	Mg	.	.	.	2														
13	Aluminum	Al	.	.	.	2	1													
14	Silicon	Si	.	.	.	2	2													
15	Phosphorus	P	.	.	.	2	3													
16	Sulfur	S	.	.	.	2	4													
17	Chlorine	Cl	.	.	.	2	5													
18	Argon	Ar	.	.	.	2	6													
19	Potassium	K	.	.	.	.	.		1											
20	Calcium	Ca	.	.	.	.	.		2											
21	Scandium	Sc	.	.	.	.	.	1	2											
22	Titanium	Ti	.	.	.	.	.	2	2											
23	Vanadium	V	.	.	.	.	.	3	2											
24	Chromium	Cr	.	.	.	.	.	5	1											
25	Manganese	Mn	.	.	.	.	.	5	2											
26	Iron	Fe	.	.	.	.	.	6	2											
27	Cobalt	Co	.	.	.	.	.	7	2											
28	Nickel	Ni	.	.	.	.	.	8	2											
29	Copper	Cu	.	.	.	.	.	10	1											
30	Zinc	Zn	.	.	.	.	.	10	2											
31	Gallium	Ga	.	.	.	.	.	10	2	1										
32	Germanium	Ge	.	.	.	.	.	10	2	2										
33	Arsenic	As	.	.	.	.	.	10	2	3										
34	Selenium	Se	.	.	.	.	.	10	2	4										
35	Bromine	Br	.	.	.	.	.	10	2	5										
36	Krypton	Kr	.	.	.	.	.	10	2	6										
37	Rubidium	Rb	.	.	.	.	.	.	.	.			1							
38	Strontium	Sr	.	.	.	.	.	.	.	.			2							
39	Yttrium	Y	.	.	.	.	.	.	.	.	1		2							
40	Zirconium	Zr	.	.	.	.	.	.	.	.	2		2							
41	Niobium	Nb	.	.	.	.	.	.	.	.	4		1							
42	Molybdenum	Mo	.	.	.	.	.	.	.	.	5		1							
43	Technetium	Tc	.	.	.	.	.	.	.	.	6		1							
44	Ruthenium	Ru	.	.	.	.	.	.	.	.	7		1							
45	Rhodium	Rh	.	.	.	.	.	.	.	.	8		1							
46	Palladium	Pd	.	.	.	.	.	.	.	.	10									
47	Silver	Ag	.	.	.	.	.	.	.	.	10		1							
48	Cadmium	Cd	.	.	.	.	.	.	.	.	10		2							
49	Indium	In	.	.	.	.	.	.	.	.	10		2	1						
50	Tin	Sn	.	.	.	.	.	.	.	.	10		2	2						
51	Antimony	Sb	.	.	.	.	.	.	.	.	10		2	3						
52	Tellurium	Te	.	.	.	.	.	.	.	.	10		2	5						
53	Iodine	I	.	.	.	.	.	.	.	.	10		2	5						
54	Xenon	Xe	.	.	.	.	.	.	.	.	10		2	6						

Electronic Structure of Selected Elements (Continued)

At. No.	Element	Sym	1s	2s	2p	3s	3p	3d	4s	4p	4d	4f	5s	5p	5d	5f	6s	6p	6d	7s
55	Cesium	Ce	.	.	.	.	.	.	.	.	.		.	.	.	.	1			
56	Barium	Ba	.	.	.	.	.	.	.	.	.		.	.	.	.	2			
57	Lantium	La	.	.	.	.	.	.	.	.	.		.	.	1	.	2			
58	Cerium	Ce	.	.	.	.	.	.	.	.	.	2	.	.	.	.	2			
59	Praseodymium	Pr	.	.	.	.	.	.	.	.	.	3	.	.	.	.	2			
60	Neodymium	Nd	.	.	.	.	.	.	.	.	.	4	.	.	.	.	2			
61	Promethium	Pm	.	.	.	.	.	.	.	.	.	5	.	.	.	.	2			
62	Samarium	Sm	.	.	.	.	.	.	.	.	.	6	.	.	.	.	2			
63	Europium	Eu	.	.	.	.	.	.	.	.	.	7	.	.	.	.	2			
64	Gadolinium	Gd	.	.	.	.	.	.	.	.	.	7	.	.	1	.	2			
65	Terbium	Tb	.	.	.	.	.	.	.	.	.	9	.	.	.	.	2			
66	Dysprosium	Dy	.	.	.	.	.	.	.	.	.	10	.	.	.	.	2			
67	Holmium	Ho	.	.	.	.	.	.	.	.	.	11	.	.	.	.	2			
68	Erbium	Er	.	.	.	.	.	.	.	.	.	12	.	.	.	.	2			
69	Thulium	Tm	.	.	.	.	.	.	.	.	.	13	.	.	.	.	2			
70	Ytterbium	Yb	.	.	.	.	.	.	.	.	.	14	.	.	.	.	2			
71	Lutetium	Lu	.	.	.	.	.	.	.	.	.	14	.	.	1	.	2			
72	Hafnium	Hf	.	.	.	.	.	.	.	.	.	14	.	.	2	.	2			
73	Tantalum	Ta	.	.	.	.	.	.	.	.	.	14	.	.	3	.	2			
74	Tungsten	W	.	.	.	.	.	.	.	.	.	14	.	.	4	.	2			
75	Rhenium	Re	.	.	.	.	.	.	.	.	.	14	.	.	5	.	2			
76	Osmium	Os	.	.	.	.	.	.	.	.	.	14	.	.	6	.	2			
77	Iridium	Ir	.	.	.	.	.	.	.	.	.	14	.	.	9	.				
78	Platinum	Pt	.	.	.	.	.	.	.	.	.	14	.	.	9	.	1			
79	Gold	Au	.	.	.	.	.	.	.	.	.	14	.	.	10	.	1			
80	Mercury	Hg	.	.	.	.	.	.	.	.	.	14	.	.	10	.	2			
81	Thallium	Tl	.	.	.	.	.	.	.	.	.	14	.	.	10	.	2	1		
82	Lead	Pb	.	.	.	.	.	.	.	.	.	14	.	.	10	.	2	2		
83	Bismuth	Bi	.	.	.	.	.	.	.	.	.	14	.	.	10	.	2	3		
84	Polonium	Po	.	.	.	.	.	.	.	.	.	14	.	.	10	.	2	4		
85	Asatine	At	.	.	.	.	.	.	.	.	.	14	.	.	10	.	2	5		
86	Radon	Rn	.	.	.	.	.	.	.	.	.	14	.	.	10	.	2	6		
87	Francium	Fr	.	.	.	.	.	.	.	.	.	.	.	.	.	.	.	.		1
88	Radium	Ra	.	.	.	.	.	.	.	.	.	.	.	.	.	.	.	.		2
89	Actinium	Ac	.	.	.	.	.	.	.	.	.	.	.	.	.	.	.	.	1	2
90	Thorium	Th	.	.	.	.	.	.	.	.	.	.	.	.	.	.	.	.	2	2
91	Protoactinium	Pa	.	.	.	.	.	.	.	.	.	.	.	.	.	2	.	.	1	2
92	Uranium	U	.	.	.	.	.	.	.	.	.	.	.	.	.	3	.	.	1	2
93	Neptunium	Np	.	.	.	.	.	.	.	.	.	.	.	.	.	4	.	.	1	2
94	Plutonium	Pu	.	.	.	.	.	.	.	.	.	.	.	.	.	6	.	.		2
95	Americium	Am	.	.	.	.	.	.	.	.	.	.	.	.	.	7	.	.		2
96	Curium	Cm	.	.	.	.	.	.	.	.	.	.	.	.	.	7	.	.	1	2
97	Berkelium	Bk	.	.	.	.	.	.	.	.	.	.	.	.	.	9	.	.		2
98	Californium	Cf	.	.	.	.	.	.	.	.	.	.	.	.	.	10	.	.		2
99	Einsteinium	Es	.	.	.	.	.	.	.	.	.	.	.	.	.	11	.	.		2
100	Fermium	Fm	.	.	.	.	.	.	.	.	.	.	.	.	.	12	.	.		2
101	Mendelevium	Md	.	.	.	.	.	.	.	.	.	.	.	.	.	13	.	.		2
102	Nobelium	No	.	.	.	.	.	.	.	.	.	.	.	.	.	14	.	.		2
103	Lawrencium	Lw	.	.	.	.	.	.	.	.	.	.	.	.	.	14	.	.	1	2

PROPERTIES OF SELECTED ELEMENTS

At. No.	Element	Sym.	Atomic Mass	Solid Density (Mg/m^3)	Crystal Structure	Melting Point (°C)
1	Hydrogen	H	1.008			-259.14
2	Helium	He	4.003			-272.2
3	Lithium	Li	6.941	0.533	bcc	180.54
4	Beryllium	Be	9.012	1.85	hcp	1278
5	Boron	B	10.81	2.47		2300
6	Carbon	C	12.01	2.27	hex.	~3550
7	Nitrogen	N	14.01			-209.86
8	Oxygen	O	16.00			-218.4
9	Fluorine	F	19.00			-219.62
10	Neon	N	20.18			-248.67
11	Sodium	Na	22.99	0.966	bcc	97.81
12	Magnesium	Mg	24.31	1.74	hcp	648.8
13	Aluminum	Al	26.98	2.7	fcc	660.37
14	Silicon	Si	28.09	2.33	dia. cub.	1410
15	Phosphorus (White)	P	30.97	1.82	ortho.	44.1
16	Sulfur	S	32.06	2.09	ortho.	112.8
17	Chlorine	Cl	35.45			-100.98
18	Argon	Ar	39.95			-189.2
19	Potassium	K	39.1	0.862	bcc	63.65
20	Calcium	Ca	40.08	1.53	fcc	839
21	Scandium	Sc	44.96	2.99	fcc	1539
22	Titanium	Ti	47.9	4.51	hcp	1660
23	Vanadium	V	50.94	6.09	bcc	1890
24	Chromium	Cr	52.00	7.19	bcc	1857

Properties of Selected Elements (Continued)

At. No.	Element	Sym.	Atomic Mass	Solid Density (Mg/m^3)	Crystal Structure	Melting Point (°C)
25	Manganese	Mn	54.94	7.47	cubic	1244
26	Iron	Fe	55.85	7.87	bcc	1535
27	Cobalt	Co	58.93	8.8	hcp	1495
28	Nickel	Ni	58.71	8.91	fcc	1453
29	Copper	Cu	63.55	8.93	fcc	1083.4
30	Zinc	Zn	65.38	7.13	hcp	419.58
31	Gallium	Ga	69.72	5.91	ortho.	29.78
32	Germanium	Ge	72.59	5.32	dia. Cub.	937.4
33	Arsenic	As	74.92	5.78	rhomb.	817
34	Selenium	Se	78.96	4.81	hex.	217
35	Bromine	Br	79.9			-7.2
36	Krypton	Kr	83.8			-156.6
37	Rubidium	Rb	85.47	1.53	bcc	38.89
38	Strontium	Sr	87.62	2.58	fcc	769
39	Yttrium	Y	88.91	4.48	hcp	1523
40	Zirconium	Zr	91.22	6.51	hcp	1852
41	Niobium	Nb	92.91	8.58	bcc	2408
42	Molybdenum	Mo	95.94	10.22	bcc	2617
43	Technetium	Tc	98.91	11.5	hcp	2172
44	Ruthenium	Ru	101.07	12.36	hcp	2310
45	Rhodium	Rh	102.91	12.42	fcc	1966
46	Palladium	Pd	106.4	12.00	fcc	1552
47	Silver	Ag	107.87	10.50	fcc	961.93
48	Cadmium	Cd	112.4	8.65	hcp	320.9
49	Indium	In	114.82	7.29	fct	156.61

Properties of Selected Elements (Continued)

At. No.	Element	Sym.	Atomic Mass	Solid Density (Mg/m³)	Crystal Structure	Melting Point (°C)
50	Tin	Sn	118.69	7.29	bct	231.9681
51	Antimony	Sb	121.75	6.69	rhomb.	630.74
52	Tellurium	Te	127.6	6.25	hex.	449.5
53	Iodine	I	126.9	4.95	ortho.	113.5
54	Xenon	Xe	131.3			-111.9
55	Cesium (-10°)	Ce	132.91	1.91	bcc	28.4
56	Barium	Ba	137.33	3.59	bcc	7.25
57	Lantium	La	138.91	6.17	hex.	920
58	Cerium	Ce	140.12	6.77	fcc	798
59	Praseodymium	Pr	140.91	6.78	hex.	931
60	Neodymium	Nd	144.24	7.00	hex.	1010
61	Promethium	Pm	(145)		hex.	~1080
62	Samarium	Sm	150.4	7.54	rhomb.	1072
63	Europium	Eu	151.96	5.25	bcc	822
64	Gadolinium	Gd	157.25	7.87	hcp	1311
65	Terbium	Tb	158.93	8.27	hcp	1360
66	Dysprosium	Dy	162.5	8.53	hcp	1409
67	Holmium	Ho	164.93	8.80	hcp	1470
68	Erbium	Er	167.26	9.04	hcp	1522
69	Thulium	Tm	168.93	9.33	hcp	1545
70	Ytterbium	Yb	173.04	6.97	fcc	824
71	Lutetium	Lu	174.97	9.84	hcp	1659
72	Hafnium	Hf	178.49	13.28	hcp	2227
73	Tantalum	Ta	180.95	16.67	bcc	2996

Properties of Selected Elements (Continued)

At. No.	Element	Sym.	Atomic Mass	Solid Density (Mg/m^3)	Crystal Structure	Melting Point (°C)
74	Tungsten	W	183.85	19.25	bcc	3410
75	Rhenium	Re	186.2	21.02	hcp	3180
76	Osmium	Os	190.2	22.58	hcp	3045
77	Iridium	Ir	192.22	22.55	fcc	2410
78	Platinum	Pt	195.09	21.44	fcc	1772
79	Gold	Au	196.97	19.28	fcc	1064.43
80	Mercury	Hg	200.59			-38.87
81	Thallium	Tl	204.37	11.87	hcp	303.5
82	Lead	Pb	207.2	11.34	fcc	327.502
83	Bismuth	Bi	208.98	9.80	rhomb.	271.3
84	Polonium	Po	(~210)	9.2	monoclinic	254
85	Asatine	At	(210)			302
86	Radon	Rn	(222)			-71
87	Francium	Fr	(223)		bcc	~27
88	Radium	Ra	226.03		bct	700
89	Actinium	Ac	(227)		fcc	1050
90	Thorium	Th	232.04	11.72	fcc	1750
91	Protoactinium	Pa	231.04		bct	<1600
92	Uranium	U	238.03	19.05	ortho.	1132
93	Neptunium	Np	237.05		ortho.	640
94	Plutonium	Pu	(244)	19.81	monoclinic	641
95	Americium	Am	(243)		hex.	994
96	Curium	Cm	(247)		hex.	1340
97	Berkelium	Bk	(247)		hex.	
98	Californium	Cf	(251)			

Properties of Selected Elements (Continued)

Density of the Elements

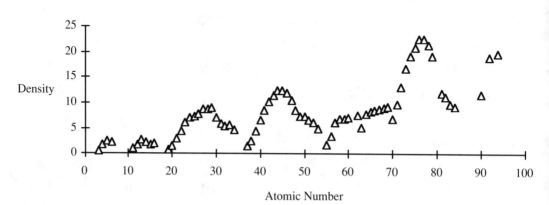

Melting Points of the Elements

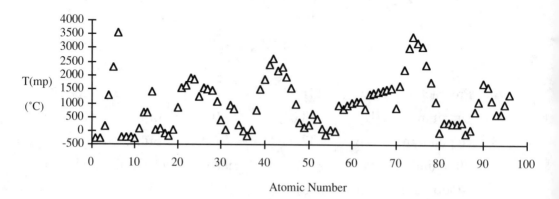

David R., *CRC Handbook of Chemistry and Physics*, CRC Press, Boca Raton, (1990).

Properties of Selected Elements (Continued)

Density of the Elements
(by group number)

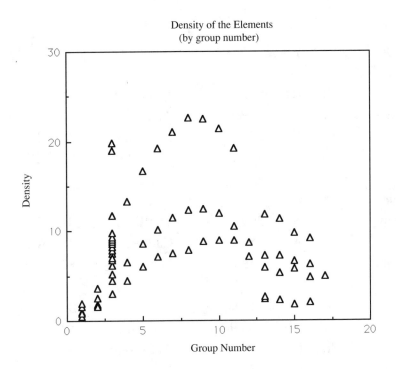

Melting Points of the Elements
(by Group Number)

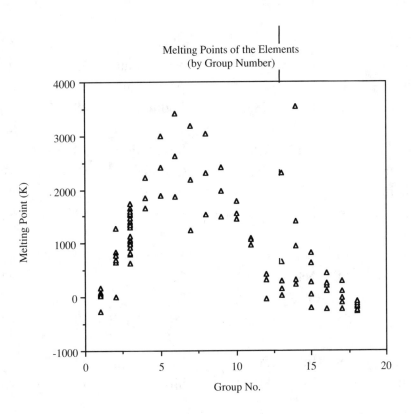

MELTING POINTS OF SELECTED ELEMENTS
Listed by Melting Point

Element	No.	At. Sym.	Melting Point (°C)	Element	No.	At. Sym.	Melting Point (°C)
Helium	2	He	-272.2	Selenium	34	Se	217
Hydrogen	1	H	-259.14	Tin	50	Sn	231.9681
Neon	10	N	-248.67	Polonium	84	Po	254
Fluorine	9	F	-219.62	Bismuth	83	Bi	271.3
Oxygen	8	O	-218.4	Asatine	85	At	302
Nitrogen	7	N	-209.86	Thallium	81	Tl	303.5
Argon	18	Ar	-189.2	Cadmium	48	Cd	320.9
Krypton	36	Kr	-156.6	Lead	82	Pb	327.502
Xenon	54	Xe	-111.9	Zinc	30	Zn	419.58
Chlorine	17	Cl	-100.98	Tellurium	52	Te	449.5
Radon	86	Rn	-71	Antimony	51	Sb	630.74
Mercury	80	Hg	-38.87	Neptunium	93	Np	640
Bromine	35	Br	-7.2	Plutonium	94	Pu	641
Barium	56	Ba	7.25	Magnesium	12	Mg	648.8
Francium	87	Fr	~27	Aluminum	13	Al	660.37
Cesium	55	Ce	28.4	Radium	88	Ra	700
Gallium	31	Ga	29.78	Strontium	38	Sr	769
Rubidium	37	Rb	38.89	Cerium	58	Ce	798
Phosphorus (White)	15	P	44.1	Arsenic	33	As	817
				Europium	63	Eu	822
Potassium	19	K	63.65	Ytterbium	70	Yb	824
Sodium	11	Na	97.81	Calcium	20	Ca	839
Sulfur	16	S	112.8	Lantium	57	La	920
Iodine	53	I	113.5	Praseodymium	59	Pr	931
Indium	49	In	156.61	Germanium	32	Ge	937.4
Lithium	3	Li	180.54	Silver	47	Ag	961.93

Melting Points of Selected Elements
Listed by Melting Point

Element	No.	At. Sym.	Melting Point (°C)	Element	No.	At. Sym.	Melting Point (°C)
Americium	95	Am	994	Titanium	22	Ti	1660
Neodymium	60	Nd	1010	Thorium	90	Th	1750
Actinium	89	Ac	1050	Platinum	78	Pt	1772
Gold	79	Au	1064.43	Zirconium	40	Zr	1852
Samarium	62	Sm	1072	Chromium	24	Cr	1857
Promethium	61	Pm	~1080	Vanadium	23	V	1890
Copper	29	Cu	1083.4	Rhodium	45	Rh	1966
Uranium	92	U	1132	Technetium	43	Tc	2172
Manganese	25	Mn	1244	Hafnium	72	Hf	2227
Beryllium	4	Be	1278	Boron	5	B	2300
Gadolinium	64	Gd	1311	Ruthenium	44	Ru	2310
Curium	96	Cm	1340	Niobium	41	Nb	2408
Terbium	65	Tb	1360	Iridium	77	Ir	2410
Dysprosium	66	Dy	1409	Molybdenum	42	Mo	2617
Silicon	14	Si	1410	Tantalum	73	Ta	2996
Nickel	28	Ni	1453	Osmium	76	Os	3045
Holmium	67	Ho	1470	Rhenium	75	Re	3180
Cobalt	27	Co	1495	Tungsten	74	W	3410
Erbium	68	Er	1522	Carbon	6	C	~3550
Yttrium	39	Y	1523				
Iron	26	Fe	1535				
Scandium	21	Sc	1539				
Thulium	69	Tm	1545				
Palladium	46	Pd	1552				
Protoactinium	91	Pa	<1600				
Lutetium	71	Lu	1659				

James F. Shackelford, *Introduction to Materials Science for Engineers, Second Edition*, Macmillian Publishing Company, New York, pp.686-688, (1985).

DENSITIES OF SELECTED ELEMENTS
Listed by Density

Element	At. No.	Sym.	Solid Density (Mg/m^3)	Element	At. No.	Sym.	Solid Density (Mg/m^3)
Lithium	3	Li	0.533	Tellurium	52	Te	6.25
Potassium	19	K	0.862	Zirconium	40	Zr	6.51
Sodium	11	Na	0.966	Antimony	51	Sb	6.69
Calcium	20	Ca	1.53	Cerium	58	Ce	6.77
Rubidium	37	Rb	1.53	Praseodymium	59	Pr	6.78
Magnesium	12	Mg	1.74	Ytterbium	70	Yb	6.97
Phosphorus (White)	15	P	1.82	Neodymium	60	Nd	7.00
				Zinc	30	Zn	7.13
Beryllium	4	Be	1.85	Chromium	24	Cr	7.19
Cesium	55	Ce	1.91	Indium	49	In	7.29
Sulfur	16	S	2.09	Tin	50	Sn	7.29
Carbon	6	C	2.27	Manganese	25	Mn	7.47
Silicon	14	Si	2.33	Samarium	62	Sm	7.54
Boron	5	B	2.47	Iron	26	Fe	7.87
Strontium	38	Sr	2.58	Gadolinium	64	Gd	7.87
Aluminum	13	Al	2.7	Terbium	65	Tb	8.27
Scandium	21	Sc	2.99	Dysprosium	66	Dy	8.53
Barium	56	Ba	3.59	Niobium	41	Nb	8.58
Yttrium	39	Y	4.48	Cadmium	48	Cd	8.65
Titanium	22	Ti	4.51	Cobalt	27	Co	8.8
Selenium	34	Se	4.81	Holmium	67	Ho	8.80
Iodine	53	I	4.95	Nickel	28	Ni	8.91
Europium	63	Eu	5.25	Copper	29	Cu	8.93
Germanium	32	Ge	5.32	Erbium	68	Er	9.04
Arsenic	33	As	5.78	Polonium	84	Po	9.2
Gallium	31	Ga	5.91	Thulium	69	Tm	9.33
Vanadium	23	V	6.09	Bismuth	83	Bi	9.80
Lantium	57	La	6.17	Lutetium	71	Lu	9.84

Densities of Selected Elements
Listed by Density

Element	At. No.	Sym.	Solid Density (Mg/m³)	Element	At. No.	Sym.	Solid Density (Mg/m³)
Molybdenum	42	Mo	10.22	Tungsten	74	W	19.25
Silver	47	Ag	10.50	Gold	79	Au	19.28
Lead	82	Pb	11.34	Plutonium	94	Pu	19.81
Technetium	43	Tc	11.5	Rhenium	75	Re	21.02
Thorium	90	Th	11.72	Platinum	78	Pt	21.44
Thallium	81	Tl	11.87	Iridium	77	Ir	22.55
Palladium	46	Pd	12.00	Osmium	76	Os	22.58
Ruthenium	44	Ru	12.36				
Rhodium	45	Rh	12.42				
Hafnium	72	Hf	13.28				
Tantalum	73	Ta	16.67				
Uranium	92	U	19.05				

James F. Shackelford, *Introduction to Materials Science for Engineers*, Second Edition, Macmillian Publishing Company, New York, (1988).

CRYSTAL STRUCTURE OF THE ELEMENTS

The Body Centered Cubic Elements

1	2	3	4	5	6	7	8	9	10	11	12	13	14	15	16	17	18
IA	IIA	IIIB	IVB	VB	VIB	VIIB	VIII	VIII	VIII	IB	IIB	IIIA	IVA	VA	VIA	VIIA	VIIIA
3 Li																	
11 Na																	
19 K				23 V	24 Cr		26 Fe										
37 Rb				41 Nb	42 Mo												
55 Cs	56 Ba			73 Ta	74 W												
87 Fr																	

Lanthanide: 63 Eu

Crystal Structure of the Elements (Continued)

The Face Centered Cubic Elements

	1 IA	2 IIA	3 IIIB	4 IVB	5 VB	6 VIB	7 VIIB	8 -----	9 VIII	10 -----	11 IB	12 IIB	13 IIIA	14 IVA	15 VA	16 VIA	17 VIIA	18 VIIA
			20 Ca	21 Sc							28 Ni	29 Cu		13 Al				70 Yb
			38 Sr			58 Ce			45 Rh		46 Pd	47 Ag						
			89 Ac	90 Th		77 Ir	78 Pt	79 Au			82 Pb							

Crystal Structure of the Elements (Continued)

The Hexagonal Close Packed Elements

1 IA	2 IIA	3 IIIB	4 IVB	5 VB	6 VIB	7 VIIB	8 VIIB	9 VIII	10 -------	11 IB	12 IIB	13 IIIA	14 IVA	15 VA	16 VIA	17 VIIA	18 VIIA
			4 Be														
			12 Mg														
		39 Y	22 Ti						27 Co			30 Zn					
			40 Zr				43 Tc	44 Ru				48 Cd					
71 Lu	72 Hf										64 Gd	65 Tb	66 Dy	67 Ho	68 Er	69 Tm	

	75 Re	76 Os			81 Tl		

Crystal Structure of the Elements (Continued)

The Hexagonal Elements

1 IA	2 IIA	3 IIIB	4 IVB	5 VB	6 VIB	7 VIIB	8 ----	9 VIII	10 ----	11 IB	12 IIB	13 IIIA	14 IVA	15 VA	16 VIA	17 VIIA	18 VIIA
														6 C			
																34 Se	
																52 Te	
				57 La		59 Pr	60 Nd	61 Pm									
								95 Am	96 Cm	97 Bk							

ATOMIC AND IONIC RADII OF THE ELEMENTS

Atomic Number	Symbol	Atomic Radius (nm)	Ion	Ionic Radius (nm)
1	H	0.046	H^-	0.154
2	He	–	–	–
3	Li	0.152	Li^+	0.078
4	Be	0.114	Be^{2+}	0.054
5	B	0.097	B^{3+}	0.02
6	C	0.077	C^{4+}	<0.02
7	N	0.071	N^{5+}	0.01– 0.2
8	O	0.060	O^{2-}	0.132
9	F	–	F^-	0.133
10	Ne	0.160	–	–
11	Na	0.186	Na^+	0.098
12	Mg	0.160	Mg^{2+}	0.078
13	Al	0.143	Al^{3+}	0.057
14	Si	0.117	Si^{4-}	0.198
			Si^{4+}	0.039
15	P	0.109	P^{5+}	0.03– 0.04
16	S	0.106	S^{2-}	0.174
			S^{6+}	0.034
17	Cl	0.107	Cl^-	0.181
18	Ar	0.192	–	–
19	K	0.231	K^+	0.133
20	Ca	0.197	Ca^{2+}	0.106
21	Sc	0.160	Sc^{2+}	0.083
22	Ti	0.147	Ti^{2+}	0.076
			Ti^{3+}	0.069
			Ti^{4+}	0.064

Atomic and Ionic Radii of the Elements (Continued)

Atomic Number	Symbol	Atomic Radius (nm)	Ion	Ionic Radius (nm)
23	V	0.132	V^{3+}	0.065
			V^{4+}	0.061
			V^{5+}	0.04
24	Cr	0.125	Cr^{3+}	0.064
			Cr^{6+}	0.03– 0.04
25	Mn	0.112	Mn^{2+}	0.091
			Mn^{3+}	0.070
			Mn^{4+}	0.052
26	Fe	0.124	Fe^{2+}	0.087
			Fe^{2+}	0.067
27	Co	0.125	Co^{2+}	0.082
			Co^{3+}	0.065
28	Ni	0.125	Ni^{2+}	0.078
29	Cu	0.128	Cu^{+}	0.096
30	Zn	0.133	Zn^{2+}	0.083
31	Ga	0.135	Ga^{3+}	0.062
32	Ge	0.122	Ge^{4+}	0.044
33	As	0.125	As^{3+}	0.069
			As^{5+}	~0.04
34	Se	0.116	Se^{2-}	0.191
			Se^{6+}	0.03–0.04
35	Br	0.119	Br^{-}	0.196
36	Kr	0.197	–	–
37	Rb	0.251	Rb^{+}	0.149
38	Sr	0.215	Sr^{2+}	0.127
39	Y	0.181	Y^{3+}	0.106

Atomic and Ionic Radii of the Elements (Continued)

Atomic Number	Symbol	Atomic Radius (nm)	Ion	Ionic Radius (nm)
40	Zr	0.158	Zr^{4+}	0.087
41	Nb	0.143	Nb^{4+}	0.074
			Nb^{5+}	0.069
42	Mo	0.136	Mo^{4+}	0.068
			Mo^{6+}	0.065
43	Tc	–	–	–
44	Ru	0.134	Ru^{4+}	0.065
45	Rh	0.134	Rh^{3+}	0.068
			Rh^{4+}	0.065
46	Pd	0.137	Pd^{2+}	0.050
47	Ag	0.144	Ag^{+}	0.113
48	Cd	0.150	Cd^{2+}	0.103
49	In	0.157	In^{3+}	0.091
50	Sn	0.158	Sn^{4-}	0.215
			Sn^{4+}	0.074
51	Sb	0.161	Sb^{3+}	0.090
52	Te	0.143	Te^{2-}	0.211
			Te^{4+}	0.089
53	I	0.136	I^{-}	0.220
			I^{5+}	0.094
54	Xe	0.218	–	–
55	Cs	0.265	Cs^{+}	0.165
56	Ba	0.217	Ba^{2+}	0.13
57	La	0.187	La^{3+}	0.122
58	Ce	0.182	Ce^{3+}	0.118
			Ce^{4+}	0.102

Atomic and Ionic Radii of the Elements (Continued)

Atomic Number	Symbol	Atomic Radius (nm)	Ion	Ionic Radius (nm)
59	Pr	0.183	Pr^{3+}	0.116
			Pr^{4+}	0.100
60	Nd	0.182	Nd^{3+}	0.115
61	Pm	–	Pm^{3+}	0.106
62	Sm	0.181	Sm^{3+}	0.113
63	Eu	0.204	Eu^{3+}	0.113
64	Gd	0.180	Gd^{3+}	0.111
65	Tb	0.177	Tb^{3+}	0.109
			Tb^{4+}	0.089
66	Dy	0.177	Dy^{3+}	0.107
67	Ho	0.176	Ho^{3+}	0.105
68	Er	0.175	Er^{3+}	0.104
69	Tm	0.174	Tm^{3+}	0.104
70	Yb	0.193	Yb^{3+}	0.100
71	Lu	0.173	Lu^{3+}	0.099
72	Hf	0.159	Hf^{4+}	0.084
73	Ta	0.147	Ta^{5+}	0.068
74	W	0.137	W^{4+}	0.068
			W^{6+}	0.065
75	Re	0.138	Re^{4+}	0.072
76	Os	0.135	Os^{4+}	0.067
77	Ir	0.135	Ir^{4+}	0.066
78	Pt	0.138	Pt^{2+}	0.052
			Pt^{4+}	0.055
79	Au	0.144	Au^{+}	0.137
80	Hg	0.150	Hg^{2+}	0.112
81	Tl	0.171	Tl^{+}	0.149
			Tl^{3+}	0.106

Atomic and Ionic Radii of the Elements (Continued)

Atomic Number	Symbol	Atomic Radius (nm)	Ion	Ionic Radius (nm)
82	Pb	0.175	Pb^{4-}	0.215
			Pb^{2+}	0.132
			Pb^{4+}	0.084
83	Bi	0.182	Bi^{3+}	0.120
84	Po	0.140	Po^{6+}	0.067
85	At	–	At^{7+}	0.062
86	Rn	–	–	–
87	Fr	–	Fr^{+}	0.180
88	Ra	–	Ra^{+}	0.152
89	Ac	–	Ac^{3+}	0.118
90	Th	0.180	Th^{4+}	0.110
91	Pa	–	–	–
92	U	0.138	U^{4+}	0.105

Source: After a tabulation by R. A. Flinn and P. K. Trojan, *Engineering Materials and Their Applications*, Houghton Mifflin Company, Boston, 1975. The ionic radii are based on the calculations of V. M. Goldschmidt, who assigned radii based on known interatomic distances in various ionic crystals.

ATOMIC RADII OF THE ELEMENTS
Listed by Value

Atomic Number	Symbol	Atomic Radius (nm)	Atomic Number	Symbol	Atomic Radius (nm)
1	H	0.046	74	W	0.137
8	O	0.060	75	Re	0.138
7	N	0.071	78	Pt	0.138
6	C	0.077	92	U	0.138
5	B	0.097	84	Po	0.140
16	S	0.106	13	Al	0.143
17	Cl	0.107	41	Nb	0.143
15	P	0.109	52	Te	0.143
25	Mn	0.112	47	Ag	0.144
4	Be	0.114	79	Au	0.144
34	Se	0.116	22	Ti	0.147
14	Si	0.117	73	Ta	0.147
35	Br	0.119	48	Cd	0.150
32	Ge	0.122	80	Hg	0.150
26	Fe	0.124	3	Li	0.152
24	Cr	0.125	49	In	0.157
27	Co	0.125	40	Zr	0.158
28	Ni	0.125	50	Sn	0.158
33	As	0.125	72	Hf	0.159
29	Cu	0.128	10	Ne	0.160
23	V	0.132	12	Mg	0.160
30	Zn	0.133	21	Sc	0.160
44	Ru	0.134	51	Sb	0.161
45	Rh	0.134	81	Tl	0.171
31	Ga	0.135	71	Lu	0.173
76	Os	0.135	69	Tm	0.174
77	Ir	0.135	68	Er	0.175
42	Mo	0.136	82	Pb	0.175
53	I	0.136	67	Ho	0.176
46	Pd	0.137	65	Tb	0.177

Atomic Radii of the Elements (Continued)
Listed by Value

Atomic Number	Symbol	Atomic Radius (nm)	Atomic Number	Symbol	Atomic Radius (nm)
66	Dy	0.177	18	Ar	0.192
64	Gd	0.180	70	Yb	0.193
90	Th	0.180	20	Ca	0.197
39	Y	0.181	36	Kr	0.197
62	Sm	0.181	63	Eu	0.204
58	Ce	0.182	38	Sr	0.215
60	Nd	0.182	56	Ba	0.217
83	Bi	0.182	54	Xe	0.218
59	Pr	0.183	19	K	0.231
11	Na	0.186	37	Rb	0.251
57	La	0.187	55	Cs	0.265

Source: After a tabulation by R. A. Flinn and P. K. Trojan, *Engineering Materials and Their Applications,* Houghton Mifflin Company, Boston, 1975. The ionic radii are based on the calculations of V. M. Goldschmidt, who assigned radii based on known interatomic distances in various ionic crystals.

IONIC RADII OF THE ELEMENTS
Listed by Value

Ion	Ionic Radius (nm)	Ion	Ionic Radius (nm)	Ion	Ionic Radius (nm)
N^{5+}	0.01 –0.2	Mo^{6+}	0.065	Sc^{2+}	0.083
C^{4+}	<0.02	Ru^{4+}	0.065	Zn^{2+}	0.083
B^{3+}	0.02	Rh^{4+}	0.065	Hf^{4+}	0.084
P^{5+}	0.03 –0.04	W^{6+}	0.065	Pb^{4+}	0.084
Cr^{6+}	0.03 –0.04	Ir^{4+}	0.066	Fe^{2+}	0.087
Se^{6+}	0.03 –0.04	Fe^{2+}	0.067	Zr^{4+}	0.087
S^{6+}	0.034	Os^{4+}	0.067	Te^{4+}	0.089
Si^{4+}	0.039	Po^{6+}	0.067	Tb^{4+}	0.089
V^{5+}	0.04	Mo^{4+}	0.068	Sb^{3+}	0.090
As^{5+}	~0.04	Rh^{3+}	0.068	Mn^{2+}	0.091
Ge^{4+}	0.044	Ta^{5+}	0.068	In^{3+}	0.091
Pd^{2+}	0.050	W^{4+}	0.068	I^{5+}	0.094
Mn^{4+}	0.052	Ti^{3+}	0.069	Cu^{+}	0.096
Pt^{2+}	0.052	As^{3+}	0.069	Na^{+}	0.098
Be^{2+}	0.054	Nb^{5+}	0.069	Lu^{3+}	0.099
Pt^{4+}	0.055	Mn^{3+}	0.070	Pr^{4+}	0.100
Al^{3+}	0.057	Re^{4+}	0.072	Yb^{3+}	0.100
V^{4+}	0.061	Nb^{4+}	0.074	Ce^{4+}	0.102
Ga^{3+}	0.062	Sn^{4+}	0.074	Cd^{2+}	0.103
At^{7+}	0.062	Ti^{2+}	0.076	Er^{3+}	0.104
Ti^{4+}	0.064	Li^{+}	0.078	Tm^{3+}	0.104
Cr^{3+}	0.064	Mg^{2+}	0.078	Ho^{3+}	0.105
V^{3+}	0.065	Ni^{2+}	0.078	U^{4+}	0.105
Co^{3+}	0.065	Co^{2+}	0.082	Ca^{2+}	0.106

Ionic Radii of the Elements
Listed by Value

Ion	Ionic Radius (nm)	Ion	Ionic Radius (nm)	Ion	Ionic Radius (nm)
Y^{3+}	0.106	Bi^{3+}	0.120	H^-	0.154
Pm^{3+}	0.106	La^{3+}	0.122	Cs^+	0.165
Tl^{3+}	0.106	Sr^{2+}	0.127	S^{2-}	0.174
Dy^{3+}	0.107	Ba^{2+}	0.13	Fr^+	0.180
Tb^{3+}	0.109	0^{2-}	0.132	Cl^-	0.181
Th^{4+}	0.110	Pb^{2+}	0.132	Se^{2-}	0.191
Gd^{3+}	0.111	F^-	0.133	Br^-	0.196
Hg^{2+}	0.112	K^+	0.133	Si^{4-}	0.198
Ag^+	0.113	Au^+	0.137	Te^{2-}	0.211
Sm^{3+}	0.113	Rb^+	0.149	Sn^{4-}	0.215
Eu^{3+}	0.113	Tl^+	0.149	Pb^{4-}	0.215
Nd^{3+}	0.115	Ra^+	0.152	I^-	0.220
Pr^{3+}	0.116				
Ce^{3+}	0.118				
Ac^{3+}	0.118				

Source: After a tabulation by R. A. Flinn and P. K. Trojan, *Engineering Materials and Their Applications*, Houghton Mifflin Company, Boston, 1975. The ionic radii are based on the calculations of V. M. Goldschmidt, who assigned radii based on known interatomic distances in various ionic crystals.

SELECTED PROPERTIES
OF SUPERCONDUCTIVE ELEMENTS

Element	$T_c(K)$	H_0(Oersteds)[a]	$\theta_D(K)$	θ_D(mJ•mole^{-1} deg•K^2)
Al	1.175	104.93	420	1.35
Be	0.026			0.21
Cd	0.518-0.52	29.6	209	0.688
Ga	1.0833	59.3	325	0.60
Ga (b)	5.90-6.2	560		
Ga (g)	7.62	950		
Ga (d)	7.85	815		
Hg (a)	4.154	411	71.9-87	1.81
Hg (b)	3.949	339	93	1.37
In	3.405	281.53	109	1.672
Ir	0.11-0.14	19	425	3.27
La (a)	4.88	808-798	142	10.0-11.3
La (b)	6.00	1096	139	11.3
Mo	0.916	90-98	460	1.83
Nb	9.25	1970	277-238	7.80
Os	0.655	65	500	2.35
Pa	1.4			
Pb	7.23	803	96.3	3.0
Re	1.697	188-211	415	2.35
Ru	0.493	66	580	3.0
Sb	2.6-2.7[b]	HF		
Sn	3.721	305	195	1.78
Ta	4.47	831	258	6.15
Tc	7.73-7.78	1410	411	4.84-6.28
Th	1.39	159.1	165	4.31
Ti	0.39	56-100	429-412	3.32
Ti	2.332-2.39	181	78.5	1.47
V	5.43-5.31	1100-1400	382	9.82
W	0.0154	1.15	550	0.90
Zn	0.875	55	319.7	0.633
Zr	0.53	47	290	2.78
Zr (w)	0.65			

[a] To convert oersteds to ampere/meters, multiply by 79.57.
[b] Metastable.

From Roberts, B. W., *Properties of Selected Superconductive Materials - 1974 Supplement,* NBS Technical Note 825, National Bureau of Standards, U.S. Government Printing Office, Washington,D.C., 1974, 10.

T_C FOR THIN FILMS
OF SUPERCONDUCTIVE ELEMENTS

Element	$T_c(K)$		Element	$T_c(K)$	
Al	1.18	−5.7	Re	~7	
Be	~.03	−10.6	Sn	3.6	-6.0
Bi	~2	-6.173	Ta	<1.7	-4.8
Cd	0.53	-0.91	Ti	1.3	
Ga	6.4	-8.56	Ti	2.64	
In	3.43	-4.5	V	5.14	-6.02
La	5.0	-6.74	W	<1.0	-4.1
Mo	3.3	-6.7	Zn	0.77	-1.48
Nb	6.2	-10.1			
Pb	~2	-7.7			

Element	$T_c(K)$	Pressure[a]
As	0.31–0.5	220–140 kbar
	0.2–0.25	~140–100 kbar
Ba II	~1.3	55 kbar
Ba III	3.05	85–88 kbar
	~5.2	> 14 0 kbar
Bi II	3.916	25 katm
	3.90	25.2 katm
	3.86	26.8 katm
Bi III	6.5 5	~37 kbar
	7.25	27–28.4 katm
Bi IV	7.0	43–62 kbar
Bi V	8.3–8.55	81 kbar
Bi VI	8.55	90–101 kbar
Ce	1.7	50kbar
Cs	~1.5	> ~125 kbar

T_C for Thin Films of Superconducting Elements (Continued)

Element	$T_c(K)$	Pressure[a]
Ga II	6.24–6.38	≥35 katm
Ga III	7.5	≥ 35 katm (P– 0)
Ge	4.85–5.4	~120 kbar
	5.35	115 kbar
La	~5.5–11.93	0–~140 kbar
P	4.7	> 100 kbar
	5.8	170 kbar
Pb II	3.55–3.6	160 kbar
Sb	3.55	85 kbar
	3.5 2	93 kbar
	3.5 3	100 kbar
	3.40	~150 kbar
Se II	6.75–6.95	~130 kbar
Si	6.7–7.1	120 kbar
Sn II	5.2	125 kbar
	4.85	160 kbar
Sn III	5.30	113 kbar
Te II	2.05	43 kbar
	3.4	50 kbar
Te III	4.28	70 kbar
Te IV	4.25	84 kbar
Ti, cub.	1.45	35 kbar
Ti, hex.	1.95	5 kbar
U	2.3	10 kbar
Y	~1.2–~2.7	120–170 kbar

[a] To convert katm to $N \cdot m^{-2}$, multiply by 1.013×10^8 ;
to convert kbar to $N \cdot m^{-2}$, multiply by 1×10^8.

From Roberts, B.W., *Properties of Selected Superconductive Materials*, 1974 Supplement, National Bureau of Standard Technical Note 825, U.S. Government Printing Office, Washington, D.C., 1974, 11.

Engineering Compounds

ENGINEERING CERAMICS

Ceramics have been an important class of materials because of their inertness and heat- resistant properties. Although these materials have been used for many years, ceramics are in a very active stage of development. Applications based on unique properties of ceramic materials range from refractory bricks and crucibles for molten metals to miniature electronic devices; from nuclear fuels and control elements to magnetic memory units. The field of application of ceramic materials is diverse; the materials are numerous.

Vast amounts of data have been generated on properties of ceramic materials, and volumes of literature have been written on the subject. In this section, data on the properties of several ceramic materials are presented in a condensed tabulation. It is beyond the scope of this work to reference all of the work done on the materials discussed. The ceramics in this section were selected to show some of the properties of conunercially available ceramic materials that make them unique for many applications. The ceramic materials described in this section are alumina, beryllia, zirconia, mullite, cordierite, silicon carbide, and silicon nitride. The properties tabulated are crystal chemical, thermodynamic, physical, mechanical, thermal, and electrical properties. The crystal chemical and thermodynamic properties are independent of processing conditions and are listed separately ftom the other properties. The physical, mechanical, thermal, and electrical properties are a function of the processing conditions. These properties can be modified through process changes to meet specifications for different applications. To show these effects, data on the physical, mechanical, Iherrnal, and electrical properties for some of the rnaterials are given for different levels of density and purity.

ENGINEERING CERAMICS (Continued)

ALUMINA (Al2O3)

Good mechanical strength, inertness, refractoriness, and availability make alumina one of the most widely used of the ceramic materials. Some applications for alumina are electrical insulators, abrasives, cutting tools, radomes, and wear–resistant parts. Alumina is used in cements, refractories, glasses, coatings, cermets, and seals. The ability to densify this material with little or no glass phase present is advantageous in developing properties approaching the theoretical limit for this material. This has resulted in new applications for alumina such as envelopes for high-pressure sodium vapor lamps and microwave windows.

BERYLLIA (BeO)

Beryllia is characterized by a higher thermal conductivity than any other ceramic. The thermal conductivity of beryllia is about ten times that of aluminana and is approximately the same as aluminum. A good mechanical strength in combination with the high thermal conductivity gives beryllia good thermal shock-resistant characteristics.

Beryllia is used in electronic components as an electrical insulator and a heat sink. It is used as crucibles for melting uranium, thorium, and beryllium. Other applications are in the nuclear industry as moderators for fast neutrons, a matrix for fuel elements, shielding, and control rod assemblies.

Beryllia is toxic in the powder form, and special precautions must be taken in fabrication ftom the powder and in grinding the finished parts.

ENGINEERING CERAMICS (Continued)

ZIRCONIA (ZrO2)

Zirconia has a melting point of 2.7000°C in a neutral or oxidizing atmosphere, and it is not wet by most steel alloys and noble metals. The high melting point and chemical inertness, combined with a low thermal conductivity, make zirconia a good material for refractory applications. It is also used as an opacifier for glazes and enamels and is a major constituent of lead-zirconate-titanate piezoelectrics.

Zirconia is monoclinic at room temperature and transforms to the denser tetragonal form at about 1,100 to 1,200°C, undergoing a disruptive volume change of about 9%. Zirconia is often stabilized with CaO, Y_2O_3 or MgO additions, which, when fired, results in a material having a cubic structure, The cubic form differs only slightly from the monoclinic and tetragonal forms and is stable above and below the transformation temperature.

MULLITE (3Al2O3 •2SiO2)

Mullite is a refractory, mixed-oxide, ceramic material which has good thermal shock resistance properties. Mullite is used for refractory liners and laboratory ware. The silica in mullite is not present as free silica and usually does not pose a contamination problem. Mullite often appears as a separate phase in classical, triaxial porcelain whiteware ceramics. The needlelike habit of the mullite crystals adds strength to the body.

CORDIERITE (2MgO • 2Al2O3 • 5SiO2)

Cordierite is a ceramic material that has low dielectric losses and is used in high frequency insulators. Cordierite also has a low coefficient of expansion, good thermal shock resistance, and is used for heating element supports and burner tips.

ENGINEERING CERAMICS (Continued)

Pure cordierite bodies in general have a short vitrification range so additions are often made to extend the range. Reactions to form cordierite are often sluggish, and precautions must be taken to fire the body high enough to complete the reactions.

SILICON CARBIDE (SiC)

Silicon carbide has a high melting point and hardness. It can be used to 1,6500C in an oxidizing atmosphere and to 2,3000C in an inert atmosphere. Its high thermal conductivity gives it high thermal shock resistance, which makes it useful in refractory furnace parts. Silicon carbide is also used as an abrasive and is currently being explored as a material for gas turbine components because of its high temperature modulus of rupture and corrosion resistance.

Silicon carbide occurs in alpha and beta forms. The alpha phase is hexagonal or rhombohedral and the beta phase is cubic. The beta phase undergoes an irreversible transformation to the alpha phase above 1,650°C. The base of the hexagonal unit cell is ~3.08 Å and the c-axis is a multiple of ~2.52 Å. Numerous stacking sequences can occur in the alpha form, giving rise to many polytypes. Some of these polytypes given in Ramsdell notation are 2H, 4H, 6H, 15R, 21 R, and 24R (etc.). In this notation, the number refers to the number of layers in the unit cell and the letter refers to either hexagonal or rhombohedral symmetry.

Silicon carbide had not been successfully fired to near theoretical density without hot pressing in the past. The problem has been that silicon carbide is unreactive up to the temperature at which it decomposes. Silicon carbide has now been successfully densified to greater than 96% theoretical density by using boron as a sintering aid. This development will promote other applications for silicon carbide which were previously limited by the restrictive geometry and cost associated with hot pressed materials.

ENGINEERING CERAMICS (Continued)

SILICON NITRIDE (Si3N4)

Silicon nitride has high hardness, wear resistance, high corrosion resistance, high thermal shock resistance, and good high temperature strength. The material is currently used for applications such as furnace refectories and supports, bearings and seals, nozzles, thermocouple sheaths, and crucibles. Silicon nitride is currently being explored as a material for gas turbine components because of its high temperature fracture energy, corrosion resistance, and high thermal shock resistance.

Silicon nitride occurs in alpha and beta forms. These phases are hexagonal and differ only slightly in lattice parameters. The alpha phase undergoes an irreversible transformation to the beta phase above 1,500 to 1,600°C.

This material, as with silicon carbide, has not been successfully fired to near theoretical density without hot pressing. A great deal of research effort is currently being given to silicon nitride because of its potential high temperature corrosion and wear resistance properties. If developments are made with respect to densification without hot pressing, as with silicon carbide, the number of applications for this material will be greatly increased. Complex components are currently fabricated using the reaction bonded approach. This consists of fabricating parts from silicon and nitriding by firing the parts in nitrogen. The bodies formed by this process are limited in that they exhibit porosities in the order of 30%. Theoretically dense bodies have been made by hot pressing with MgO as a sintering aid.

Practical Handbook of Materials Science, Charles T. Lynch, Ed., CRC Press, Boca Raton, Fla, (1989)

REFRACTORIES, CERAMICS, AND SALTS

Borides

Compositions and T_{mp} (K)

Group	Element	T_{mp} (K)
IIA	BeB_2	>2243
IIA	SrB_6	2508
IIA	BaB_4	2543
IVB	TiB_2	3253
IVB	ZrB_2	3313
VB	VB_2	2373
VB	NbB_2	>2270
VB	TaB	>2270
VIB	CrB_2	2123
VIB	MoB	2625
VIB	WB	3133
	CeB_6	2463
	ThB_4	>2270
	UB_2	>1770

REFRACTORIES, CERAMICS, AND SALTS (Continued)

Bromides

Compositions and T_{mp} (K)

1 IA	2 IIA	3 IIIB	4 IVB	5 VB	6 VIB	7 VIIB	8 ------	9 VIII	10 ------	11 IB	12 IIB	13 IIIA	14 IVA	15 VA	16 VIA	17 VIIA	18 VIIA
LiBr 823	BeBr$_2$ 793											BBr$_3$ 227					
NaBr 1023	MgBr$_2$ 984											AlBr$_3$ 371					
KBr 1008	CaBr$_2$ 1003		TiBr$_4$ 312				FeBr$_2$ 955		NiBr$_2$ 1236	CuBr 777	ZnBr$_2$ 667						
	SrBr$_2$ 916		ZrBr$_2$ >625							AgBr 703	CdBr$_2$ 841	InBr$_3$ 709	SnBr$_2$ 488	SbBr$_3$ 370	TeBr$_2$ 612		
	BaBr$_2$ 1123			TaBr$_5$ 538					PtBr$_2$ 523				PbBr$_2$ 643	BiBr$_3$ 491			

					ThBr$_4$ 883	UBr$_4$ 789	

REFRACTORIES, CERAMICS, AND SALTS (Continued)

Carbides
Compositions and T_{mp} (K)

1 IA	2 IIA	3 IIIB	4 IVB	5 VB	6 VIB	7 VIIB	8 VIII	9 VIII	10 VIII	11 IB	12 IIB	13 IIIA	14 IVA	15 VA	16 VIA	17 VIIA	18 VIIIA
	Be_2C >2375											B_4C 2720	SiC 2970				
NaC_2 973												Al_4C_3 2000					
			TiC 3433	VC 3600	Cr_3C_2 2168		Fe_3C 2110										
	SrC_2 >1970		ZrC 3533	NbC 3770	Mo_2C 2963												
				TaC 3813	WC 2900												

Actinide series:

ThC 2898		UC 2863		

REFRACTORIES, CERAMICS, AND SALTS (Continued)

Chlorides

Compositions and T_{mp} (K)

1 IA	2 IIA	3 IIIB	4 IVB	5 VB	6 VIB	7 VIIB	8 VIII	9 VIII	10 VIII	11 IB	12 IIB	13 IIIA	14 IVA	15 VA	16 VIA	17 VIIA	18 VIIA
LiCl 883	BeCl$_2$ 713											BCl$_3$ 166					
NaCl 1073	MgCl$_2$ 987											AlCl$_3$ 465					
KCL 1043	CaCl$_2$ 1055		TiCl$_4$ 250	VCl$_4$ 245		MnCl$_2$ 923	FeCl$_2$ 945		NiCl$_3$ 1274	CuCl 695	ZnCl$_2$ 548						
	SrCl$_2$ 1148		ZrCl$_2$ 623						PtCl$_2$ 854	AgCl 728	CdCl$_2$ 841	InCl 498	SnCl$_2$ 581	SbCl$_3$ 346	TeCl$_2$ 448		
	BaCl$_2$ 1235			TaCl$_5$ 489	WCl$_6$ 548								PbCl$_2$ 771	BiCl$_3$ 507			

CeCl$_3$ 1095				
ThCl$_4$ 1043		UCl$_4$ 843		

REFRACTORIES, CERAMICS, AND SALTS (Continued)

Fluorides

Compositions and T_{mp} (K)

1 IA	2 IIA	3 IIIB	4 IVB	5 VB	6 VIB	7 VIIB	8 ------	9 VIII	10 ------	11 IB	12 IIB	13 IIIA	14 IVA	15 VA	16 VIA	17 VIIA	18 VIIA
LiF 1119	BeF$_2$ 813											BF$_3$ 146					
NaF 1267	MgF$_2$ 1535											AlF$_3$ 1564	SiF$_4$ 183				
KF 1131	CaF$_2$ 1675		TiF$_3$ 1475	VF$_3$ >1075		MnF$_2$ 1129	FeF$_3$ >1275		NiF$_2$ 1273	CuF$_2$ 1129	ZnF$_2$ 1145						
	SrF$_2$ 1736		ZrF$_4$ 873		MoF$_6$ 290					AgF 708	CdF$_2$ 1373	InF$_3$ 1443	SnF$_4$ 978	SbF$_3$ 565			
	BaF$_2$ 1627			TaF$_5$ 370									PbF$_2$ 1095	BiF$_3$ 1000			

CeF$_2$ 1710					
ThF$_4$ 1375		UF$_4$ 1233			

REFRACTORIES, CERAMICS, AND SALTS (Continued)

Iodides

Compositions and T_{mp} (K)

1 IA	2 IIA	3 IIIB	4 IVB	5 VB	6 VIB	7 VIIB	8 ------	9	10 VIII	11 IB	12 IIB	13 IIIA	14 IVA	15 VA	16 VIA	17 VIIA	18 VIIA
LiI 722	BeI$_2$ 783																
NaI 935	MgI$_2$ <910											AlI 464					
KI 958	CaI$_2$ 848		TiI$_2$ 873	FI$_2$ 1048					NiI$_2$ 1070	CuI 878	ZnI2 719						
	SrI$_2$ 593		ZrI$_4$ 772		MoI4 373					AgI 831	CdI$_2$ 423	InI$_3$ 483	SnI$_2$ 788	SbI$_3$ 443			
	BaI$_2$ 1013								PtI$_2$ 633				PbI$_2$ 675	BiI$_3$ 681			

	CeI$_3$ 1025		UI$_4$ 779	

REFRACTORIES, CERAMICS, AND SALTS (Continued)

Nitrates

Compositions and T_{mp} (K)

1 IA	2 IIA	3 IIIB	4 IVB	5 VB	6 VIB	7 VIIB	8 ------	9 VIII	10 ------	11 IB	12 IIB	13 IIIA	14 IVA	15 VA	16 VIA	17 VIIA	18 VIIA
$LiNO_3$ 527																	
$NaNO_3$ 583																	
KNO_3 610	$Ca(NO_3)_2$ 623																
	$Sr(NO_3)_2$ 643									$AgNO_3$ 483	$Cd(NO_3)_2$ 834						
	$Ba(NO_3)_2$ 865												$Pb(NO_3)_2$ 743				

REFRACTORIES, CERAMICS, AND SALTS (Continued)

Nitrides

Compositions and T_{mp} (K)

	1 IA	2 IIA	3 IIIB	4 IVB	5 VB	6 VIB	7 VIIB	8 VIII	9 VIII	10 VIII	11 IB	12 IIB	13 IIIA	14 IVA	15 VA	16 VIA	17 VIIA	18 VIIA
	Li_3N 1118	Be_3N_2 2513											BN 3000					
	Na_2N 573												AlN >2475	Si_3N_4 2715				
		Ca_3N_2 1468		TiN 3200	VN 2593	CrN 1770					Cu_3N 573							
				ZrN 3250	NbN 2323													
					Ta_2N 3360													

			3	4	5 VB											
			ThN 2903		UN 3123											

REFRACTORIES, CERAMICS, AND SALTS (Continued)

Oxides

Compositions and T_{mp} (K)

1	2	3	4	5	6	7	8	9	10	11	12	13	14	15	16	17	18
IA	IIA	IIIB	IVB	VB	VIB	VIIB	--------	VIII	--------	IB	IIB	IIIA	IVA	VA	VIA	VIIA	VIIIA
Li_2O >1975	BeO 2725											B_2O_3 723					
	MgO 3098											Al_2O_3 2322	SiO_2 1978				
K_2O 703	CaO 3183		TiO_2 2113	V_2O_5 947	Cr_2O_3 >2603	MnO 1840	Fe_2O_3 1864		NiO 2257	Cu_2O 1508	ZnO 2248						
	SrO 2933		ZrO_2 3123	Nb_2O_5 1764	MoO_3 1068					Ag_2O 573	CdO 1773	In_2O_3 2183	SnO 1353	Sb_2O_3 928	TeO_2 1006		
	BaO 2283			Ta_2O_5 2100	WO_3 1744								PbO 1159	B_2O_3 1098			

CeO_2 >2873	UO_2 3151
ThO_2 3493	

REFRACTORIES, CERAMICS, AND SALTS (Continued)

Silicides

Compositions and T_{mp} (K)

Group	Element	T_{mp} (K)
IIA	Mg_2Si	1375
IVB	$TiSi_2$	1813
VB	VSi_2	2023
VB	$NbSi_2$	2203
VB	$TaSi_2$	2670
VIB	$CrSi_2$	1843
VIB	$MoSi_2$	2553
VIB	WSi_2	2320
IB	Cu_4Si	1123
	USi_2	1970

REFRACTORIES, CERAMICS, AND SALTS (Continued)

Sulfates

Compositions and T_{mp} (K)

1 IA	2 IIA	3 IIIB	4 IVB	5 VB	6 VIB	7 VIIB	8 ----	9 VIII	10 ----	11 IB	12 IIB	13 IIIA	14 IVA	15 VA	16 VIA	17 VIIA	18 VIIA
Li_2SO_4 1132	$BeSO_4$ 848																
Na_2SO_4 1157	$MgSO_4$ 1397											$Al_2(SO_4)_3$ 1043					
K_2SO_4 1342	$CaSO_4$ 1723						$Fe_2(SO_4)_3$ 753		$NiSO_4$ 1121		$ZnSO_4$ 873						
	$SrSO_4$ 1878		$Zr(SO_4)_2$ 683							Ag_2SO_4 933	$CdSO_4$ 1273		$SnSO_4$ >635				
	$BaSO_4$ 1853												$PbSO_4$ 1443	$Bi(SO_4)_3$ 678			

$Ce(SO_4)_2$ 468

REFRACTORIES, CERAMICS, AND SALTS (Continued)

Sulfides

Compositions and T_{mp} (K)

1 IA	2 IIA	3 IIIB	4 IVB	5 VB	6 VIB	7 VIIB	8 -----	9 VIII	10 -----	11 IB	12 IIB	13 IIIA	14 IVA	15 VA	16 VIA	17 VIIA	18 VIIA
Li₂S 1198												BS₄ 663					
Na₂S 1453	MgS >2275											Al₂S₃ 1373					
K₂S 1113	SrS >2275		ZrS₂ 1823	V₂S₃ >875	MoS₂ 1458		FeS 1468		NiS 1070	Cu₂S 1400	CdS 2023	In₂S₃ 1323	SnS 1153	SbS₃ 820			
	BaS 1473			CeS 2400	TaS₄ >1575				WS₂ 1523	Ag₂S 1098			PtS₂ 508	PbS 1387			

3	4	5
Bi₂S₃ 1020		
ThS₂ 2198		US₂ >1375

(Group labels as printed in the chart: Li₂S = Li_2S 1198; Na₂S = Na_2S 1453; MgS >2275; Al₂S₃ = Al_2S_3 1373; K₂S = K_2S 1113; SrS >2275; BaS 1473; ZrS₂ = ZrS_2 1823; V₂S₃ = V_2S_3 >875; CeS 2400; MoS₂ = MoS_2 1458; TaS₄ = TaS_4 >1575; FeS 1468; NiS 1070; WS₂ = WS_2 1523; Cu₂S = Cu_2S 1400; Ag₂S = Ag_2S 1098; CdS 2023; BS₄ = BS_4 663; In₂S₃ = In_2S_3 1323; SnS 1153; PtS₂ = PtS_2 508; SbS₃ = SbS_3 820; PbS 1387; Bi₂S₃ = Bi_2S_3 1020; ThS₂ = ThS_2 2198; US₂ = US_2 >1375.)

TYPE II SUPERCONDUCTING COMPOUNDS: CRITICAL TEMPERATURE AND CRYSTAL STRUCTURE DATA

Compound	T_c, K	Crystal Structure
$Ag_xAl_yAn_{1-x-y}$	0.5–0.845	
$Ag_7BF_4O_8$	0.15	Cubic
$AgBi_2$	3.0–2.78	
$Ag_7F_{0.25}N_{0.75}O_{10.25}$	0.85–0.90	
Ag_7FO_8	0.3	Cubic
Ag_2F	0.066	
$Ag_{0.8-0.3}Ga_{0.2-0.7}$	6.5–8	
Ag_4Ge	0.85	Hex., c.p.
$Ag_{0.438}Hg_{0.562}$	0.64	$D8_2$
$AgIn_2$	~2.4	C16
$Ag_{0.1}In_{0.9}Te$ ($n = 1.40 \times 10^{22}$)	1.20–1.89	B1
$Ag_{0.2}In_{0.8}Te$ ($n = 1.07 \times 10^{22}$)	0.77–1.00	B1
$AgLa$ (9.5 kbar)	1.2	B2
Ag_7NO_{11}	1.04	Cubic
Ag_xPb_{1-x}	7.2 max.	
Ag_xSn_{1-x} (film)	2.0–3.8	
Ag_xSn_{1-x}	1.5–3.7	
$AgTe_3$	2.6	Cubic
$AgTh_2$	2.26	C16
$Ag_{0.03}Tl_{0.97}$	2.67	
$Ag_{0.94}Tl_{0.06}$	2.32	
Ag_xZn_{1-x}	0.5–0.845	
Al (film)	1.3–2.31	
Al (l to 21 katm)	1.170–0.687	A1

Type II Superconducting Compounds:
Critical Temperature and Crystal Structure Data (Continued)

Compound	T_c, K	Crystal Structure
$AlAu_4$	0.4–0.7	Like A13
Al_2CMo_3	10.0	A13
Al_2CMo_3	9.8–10.2	A13 + trace 2nd phase
Al_2CaSi	5.8	
$Al_{0.131}Cr_{0.088}V_{0.781}$	1.46	Cubic
$AlGe_2$	1.75	
$Al_{0.5}Ge_{0.5}Nb$	12.6	A15
$Al_{\sim0.8}Ge_{\sim0.2}Nb_3$	20.7	A15
$AlLa_3$	5.57	DO_{19}
Al_2La	3.23	C15
Al_3Mg_2	0.84	Cubic, f.c.
$AlMo_3$	0.58	A15
$AlMo_6Pd$	2. 1	
AlN	1.55	B4
Al_2NNb_3	1.3	A13
$AlNo_3$	18.0	A15
Al_xNb_{1-x}	<4.2–13.5	$D8_b$
Al_xNb_{1-x}	12–17.5	A15
$Al_{0.27}Nb_{0.73-0.48}V_{0-0.25}$	14.5–17.5	A15
$AlNb_xV_{1-x}$	<4.2–13.5	
$AlOs$	0.39	B2
Al_3Os	5.90	
$AlPb$ (films)	12–7	
Al_2Pt	0.48–0.55	Cl

Type II Superconducting Compounds:
Critical Temperature and Crystal Structure Data (Continued)

Compound	T_c, K	Crystal Structure
Al_5Re_{24}	3.35	A12
Al_3Th	0.75	DO_{19}
$Al_xTi_yV_{1-y}$	2.05–3.62	Cubic
$Al_{0.108}V_{0.897}$	1.82	Cubic
Al_xZn_{1-x}	0.5–0.845	
$AlZr_3$	0.73	$L1_2$
AsBiPb	9.0	
AsBiPbSb	9.0	
$As_{0.33}InTe_{0.67}$ ($n = 1.24 \times 10^{22}$)	0.85–1.15	B1
$As_{0.5}InTe_{0.5}$ ($n = 0.97 \times 10^{22}$)	0.44–0.62	B1
$As_{0.50}Ni_{0.06}Pd_{0.44}$	1.39	C2
AsPb	8.4	
$AsPd_2$ (low-temperature phase)	0.60	Hexagonal
$AsPd_2$ (high-temp phase)	1.70	C22
$AsPd_5$	0.46	Complex
AsRh	0.58	B31
$AsRh_{1.4-1.6}$	<0.03–0.56	Hexagonal
AsSn	4.10	
AsSn ($n = 2.14 \times 10^{22}$)	3.41–3.65	B1
$As_{\sim 2}Sn_{\sim 3}$	3.5–3.6, 1.21–1.17	
As_3Sn_4 ($n = 0.56 \times 10^{22}$)	1.16–1.19	Rhombohedral
Au_5Ba	0.4–0.7	$D2_d$
AuBe	2.64	B20
Au_2Bi	1.80	C15

Type II Superconducting Compounds:
Critical Temperature and Crystal Structure Data (Continued)

Compound	T_c, K	Crystal Structure
Au_5Ca	0.34–0.38	$C15_b$
$AuGa$	1.2	B31
$Au_{0.40-0.92}Ge_{0.60-0.08}$	<0.32–1.63	Complex
$AuIn$	0.4–0.6	Complex
$AuLu$	<0.35	B2
$AuNb_3$	11.5	A15
$AuNb_3$	1.2	A2
$Au_{0.03}Nb_{1-0.7}$	1.1–11.0	
$Au_{0.2-0.98}Nb_3Rh_{0.98-0.2}$	2.53–10.9	A15
$AuNb_{3(1-x)}V_{3x}$	1.5–11.0	A15
$AuPb_2$	3.15	
$AuPb_2$ (film)	4.3	
$AuPb_3$	4.40	
$AuPb_3$ (film)	4.25	
Au_2Pb	1.18, 6–7	C15
$AuSb_2$	0.58	C2
$AuSn$	1.25	$B8_1$
$Au_xSn_{(1-x)}$ (film)	2.0–3.8	
Au_5Sn	0.7–1.1	A3
Au_3Te_5	1.62	Cubic
$AuTh_2$	3.08	C16
$AuTl$	1.92	
AuV_3	0.74	A15
$Au_xZn(1-x)$	0.50–0.845	

Type II Superconducting Compounds:
Critical Temperature and Crystal Structure Data (Continued)

Compound	T_c, K	Crystal Structure
$AuZn_3$	1.21	Cubic
Au_xZr_y	1.7–2.8	A3
$AuZr_3$	0.92	A15
$BCMo_2$	5.4–7.2	Orthorhombic
$B_{0.03}C_{0.51}Mo_{0.47}$	12.5	
$BCMo_2$	5.3–7.0	Orthorhombic
BHf	3.1	Cubic
B_6La	5.7	
$B_{12}Lu$	0.48	
BMo	0.5 (extrapolated)	
BMo_2	4.74	C16
BNb	8.25	B_f
BRe_2	2.80, 4.6	
$B_{0.3}Ru_{0.7}$	2.58	$D10_2$
$B_{12}Sc$	0.39	
BTa	4.0	B_f
B_6Th	0.74	
BW_2	3.1	C16
B_6Y	6.5–7.1	
$B_{12}Y$	4.7	
BZr	3.4	Cubic
$B_{12}Zr$	5.82	
$BaBi_3$	5.69	Tetragonal
$Ba_xO_3Sr_{1-x}Ti$ ($n = 4.2–11 \times 10^{19}$)	<0.1–0.55	

Type II Superconducting Compounds:
Critical Temperature and Crystal Structure Data (Continued)

Compound	T_c, K	Crystal Structure
$BaO_{0.13}O_3W$	1.9	Tetragonal
$BaO_{14}O_3W$	< 1.25–2.2	Hexagonal
$BaRh_2$	6.0	C15
$Be_{22}Mo$	2.51	Cubic, like $Be_{22}Re$
$Be_8Nb_5Zr_2$	5.2	
$Be_{0.98-0.92}Re_{0.02-0.08}$ (quenched)	9.5–9.75	Cubic
$Be_{0.957}Re_{0.043}$	9.62	Cubic, like $Be_{22}Re$
$BeTc$	5.21	Cubic
$Be_{22}W$	4.12	Cubic, like $Be_{22}Re$
$Be_{23}W$	4.1	Tetragonal
Bi_3Ca	2.0	
$Bi_{0.5}Cd_{0.13}Pb_{0.25}Sn_{0.12}$ (weight fractions)	8.2	
$BiCo$	0.42–0.49	
Bi_2Cs	4.75	C15
Bi_xCu_{1-x} (electrodeposited)	2.2	
$BiCu$	1.33	140
$Bi_{0.019}In_{0.981}$	3.86	
$Bi_{0.05}In_{0.95}$	4.65	α–phase
$Bi_{0.010}In_{0.90}$	5.05	β–phase
$Bi_{0.15-0.30}In_{0.85-0.70}$	5.3–5.4	α– and β–phases
$Bi_{0.34-0.48}In_{0.66-0.52}$	4.0–4.1	
Bi_3In_5	4.1	
$BiIn_2$	5.65	β–phase

Type II Superconducting Compounds:
Critical Temperature and Crystal Structure Data (Continued)

Compound	T_c, K	Crystal Structure
Bi_2Ir	1.7–2.3	
Bi_2Ir (quenched)	3.0–3.96	
BiK	3.6	
Bi_2K	3.58	C15
BiLi	2.47	$L1_0$ α–phase
$Bi_{4-9}Mg$	0.7–~ 1.0	
Bi_3Mo	3–3.7	
BiNa	2.25	$L1_0$
BiNb (high presure and temperature)	3.05	A15
BiNi	4.25	$B8_1$
Bi_3Ni	4.06	Orthorhombic
$Bi_{1-0}Pb_{0-1}$	7.26–9.14	
$Bi_{1-0}Pb_{0-1}$ (film)	7.25–8.67	
$Bi_{0.05-0.49}Pb_{0.95-0.60}$	7.35–8.4	Hexagonal c.p., to ε-phase
BiPbSb	8.9	
$Bi_{0.5}Pb_{0.31}Sn_{0.19}$ (weight fractions)	8.5	
$Bi_{0.5}Pb_{0.25}Sn_{0.25}$	8.5	
$BiPd_2$	4.0	
$Bi_{0.4}Pd_{0.6}$	3.7–4	Hexagonal, ordered
BiPd	3.7	Orthorhombic
Bi_2Pd	1.70	Monoclinic, α-phase
Bi_2Pd	4.25	Tetragonal, β-phase
BiPdSe	1.0	C2
BiPdTe	1.2	C2

Type II Superconducting Compounds:
Critical Temperature and Crystal Structure Data (Continued)

Compound	T_c, K	Crystal Structure
BiPt	1.21	$B8_1$
BiPtSe	1.45	C2
BiPtTe	1.15	C2
Bi_2Pt	0.155	Hexagonal
Bi_2Rb	4.25	C15
$BiRe_2$	1.9–2.2	
BiRh	2.06	$B8_1$
Bi_3Rh	3.2	Orthorhombic, like NiB_3
Bi_4Rh	2.7	Hexagonal
Bi_3Sn	3.6–3.8	
BiSn	3.8	
Bi_xSn_y	3.85–4.18	
Bi_3Sr	5.62	L12
Bi_3Te	0.75–1.0	
Bi_5Tl_3	6.4	
$Bi_{0.26}Tl_{0.74}$	4.4	Cubic, disordered
$Bi_{0.26}Tl_{0.74}$	4.15	L12, ordered?
Bi_2Y_3	2.25	
Bi_3Zn	0.8–0.9	
$Bi_{0.3}Zr_{0.7}$	1.51	
$BiZr_3$	2.4–2.8	
CCs_x	0.020–0.135	Hexagonal
C_8K (gold)	0.55	
$CGaMo_2$	3.7–4.1	Hexagonal, H-phase

Type II Superconducting Compounds:
Critical Temperature and Crystal Structure Data (Continued)

Compound	T_c, K	Crystal Structure
$CHf_{0.5}Mo_{0.5}$	3.4	B1
$CHf_{0.3}Mo_{0.7}$	5.5	B1
$CHf_{0.25}Mo_{0.75}$	6–6	B1
$CHf_{0.7}Nb_{0.3}$	6.1	B1
$CHf_{0.6}Nb_{0.4}$	4.5	B1
$CHf_{0.5}Nb_{0.5}$	4.8	B1
$CHf_{0.4}Nb_{0.6}$	5.6	B1
$CHf_{0.25}Nb_{0.75}$	7.0	B1
$CHf_{0.2}Nb_{0.8}$	7.8	B1
$CHf_{0.9-0.1}Ta_{1-0.9}$	5.0–9.0	B1
CK (excess K)	0.55	Hexagonal
C_8K	0.39	Hexagonal
$Co_{0.40-0.44}Mo_{0.60-0.56}$	9–13	
CMo	6.5, 9.26	
CMo_2	12.2	Orthorhombic
$C_{0.44}Mo_{0.56}$	1.3	B1
$C_{0.5}Mo_xNb_{1-x}$	10.8–12.5	B1
$C_{0.6}Mo_{48}Si_3$	7.6	$D8_8$
$CMo_{0.2}Ta_{0.8}$	7.5	B1
$CMo_{0.5}Ta_{0.5}$	7.7	B1
$CMo_{0.75}Ta_{0.25}$	8.5	B1
$CMo_{0.8}Ta_{0.2}$	8.7	B1
$CMo_{0.85}Ta_{0.15}$	8.9	B1
CMo_xTi_{1-x}	10.2 max.	B1

Type II Superconducting Compounds:
Critical Temperature and Crystal Structure Data (Continued)

Compound	T_c, K	Crystal Structure
$CMo_{0.83}Ti_{0.17}$	10.2	B1
CMo_xV_{1-x}	2.9–9.3	B1
CMo_xZr_{1-x}	3.8–9.5	B1
$C_{0.1-0.9}N_{0.9-0.1}Nb$	8.5–17.9	
$C_{0.-0.38}N_{1-0.62}Ta$	10.0–11.3	
CNb (whiskers)	7.5–10.5	
$C_{0.984}Nb$	9.8	B1
CNb (extrapolated)	~ 14	
$C_{0.7-1.0}Nb_{0.3-0}$	6–11	B1
CNb_2	9.1	
CNb_xTa_{1-x}	8.2–13.9	
CNb_xTi_{1-x}	<4.2–8.8	B1
$CNb_{0.6-0.9}W_{0.4-0.1}$	12.5–11.6	B1
$CNb_{0.6-0.9}Zr_{0.4-0.1}$	4.2–8.4	B1
CRb_x (gold)	0.023–0.151	Hexagonal
$CRe_{0.01-0.08}W$	1.3–5.0	
$CRe_{0.06}W$	5.0	
CTa	~11 (extrapolated)	
$C_{0.987}Ta$	9.7	
$C_{0.848-0.987}Ta$	2.04–9.7	
CTa (film)	5.09	B1
CTa_2	3.26	L′3
$CTa_{0.4}Ti_{0.6}$	4.8	B1
$CTa_{1-0.4}W_{0-0.6}$	8.5–10.5	B1

Type II Superconducting Compounds:
Critical Temperature and Crystal Structure Data (Continued)

Compound	T_c, K	Crystal Structure
$CTa_{0.2-0.9}Zr_{0.8-0.1}$	4.6–8.3	B1
CTc (excess C)	3.85	Cubic
$CTi_{0.5-0.7}W_{0.5-0.3}$	6.7–2.1	B1
CW	1.0	
CW_2	2.74	L´3
CW_2	5.2	Cubic, f.c.
$CaIr_2$	6.15	C15
$Ca_xO_3Sr_{1-x}Ti$ ($n = 3.7–11.0 \times 10^{19}$)	<0.1–0.55	
$Ca_{0.1}O_3W$	1.4–3.4	Hexagonal
CaPb	7.0	
$CaRh_2$	6.40	C15
$Cd_{0.3-0.5}Hg_{0.7-0.5}$	1.70–1.92	
CdHg	1.77, 2.15	Tetragonal
$Cd_{0.0075-0.05}In_{1-x}$	3.24–3.36	Tetragonal
$Cd_{0.97}Pb_{0.03}$	4.2	
CdSn	3.65	
$Cd_{0.97}Tl_{0.03}$	2–3	
$Cd_{0.18}Tl_{0.82}$	2.54	
$CeCO_2$	0.84	C15
$CeCo_{1.67}Ni_{0.33}$	0.46	C15
$CeCo_{1.67}Rh_{0.33}$	0.47	C15
$Ce_xGd_{1-x}Ru_2$	3.2–5.2	C15
$CeIr_3$	3.34	
CeI_5	1.82	

Type II Superconducting Compounds:
Critical Temperature and Crystal Structure Data (Continued)

Compound	T_c, K	Crystal Structure
$Ce_{0.005}La_{0.995}$	4.6	
Ce_xLa_{1-x}	1.3–6.3	
$Ce_xPr_{1-x}Ru_2$	1.4–5.3	C15
Ce_xPt_{1-x}	0.7–1.55	
$CeRu_2$	6.0	C15
$Co_xFe_{1-x}Si_2$	1.4 max.	Cl
$CoHf_2$	0.56	E93
$CoLa_3$	4.28	
$CoLu_3$	~ 0.35	
$Co_{0-0.01}Mo_{0.8}RC_{0.02}$	2–10	
$Co_{0.02-0.10}Nb_3Rh_{0.098-0.90}$	2.28–1.90	A15
$Co_xNi_{1-x}Si_2$	1.4 max.	Cl
$Co_{0.5}Rh_{0.5}Si_2$	2.5	
$Co_xRh_{1-x}Si_2$	3.65 max.	
$Co_{\sim0.3}Sc_{\sim0.7}$	~0.35	
$CoSi_2$	1.40, 1.22	Cl
Co_3Th_7	1.83	D102
Co_xTi_{1-x}	2.8 max.	Co in α-Ti
Co_xTi_{1-x}	3.8 max.	Co in β-Ti
$CoTi_2$	3.44	E93
$CoTi$	0.71	A2
CoU	1.7	B2, distorted
CoU_6	2.29	D2$_c$
$Co_{0.28}Y_{0.072}$	0.34	

Type II Superconducting Compounds:
Critical Temperature and Crystal Structure Data (Continued)

Compound	T_c, K	Crystal Structure
CoY_3	<0.34	
$CoZr_2$	6.3	C16
$Co_{0.1}Zr_{0.9}$	3.9	A3
$Cr_{0.6}Ir_{0.4}$	0.4	Hexagonal, c.p.
$Cr_{0.65}Ir_{0.35}$	0.59	Hexagonal, c.p.
$Cr_{0.7}Ir_{0.3}$	0.76	Hexagonal, c.p.
$Cr_{0.72}Ir_{0.28}$	0.83	
Cr_3Ir	0.45	A15
$Cr_{0-0.1}Nb_{1-0.9}$	4.6–9.2	A2
$Cr_{0.80}Os_{0.20}$	2.5	Cubic
Cr_xRe_{1-x}	1.2–5.2	
$Cr_{0.40}Re_{0.060}$	2.15	D8$_h$
$Cr_{0.8-0.6}Rh_{0.2-0.4}$	0.5	
Cr_3Ru (annealed)	3.3	A15
Cr_2Ru	2.02	D8$_h$
$Cr_{0.1-0.5}Ru_{0.9-0.5}$	0.34–1.65	A3
Cr_xTi_{1-x}	3.6 max.	Cr in α-Ti
Cr_xTi_{1-x}	4.2 max.	Cr in β-Ti
$Cr_{0.1}Ti_{0.3}V_{0.6}$	5.6	
$Cr_{0.0175}U_{0.9825}$	0.75	β-phase
$Cs_{0.32}O_3W$	1.12	Hexagonal
$Cu_{0.15}In_{0.85}$ (film)	3.75	
$Cu_{0.04-0.09}In_{1-x}$	4.4	
$CuLa$	5.85	

Type II Superconducting Compounds:
Critical Temperature and Crystal Structure Data (Continued)

Compound	T_c, K	Crystal Structure
Cu_xPb_{1-x}	5.7–7.7	
CuS	1.62	B18
CuS_2	1.48–1.53	C18
CuSSe	1.5–2.0	C18
$CuSe_2$	2.3–2.43	C18
CuSeTe	1.6–2.0	C18
Cu_xSn_{1-x}	3.2–3.7	
Cu_xSn_{1-x} (film, made at 10°K)	3.6–7	
Cu_xSn_{1-x} (film, made at 300°K)	2.8–3.7	
CuTh2	3.49	C16
Cu0–0.27V	3.9–5.3	A2
Cu_xZn_{1-x}	0.5–0.845	
Er_xLa_{1-x}	1.4–6.3	
$Fe_{0-0.5}Mo_{0.8}Re_{0.2}$	1–10	
$Fe_{0.05}Ni_{0.05}Zr_{0.9}$	~ 3.9	
Fe_3Th_7	1.86	D10
Fe_xTi_{1-x}	3.2 max.	Fe in α-Ti
Fe_xTi_{1-x}	3.7 max.	Fe in β-Ti
$Fe_xTi_{0.6}V_{1-x}$	6.8 max.	
FeU_6	3.86	D2
$Fe_{0.1}Zr_{0.9}$	1.0	A3
$Ga_{0.5}Ge_{0.5}Nb_3$	7.3	A15
$GaLa_3$	5.84	

Type II Superconducting Compounds:
Critical Temperature and Crystal Structure Data (Continued)

Compound	T_c, K	Crystal Structure
Ga_2Mo	9.5	
$GaMo_3$	0.76	A15
Ga_4Mo	9.8	
GaN (black)	5.85	B4
$GaNb_3$	14.5	A15
$Ga_xNb_3Sn_{1-x}$	14–18.37	A15
$Ga_{0.07}Pt_{0.03}$	2.9	Cl
GaPt	1.74	B20
GaSb (120 kbar, 77°K, annealed)	4.24	A5
GaSb (unannealed)	~ 5.9	
$Ga_{0-1}Sn_{1-0}$ (quenched)	3.47–4.18	
$Ga_{0-1}Sn_{1-0}$ (annealed)	2.6–3.85	
Ga_5V_2	3.55	Tetragonal, Mn_2Hg_5 type
GaV_3	16.8	A15
$GaV_{2.1-3.5}$	6.3–14.45	A15
$GaV_{4.5}$	9. 1 5	
Ga_3Zr	1.38	
Gd_xLa_{1-x}	< 1.0–5.5	
$Gd_xOs_2Y_{1-x}$	1.4–4.7	
$Gd_xRu_2Th_{1-x}$	3.6 max.	C15
GeIr	4.7	B31
Ge_2La	1.49, 2.2	Orthorhombic, distorted $ThSi_2$-type
$GeMo_3$	1.43	A15

Type II Superconducting Compounds:
Critical Temperature and Crystal Structure Data (Continued)

Compound	T_c, K	Crystal Structure
$GeNb_2$	1.9	
$GeNb_3$ (quenched)	6–17	A15
$Ge_{0.29}Nb_{0.71}$	6	A15
$Ge_xNb_3Sn_{1-x}$	17.6–18.0	A15
$Ge_{0.5}Nb_{35}Sn_{0.5}$	11.3	
$GePt$	0.40	B31
Ge_3Rh_5	2.12	Orthorhombic, related to $InNi_2$
Ge_2Sc	1.3	
Ge_3Te_4 ($n = 1.06 \times 10^{22}$)	1.55–1.80	Rhombohedral
Ge_xTe_{1-x} ($n = 8.5$–64×10^{20})	0.07–0.41	B1
GeV_3	6.01	A15
Ge_2Y	3.80	C_c
$Ge_{0.62}Y$	2.4	
$H_{0.33}Nb_{0.67}$	7.28	Cubic, b.c.
$H_{0.1}Nb_{0.9}$	7.38	Cubic, b.c.
$H_{0.05}Nb_{0.85}$	7.83	Cubic, b.c.
$H_{0.12}Nb_{0.88}$	2.81	Cubic, b.c.
$H_{0.08}Nb_{0.12}$	3.26	Cubic, b.c.
$H_{0.04}Nb_{0.96}$	3.62	Cubic, b.c.
$HfN_{0.989}$	6.6	B1
$Hf_{0-0.5}Nb_{1-0.5}$	8.3–9.5	A2
$Hf_{0.75}Nb_{0.25}$	>4.2	
$HfOs_2$	2.69	C14
$HfRe_2$	4.80	C14

Type II Superconducting Compounds:
Critical Temperature and Crystal Structure Data (Continued)

Compound	T_c, K	Crystal Structure
$Hf_{0.14}Re_{0.86}$	5.86	A12
$Hf_{0.99-0.96}Rh_{0.01-0.04}$	0.85–1.51	
$Hf_{0-0.55}Ta_{1-0.45}$	4.4–6.5	A2
HfV_2	8.9–9.6	C15
Hg_xIn_{1-x}	3.14–4.55	
$HgIn$	3.81	
Hg_2K	1.20	Orthorhombic
Hg_3K	3.18	
Hg_4K	3.27	
Hg_8K	3.42	
Hg_3Li	1.7	Hexagonal
Hg_2Na	1.62	Hexagonal
Hg_4Na	3.05	
Hg_xPb_{1-x}	4.14–7.26	
$HgSn$	4.2	
Hg_xTl_{1-x}	2.30–4.109	
Hg_5Tl_2	3.86	
Ho_xLa_{1-x}	1.3–6.3	
$InLa_3$	9.83, 10.4	$L1_2$
$InLa_3$ (0–35 kbar)	9.75–10.55	
$In_{1-0.86}Mg_{0-0.14}$	3.395–3.363	
$InNb_3$ (high pressure and temp.)	4–8, 9.2	A15
$In_{0-0.3}Nb_3Sn_{1-0.07}$	18.0–18.19	A15
$In_{0.5}Nb_3Zr_{0.5}$	6.4	

Type II Superconducting Compounds:
Critical Temperature and Crystal Structure Data (Continued)

Compound	T_c, K	Crystal Structure
$In_{0.11}O_3W$	< 1.25–2.8	Hexagonal
$In_{0.95–0.85}Pb_{0.05–0.15}$	3.6–5.05	
$In_{0.98–0.91}Pb_{0.02–0.09}$	3.45–4.2	
InPb	6.65	
InPd	0.7	B2
InSb		
(quenched, from 170 kbar into liquid N_2)	4.8	Like AS
InSb	2.1	
$(InSb)_{0.95–0.10}Sn_{0.05–0.90}$	3.8–5.1	
(various heat treatments)		
$(InSb)_{0–0.7}Sn_{1–0.93}$	3.67–3.74	
In_3Sn	~ 5.5	
$In_xSn_{1–x}$	3.4–7.3	
$In_{0.82–1}Te$ (n = 0.83–1.71 x 10^{22})	1.02–3.45	B1
$In_{1.000}Te_{1.002}$	3.5–3.7	B1
In_3Te_4 (n = 0.47 x 10^{22})	1.15–1.25	Rhombohedral
$In_xTl_{1–x}$	2.7–3.374	
$In_{0.8}Tl_{0.2}$	3.223	
$In_{0.62}Tl_{0.38}$	2.760	
$In_{0.78–0.69}Tl_{0.22-0.31}$	3.18–3.32	Tetragonal
$In_{0.69–0.62}Tl_{0.31–0.38}$	2.98–3.3	Cubic, f.c.
Ir_2La	0.48	C15
Ir_3La	2.32	$D10_2$
Ir_3La_7	2.24	$D10_2$
Ir_5La	2.13	

Type II Superconducting Compounds:
Critical Temperature and Crystal Structure Data (Continued)

Compound	T_c, K	Crystal Structure
Ir_2Lu	2.47	C15
Ir_3Lu	2.89	C15
IrMo	< 1.0	A3
$IrMo_3$	8.8	A15
$IrMo_3$	6.8	D8b
$IrNb_3$	1.9	A15
$Ir_{0.4}Nb_{0.6}$	9.8	$D8_b$
$Ir_{0.37}Nb_{0.63}$	2.32	$D8_b$
IrNb	7.9	$D8_b$
$Ir_{0.2}Nb_3Rh_{0.98}$	2.43	A15
$Ir_{0.05}Nb_3Rh_{0.95}$	2.38	A15
$Ir_{0.287}O_{0.14}Ti_{0.573}$	5.5	$E9_3$
$Ir_{0.265}O_{0.035}Ti_{0.65}$	2.30	$E9_3$
Ir_xOs_{1-x}	0.3–0.98 (max.)–0.6	
IrOsY	2.6	C15
$Ir_{1.5}Os_{0.5}$	2.4	C14
Ir_2Sc	2.07	C15
$Ir_{2.5}SC$	2.46	C15
$IrSn_2$	0.65–0.78	C1
Ir_2Sr	5.70	C15
$Ir_{0.5}Te_{0.5}$	~3	
$IrTe_3$	1.18	C2
IrTh	<0.37	B_f
Ir_2Th	6.50	C15

Type II Superconducting Compounds:
Critical Temperature and Crystal Structure Data (Continued)

Compound	T_c, K	Crystal Structure
Ir_3Th	4.71	
Ir_3Th_7	1.52	$D10_2$
Ir_5Th	3.93	$D2_d$
$IrTi_3$	5.40	A15
IrV_2	1.39	A15
IrW_3	3.82	
$Ir_{0.28}W_{0.72}$	4.49	
Ir_2Y	2.18, 1.38	C15
$Ir_{0.69}Y_{0.31}$	1.98, 1.44	C15
$Ir_{0.70}Y_{0.30}$	2.16	C15
Ir_2Y	1.09	C15
Ir_2Y_3	1.61	
Ir_xY_{1-x}	0.3–3.7	
Ir_2Zr	4.10	C15
$Ir_{0.1}Zr_{0.9}$	5.5	A3
$K_{0.27-0.031}O_3W$	0.50	Hexagonal
$K_{0.40-0.57}O_3W$	1.5	Tetragonal
$La_{0.55}Lu_{0.45}$	2.2	Hexagonal, La type
$La_{0.8}Lu_{0.2}$	3.4	Hexagonal, La type
$LaMg_2$	1.05	C15
LaN	1.35	
$LaOs_2$	6.5	C15
$LaPt_2$	0.46	C15
$La_{0.28}Pt_{0.72}$	0.54	

Type II Superconducting Compounds:
Critical Temperature and Crystal Structure Data (Continued)

Compound	T_c, K	Crystal Structure
$LaRh_3$	2.60	
$LaRh_5$	1.62	
La_7Rh_3	2.58	$D10_2$
$LaRu_2$	1.63	C15
La_3S_4	6.5	$D7_3$
La_3Se_4	8.6	$D7_3$
$LaSi_2$	2.3	C_c
La_xY_{1-x}	1.7–5.4	
LaZn	1.04	B2
LiPb	7.2	
$LuOs_2$	3.49	C14
$Lu_{0.275}Rh_{0.725}$	1.27	C15
$LuRh_5$	0.49	
$LuRu_2$	0.86	C14
$Mg_{\sim0.047}Tl_{\sim0.53}$	2.75	B2
Mg_2Nb	5.6	
Mn_xTi_{1-x}	2.3 max.	Mn in α-Ti
Mn_xTi_{1-x}	1.1–3.0	Mn in β-Ti
MnU_6	2.32	$D2_c$
MoN	12	Hexagonal
Mo_2N	5.0	Cubic f c.
Mo_xNb_{1-x}	0.016–9.2	
Mo_3Os	7.2	A15
$Mo_{0.62}Os_{0.38}$	5.65	$D8_b$

Type II Superconducting Compounds:
Critical Temperature and Crystal Structure Data (Continued)

Compound	T_c, K	Crystal Structure
Mo_3P	5.31	DO_e
$Mo_{0.5}Pd_{0.05}$	3.52	A3
Mo_3Re	10.0	
Mo_xRe_{1-x}	1.2–12.2	
$MoRe_3$	9.25–9.89	A12
$Mo_{0.42}Re_{0.58}$	6.35	$D8_b$
$Mo_{0.52}Re_{0.48}$	11.1	
$Mo_{0.57}Re_{0.43}$	14.0	
$Mo_{\sim0.60}Re_{0.395}$	10.6	
$MoRh$	1.97	A3
Mo_xRh_{1-x}	1.5–8.2	Cubic, b.c.
$MoRu$	9.5–10.5	A3
$Mo_{0.61}Ru_{0.39}$	7.18	$D8_b$
$Mo_{0.2}Ru_{0.8}$	1.66	A3
Mo_3Sb_4	2.1	
Mo_3Si	1.30	A15
$MoSi_{0.7}$	1.34	
Mo_xSiV_{3-x}	4.54–16.0	A15
Mo_xTc_{1-x}	10.8–15.8	
$Mo_{0.16}Ti_{0.84}$	4.18, 4.25	
$Mo_{0.913}Ti_{0.087}$	2.95	
$Mo_{0.04}Ti_{0.96}$	2.0	Cubic
$Mo_{0.25}Ti_{0.95}$	1.8	
Mo_xU_{1-x}	0.7–2.1	

Type II Superconducting Compounds:
Critical Temperature and Crystal Structure Data (Continued)

Compound	T_c, K	Crystal Structure
Mo_xV_{1-x}	$0 - \sim 5.3$	
Mo_2Zr	4.27–4.75	C15
NNb (whiskers)	10–14.5	
NNb (diffusion wires)	16.10	
NNb (film)	6.9	B1
$N_{0.988}Nb$	14.9	B1
$N_{0.824-0.988}Nb$	14.4–15.3	B1
$N_{0.70-0.795}Nb$	11.3–12.9	Cubic and tetragonal
NNb_xO_y	13.5–17.0	B1
NNb_xO_y	6.0–11	
$N_{100-42w/o}Nb_{0-58w/o}Ti$	15–16.8	
$N_{100-75w/o}Nb_{0-25w/o}Zr$	12.5–16.35	
NNb_xZr_{1-x}	9.8–13.8	B1
$N_{0.93}Nb_{0.85}Zr_{0.15}$	13.8	B1
$N_xO_yTi_z$	2.9–5.6	Cubic
$N_xO_yV_z$	5.8–8.2	Cubic
$N_{0.34}Re$	4–5	Cubic, f.c.
NTa (extrapolated value)	12–14	B1
NTa (film)	4.84	B1
$N_{0.6-0.987}Ti$	< 1.17–5.8	B1
$N_{0.82-0.99}V$	2.9–7.9	B1
NZr	9.8	B1
$N_{0.906-0.984}Zr$	3.0–9.5	B1
$Na_{0.28-0.35}O_3W$	0.56	Tetragonal

Type II Superconducting Compounds:
Critical Temperature and Crystal Structure Data (Continued)

Compound	T_c, K	Crystal Structure
$Na_{0.28}Pb_{0.72}$	7.2	
NbO	1.25	
$NbOs_2$	2.52	A12
Nb_3Os	1.05	A15
$Nb_{0.6}Os_{0.4}$	1.89, 1.78	$D8_b$
$Nb_3Os_{0.02-0.10}Rh_{0.98-0.90}$	2.42–2.30	A15
$Nb_{0.6}Pd_{0.4}$	1.60	$D8_f$ plus cubic
$Nb_3Pd_{0.02-0.10}Rh_{0.98-0.90}$	2.49–2.55	A15
$Nb_{0.62}Pt_{0.38}$	4.21	$D8_b$
Nb_3Pt	10.9	A15
$NbsPt_3$	3.73	$D8_b$
$Nb_3Pt_{0.02-0.98}Rh_{0.98-0.02}$	2.52–9.6	A15
$Nb_{0.38-0.18}Re_{0.62-0.82}$	2.43–9.70	A12
Nb_3Rh	2.64	A15
$Nb_{0.60}Rh_{0.40}$	4.21	$D8_b$ plus other
$Nb_3Rh_{0.98-0.90}Ru_{0.02-0.10}$	2.42–2.44	A15
Nb_xRu_{1-x}	1.2–4.8	
NbS_2	6.163	Hexagonal, $NbSe_2$ type
NbS_2	5.0–5.5	Hexagonal, three-layer type
$Nb_3Sb_{0-0.7}Sn_{1-0.3}$	6.8–18	A15
$NbSe_2$	5.15–5.62	Hexagonal, NbS_2 type
$Nb_{1-0.05}Se_2$	2.2–7.0	Hexagonal, NbS_2 type
Nb_3Si	1.5	$L1_2$
Nb_3SiSnV_3	4.0	

Type II Superconducting Compounds:
Critical Temperature and Crystal Structure Data (Continued)

Compound	T_c, K	Crystal Structure
Nb_3Sn	18.05	A15
$Nb_{0.8}Sn_{0.2}$	18.18, 18.5	A15
Nb_xSn_{1-x} (film)	2.6–18.5	
$NbSn_2$	2.60	Orthorhombic
Nb_3Sn_2	16.6	Tetragonal
$NbSnTa_2$	10.8	A15
Nb_2SnTa	16.4	A15
$Nb_{2.5}SnTa0.5$	17.6	A15
$Nb_{2.75}SnTa_{0.25}$	17.8	A15
$Nb_{3x}SnTa_{3(1-x)}$	6.0–18.0	
$NbSnTaV$	6.2	A15
$Nb_2SnTa_{0.5}V_{0.5}$	12.2	A15
$NbSnV_2$	5.5	A15
Nb_2SnV	9.8	A15
$Nb_{2.5}SnV_{0.5}$	14.2	A15
Nb_xTa_{1-x}	4.4–9.2	A2
$NbTc_3$	10.5	A12
Nb_xTi_{1-x}	0.6–9.8	
$Nh_{0.6}Ti_{0.4}$	9.8	
Nb_xU_{1-x}	1.95 max.	
$Nb_{0.88}V_{0.12}$	5.7	A2
$Nb_{0.75}Zr_{0.25}$	10.8	
$Nb_{0.66}Zr_{0.33}$	10.8	
$Ni_{0.3}Th_{0.7}$	1.98	$D10_2$

Type II Superconducting Compounds:
Critical Temperature and Crystal Structure Data (Continued)

Compound	T_c, K	Crystal Structure
$NiZr_2$	1.52	
$Ni_{0.1}Zr_{0.9}$	1.5	A3
$O_3Rb_{0.27-0.29}W$	1.98	Hexagonal
O_3SrTi (n =1.7-12.0 x 10^{19})	0.12–0.37	
O_3SrTi (n = 10^{18}–10^{21})	0.05-0.47	
O_3SrTi (n = ~10^{20})	0.47	
OTi	0.58	
$O_3Sr_{0.08}W$	2.4	Hexagonal
$O_3Tl_{0.30}W$	2.0 – ~2.14	Hexagonal
OV_3Zr_3	7.5	E93
OW_3 (film)	3.35, 1.1	A15
OsReY	2.0	C14
Os_2Sc	4.6	C14
OsTa	1.95	A12
Os_3Th_7	1.51	D102
Os_xW_{1-x}	0.9–4.1	
OsW_3	~ 3	
Os_2Y	4.7	C14
Os_2Zr	3.0	C14
Os_xZr_{1-x}	1.50–5.6	
PPb	7.8	
$PPd_{3.0-3.2}$	< 0.35–0.7	DO_{11}
P_3Pd_7 (high temperature)	1.0	Rhombohedral
P_3Pd_7 (low temp.)	0.70	Complex

Type II Superconducting Compounds:
Critical Temperature and Crystal Structure Data (Continued)

Compound	T_c, K	Crystal Structure
PRh	1.22	
PRh$_2$	1.3	Cl
PW$_3$	2.26	DO
Pb$_2$Pd	2.95	C16
Pb$_4$Pt	2.80	Related to C16
Pb$_2$Rh	2.66	C16
PbSb	6.6	
PbTe (plus 0.1 *w/o* Pb)	5.19	
PbTe (plus 0.1 *w/o* Tl)	5.24–5.27	
PbTl$_{0.27}$	6.43	
PbTl$_{0.17}$	6.73	
PbTl$_{0.12}$	6.88	
PbTl$_{0.075}$	6.98	
PbTl$_{0.04}$	7.06	
Pb$_{1-0.26}$Tl$_{0-0.74}$	7.20–3.68	
PbTl$_2$	3.75–4.1	
Pb$_3$Zr$_5$	4.60	D88
PbZr$_3$	0.76	A15
Pd$_{0.9}$Pt$_{0.1}$Te$_2$	1.65	C6
Pd$_{0.05}$Ru$_{0.05}$Zr$_{0.9}$	~ 9	
PdSb$_2$	1.25	C2
PdSb	1.50	B8
PdSbSe	1.0	C2

Type II Superconducting Compounds:
Critical Temperature and Crystal Structure Data (Continued)

Compound	T_c, K	Crystal Structure
PdSbTe	1.2	C2
Pd_4Se	0.42	Tetragonal
$Pd_{6-7}Se$	0.66	Like Pd_4Te
$Pd_{2.3}Se$	2.3	
Pd_xSe_{1-x}	2.5 max.	
PdSi	0.93	B31
PdSn	0.41	B31
$PdSn_2$	3.34	
Pd_2Sn	0.41	C37
Pd_3Sn_2	0.47–0.64	$B8_2$
PdTe	2.3, 3.85	B8
$PdTe_{1.02-1.03}$	2.56–1.88	B8
$PdTe_2$	1.69	C6
$PdTe_{2.1}$	1.89	C6
$PdTe_{2.3}$	1.85	C6
$Pd_{1.1}Te$	4.07	$B8_1$
$PdTh_2$	0.85	C16
$Pd_{0.1}Zr_{0.9}$	7.5	A3
PtSb	2.1	B8
PtSi	0.88	B31
PtSn	0.37	$B8_1$
PtTe	0.59	Orthorhombic
PtTh	0.44	B1
Pt_3Th_7	0.98	$D10_2$

Type II Superconducting Compounds:
Critical Temperature and Crystal Structure Data (Continued)

Compound	T_c, K	Crystal Structure
Pt_5Th	3.13	
$PtTi_3$	0.58	A15
$Pt_{0.02}U_{0.98}$	0.87	α-phase
$PtV_{2.5}$	1.36	A15
PtV_3	2.87–3.20	A15
$PtV_{3.5}$	1.26	A15
$Pt_{0.5}W_{0.5}$	1.45	A1
Pt_xW_{1-x}	0.4–2.7	
Pt_2Y_3	0.90	
Pt_2Y	1.57, 1.70	C15
Pt_3Y_7	0.82	$D10_2$
$PtZr$	3.0	A3
$Re_{0.64}Ta_{0.36}$	1.46	A12
$Re_{24}Ti_5$	6.60	A12
Re_xTi_{1-x}	6.6 max.	
$Re_{0.76}V_{0.24}$	4.52	$D8_b$
$Re_{0.92}V_{0.08}$	6.8	A3
$Re_{0.6}W_{0.4}$	6.0	
$Re_{0.5}W_{0.5}$	5.12	D8
Re_2Y	1.83	C14
Re_2Zr	5.9	C14
Re_6Zr	7.40	A12
$Rh_{17}S_{15}$	5.8	Cubic
$Rh_{0.24}Sc_{0.76}$	0.88, 0.92	

Type II Superconducting Compounds:
Critical Temperature and Crystal Structure Data (Continued)

Compound	T_c, K	Crystal Structure
Rh_xSe_{1-x}	6.0 max.	
Rh_2Sr	6.2	C15
$Rh_{0.4}Ta_{0.6}$	2.35	D8$_b$
$RhTe_2$	1.51	C2
$Rh_{0.67}Te_{0.33}$	0.49	
Rh_xTe_{1-x}	1.51 max.	
RhTh	0.36	B$_f$
Rh_3Th_7	2.15	D10$_2$
Rh_5Th	1.07	
Rh_xTi_{1-x}	2.25	3.95
$Rh_{0.02}U_{0.98}$	0.96	
RhV_3	0.38	A15
RhW	~3.4	A3
RhY_3	0.65	
Rh_2Y_3	1.48	
Rh_3Y	1.07	C15
Rh_5Y	0.56	
$RhZr_2$	10.8	C16
$Rh_{0.005}Zr$ (annealed)	5.8	
$Rh_{0-0.45}Zr_{1-0.55}$	2.1–10.8	
$Rh_{0.1}Zr_{0.9}$	9.0	Hexagonal, c.p.
Ru_2Sc	1.67	C14
Ru_2Th	3.56	C15
RuTi	1.07	B2

Type II Superconducting Compounds:
Critical Temperature and Crystal Structure Data (Continued)

Compound	T_c, K	Crystal Structure
$Ru_{0.05}Ti_{0.95}$	2.5	
$Ru_{0.1}Ti_{0.9}$	3.5	
$Ru_xTi_{0.6}V_y$	6.6 max.	
$Ru_{0.45}V_{0.55}$	4.0	B2
RuW	7.5	A3
Ru_2Y	1.52	C14
Ru_2Zr	1.84	C14
$Ru_{0.1}Zr_{0.9}$	5.7	A3
$SbSn$	1.30–1.42	B1 or distorted
$SbSn$	1.42–2.37	B1
$SbTi_3$	5.8	A15
Sb_2Tl_7	5.2	
$Sb_{0.01-0.03}V_{0.99-0.97}$	3.76–2.63	A2
SbV_3	0.80	A15
Si_2Th	3.2	C_c, α-phase
Si_2Th	2.4	C32, β-phase
SiV_3	17.1	A15
$Si_{0.9}V_3Al_{0.1}$	14.05	A15
$Si_{0.9}V_3B_{0.1}$	15.8	A15
$Si_{0.9}V_3C_{0.1}$	16.4	A15
$SiV_{2.7}Cr_{0.3}$	11.3	A15
$Si_{0.9}V_3Ge_{0.1}$	14.0	A15
$SiV_{2.7}Mo_{0.3}$	11.7	A15
$SiV_{2.7}Nb_{0.3}$	12.8	A15

Type II Superconducting Compounds:
Critical Temperature and Crystal Structure Data (Continued)

Compound	T_c, K	Crystal Structure
$SiV_{2.7}Ru_{0.3}$	2.9	A15
$SiV_{2.7}Ti_{0.3}$	10.9	A15
$SiV_{2.7}Zr_{0.3}$	13.2	A15
Si_2W_3	2.8, 2.84	
$Sn_{0.174-0.104}Ta_{0.826-0.896}$	6.5–<4.2	A15
$SnTa_3$	8.35	A15, highly ordered
$SnTa_3$	6.2	A15, partially ordered
$SnTaV_2$	2.8	A15
$SnTa_2V$	3.7	A15
Sn_xTe_{1-x} ($n = 10.5$-20×10^{20})	0.07–0.22	B1
Sn_xTl_{1-x}	2.37–5.2	
SnV_3	3.8	A15
$Sn_{0.02-0.057}V_{0.98-0.943}$	2.87–~1.6	A2
$Ta_{0.225}Ti_{0.975}$	1.3	Hexagonal
$Ta_{0.5}Ti_{0.95}$	2.9	Hexagonal
$Ta_{0.05-0.75}V_{0.95-0.25}$	4.30–2.65	A2
$Ta_{0.8-1}W_{0.2-0}$	1.2–4.4	A2
$Tc_{0.1-0.4}W_{0.9-0.6}$	1.25–7.18	Cubic
$Tc_{0.50}W_{0.50}$	7.52	α plus σ
$Tc_{0.60}W_{0.40}$	7.88	σ plus α
Tc_6Zr	9.7	A12
$Th_{0-0.55}Y_{1-0.45}$	1.2–1.8	
$Ti_{0.70}V_{0.30}$	6.14	Cubic
Ti_xV_{1-x}	0.2–7.5	

Type II Superconducting Compounds:
Critical Temperature and Crystal Structure Data (Continued)

Compound	T_c, K	Crystal Structure
$Ti_{0.5}Zr_{0.5}$ (annealed)	1.23	
$Ti_{0.5}Zr_{0.5}$ (quenched)	2.0	
V_2Zr	8.80	C15
$V_{0.26}Zr_{0.74}$	~ 5.9	
W_2Zr	2.16	C15

n = number of normal carriers per cubic centimeter for semiconductor superconductors.

All compositions are denoted on an atomic basis, i.e., AB, AB_2, or AB_3 for compounds, unless noted. Solid solutions or odd compositions may be denoted as A_xB_{1-x} or A_zB. A series of three or more alloys is indicated as A_zB_{1-x} or by actual indication of the atomic fraction range, such as $A_{0-0.6}B_{1-0.4}$. The critical temperature of such a series of alloys is denoted by a range of values or possibly the maximum value.

The selection of the critical temperature from a transition in the effective permeability, or the change in resistance, or possibly the incremental changes in frequency observed by certain techniques is not often obvious from the literature. Most authors choose the mid-point of such curves as the probable critical temperature of the idealized material, while others will choose the highest temperature at which a deviation from the normal-state property is observed. In view of the previous discussion concerning the variability of the superconductive properties as a function of purity and other metallurgical aspects, it is recommended that appropriate literature be checked to determine the most probable critical temperature or critical field of a given alloy.

Data Compiled by J.S. Park from: *CRC Handbook of Material Science,* Vol. 3, Charles T. Lynch, Ed., CRC Press, Cleveland, (1974)

HIGH TEMPERATURE SUPERCONDUCTING COMPOUNDS: CRITICAL TEMPERATURE AND CRYSTAL STRUCTURE DATA

Compound	T_c, K	Crystal Structure
La-Ba -Cu-O System[a]		
$Ba_xLa_{5-x}Cu_5O_{5(3-y)}$	30-35	D_{4h}
$(La_{0.9}Ba_{0.1})_2CuO_{4-y}$, at 1Gpa	52.5	D_{4h}
$(La_{1-x}Sr_x)_2CuO_4$	34-37	tetragonal
$(La_{0.9}Sr_{0.1})_2CuO_4$	30-36	D_{4h}
$(La_{0.925}Sr_{0.075})_2CuO_4$	33-38.5	tetragonal
$(La_{0.925}Ba_{0.075})_2CuO_4$	~ 30	tetragonal
X-Ba-Cu-O System[a]		
$YBa_2Cu_3O_{7-x}$	93	D_{2v} (a=3.820Å, b=3.893Å, c=11.688Å
$LaBa_2Cu_3O_{7-x}$	59.2	orthohombic
$NdBa_2Cu_3O_{7-x}$	78.3	orthohombic
$SmBa_2Cu_3O_{7-x}$	88.6	orthohombic
$EuBa_2Cu_3O_{7-x}$	91.1	orthohombic
$GdBa_2Cu_3O_{7-x}$	90.9	orthohombic
$DyBa_2Cu_3O_{7-x}$	91.8	orthohombic
$HoBa_2Cu_3O_{7-x}$	91.1	orthohombic
$ErBa_2Cu_3O_{7-x}$	90.7	orthohombic
$TmBa_2Cu_3O_{7-x}$	90.5	orthohombic
$YbBa_2Cu_3O_{7-x}$	89.3	orthohombic
$LuBa_2Cu_3O_{7-x}$	72.6	orthohombic

High Temperature Superconducting Compounds:
Critical Temperature and Crystal Structure Data (Continued)

Compound	T_c, K	Crystal Structure
$Y_{0.75}Sc_{0.25}Ba_2Cu_3O_{7-x}$	91	orthohombic
$Y_{0.5}Sc_{0.5}Ba_2Cu_3O_{7-x}$	90	orthohombic
$Y_{0.5}La_{0.5}Ba_2Cu_3O_{7-x}$	87	orthohombic
$Y_{0.25}Eu_{0.75}Ba_2Cu_3O_{7-x}$	95	orthohombic
$Y_{0.1}Eu_{0.9}Ba_2Cu_3O_{7-x}$	94.5	orthohombic
$Pr_{0.1}Eu_{0.9}Ba_2Cu_3O_{7-x}$	82	orthohombic
$Eu_{0.75}Sc_{0.25}Ba_2Cu_3O_{7-x}$	93	orthohombic
$Y(Ba_{0.75}Sr_{0.25})_2Cu_3O_{7-x}$	87	orthohombic
$LaBa_2Cu_3O_{7-x}$	60	orthohombic
$La(Ba_{0.5}Ca_{0.5})_2Cu_3O_{7-x}$	79	orthohombic

Bi-Sr-Ca-Cu-O System[a,b]

Compound	T_c, K	Crystal Structure
$Bi_2Sr_2CuO_6$	10-22	tetragonal
$Bi_2Sr_2CaCuO_8$	~ 85	D_{4h}
$Bi_{2-x}Pb_xSr_2Ca_{1-x}Y_xCu_2O_8$	~ 85	tetragonal
$Bi_{2-x}Pb_xSr_2Ca_2Cu_3O_{10}$	~ 110	tetragonal

Tl-Ba-Ca-Cu-O System[a,b]

Compound	T_c, K	Crystal Structure
$TlBa_2CaCu_2O_7$	~ 60	tetragonal
$TlBa_2Ca_2Cu_3O_9$	~ 120	tetragonal
$TlBa_2Ca_3Cu_4O_{11}$	~ 108	tetragonal
$TlSr_2CaCu_2O_7$	~ 50	tetragonal

High Temperature Superconducting Compounds:
Critical Temperature and Crystal Structure Data (Continued)

Compound	T_c, K	Crystal Structure
$Tl_{0.5}Pb_{0.5}Sr_2CaCu_2O_7$	~ 85	tetragonal
$Tl_{0.5}Pb_{0.5}Sr_2Ca_2Cu_3O_9$	~ 120	tetragonal
$Tl_2Ba_2CuO_6$	~ 30	tetragonal
$Tl_2Ba_2CaCu_2O_8$	~105	D_{4h}
$Tl_2Ba_2Ca_2Cu_3O_{10}$	~125	tetragonal
$Tl_2Ba_2Ca_3Cu_4O_{12}$	~115	tetragonal

Nd-Ce-Sr-Cu-O System[c]

Compound	T_c, K	Crystal Structure
$Nd_{2-x-z}Ce_xSr_zCuO_{4-y}$	~ 30	tetragonal
$Nd_2Ce_{0.5}Sr_{0.5}Cu_{1.2}O_y$	~ 20	tetragonal
$Nd_{2-x}Ce_xCuO_y$	~ 24	tetragonal

Compiled by J.S. Park.

[a] *Source:* data from *Copper Oxide Superconductors*, edited by Charles P. Poole, Jr. *et al*, John Wiley & Sons, New York, 1988.

[b] *Source:* data from "Structural Problems and Non Stoichiometry in Thallium, Bismuth, and Lead Oxides," B. Raveau, et al in *High Temperature Superconductors: Relationships Between Properties, Structure, and Solid State Chemistry* edited by James D. Jorgensen *et al*, Materials Research Society Symposium Proceedings, Vol. 156, Pittsburgh, PA, 1989.

[c] *Source:* data from "A New Class of Oxide Superconductor (Nd, Ce, Sr)$_2$CuO$_4$" E. Takayama-Muromachi, in *High Temperature Superconductors: Relationships Between Properties, Structure, and Solid State Chemistry* edited by James D. Jorgensen *et al*, Materials Research Society Symposium Proceedings, Vol. 156, Pittsburgh, PA, 1989.

CRYSTAL STRUCTURE TYPES

"Structk-turbericht" type	Example	Class
A1	Cu	Cubic, fc.
A2	W	Cubic b c.
A3	Mg	Hexagonal, close packed
A4	Diamond	Cubic, f.c.
A5	White Sn	Tetragonal, b.c.
A6	In	Tetragonal, b.c. (f.c. cell usually used)
A7	As	Rhombohedral
A8	Se	Trigonal
A10	Hg	Rhombohedral
A12	α-Mn	Cubic, b.c.
A13	β-Mn	Cubic
A15	"β-W" (WO_3)	Cubic
B1	NaCl	Cubic, fc.
B2	CsCl	Cubic
B3	ZnS	Cubic
B4	ZnS	Hexagonal
$B8_1$	NiAs	Hexagonal
$B8_2$	Ni_2In	Hexagonal
B10	PbO	Tetragonal
B11	γ-CuTi	Tetragonal
B17	PtS	Tetragonal
B18	CuS	Hexagonal
B20	FeSi	Cubic
B27	FeB	Orthorhombic
B31	MnP	Orthorhombic
B32	NaTl	Cubic, f.c.
B34	PdS	Tetragonal
B_f	δ-CrB	Orthorhombic

Crystal Structure Types (Continued)

"Structk- turbericht" type	Example	Class
B_g	MoB	Tetragonal, b.c.
B_h	WC	Hexagonal
B_i	γ-MoC	Hexagonal
Cl	CaF_2	Cubic, f.c.
Cl_b	MgAgAs	Cubic, f.c.
C2	FeS_2	Cubic
C6	CdI_2	Trigonal
$C11_b$	$MoSi_2$	Tetragonal, b.c.
C12	$CaSi_2$	Rhombohedral
C14	$MgZn_2$	Hexagonal
C15	Cu2Mg	Cubic, f.c.
$C15_b$	$AuBe_5$	Cubic
C16	$CuAl_2$	Tetragonal, b.c.
C18	FeS_2	Orthorhombic
C22	Fe_2P	Trigonal
C23	$PbCl_2$	Orthorhombic
C32	AlB_2	Hexagonal
C36	$MgNi_2$	Hexagonal
C37	Co_2Si	Orthorhombic
C49	$ZrSi_2$	Orthorhombic
C54	$TiSi_2$	Orthorhombic
C_c	Si_2Th	Tetragonal, b.c.
DO_3	BiF_3	Cubic, f.c.
DO_{11}	Fe_3C	Orthorhombic
DO_{18}	Na_3As	Hexagonal
DO_{l9}	Ni_3Sn	Hexagonal
DO_{20}	$NiAl_3$	Orthorhombic
DO_{22}	$TiAl_3$	Tetragonal

Crystal Structure Types (Continued)

"Structk-turbericht" type	Example	Class
DO_e	Ni_3P	Tetragonal, b.c.
$D1_3$	Al_4Ba	Tetragonal, b.c.
$D1_2$	$PtSn_4$	Orthorhombic
$D2_1$	CaB_6	Cubic
$D2_c$	$MnU6$	Tetragonal, b.c.
$D2_d$	$CaZn_5$	Hexagonal
$D5_2$	La_2O_3	Trigonal
$D5_8$	Sb_2S_3	Orthorhombic
$D7_3$	Th_3P_4	Cubic, b.c.
$D7c$	Ta_3B_4	Orthorhombic
$D8_1$	Fe_3Zn_{10}	Cubic, b.c.
$D8_2$	Cu_5Zn_8	Cubic, b.c.
$D8_3$	Cu_5Al_4	Cubic
$D8_8$	Mn_5Si_3	Hexagonal
$D8_b$	$CrFe$	Tetragonal
$D8_i$	Mo_2B_5	Rhombohedral
$D10_2$	Fe_3Th_7	Hexagonal
$E2_1$	$CaTiO_3$	Cubic
$E9_3$	Fe_3W_3C	Cubic, f.c.
$L1_0$	$CuAu$	Tetragonal
$L1_2$	Cu_3Au	Cubic
$L2_b$	ThH_2	Tetragonal, b.c.
$L3$	Fe_2N	Hexagonal

Source: CRC Handbook of Materials Science, Charles T. Lynch, Ed., CRC Press, Cleveland, (1974)

CRITICAL TEMPERATURE DATA FOR TYPE II SUPERCONDUCTING COMPOUNDS

Compound	T_c, K	Compound	T_c, K
Ag_2F	0.066	PdSn	0.41
CCs_x	0.135	Pd_2Sn	0.41
$Ag_7BF_4O_8$	0.15	Pd_4Se	0.42
CRb_x (gold)	0.151	PtTh	0.44
Bi_2Pt	0.155	Cr_3Ir	0.45
Sn_xTe_{1-x}	0.22	$AsPd_5$	0.46
Ag_7FO_8	0.3	$CeCo_{1.67}Ni_{0.33}$	0.46
$Co_{0.28}Y_{0.072}$	0.34	$LaPt_2$	0.46
CoY_3	0.34	O_3SrTi	0.47
$CoLu_3$	0.35	$CeCo_{1.67}Rh_{0.33}$	0.47
$Co_{\sim0.3}Sc_{\sim0.7}$	0.35	O_3SrTi	0.47
AuLu	0.35	$B_{12}Lu$	0.48
RhTh	0.36	Ir_2La	0.48
O_3SrTi	0.37	BiCo	0.49
PtSn	0.37	$LuRh_5$	0.49
IrTh	0.37	$Rh_{0.67}Te_{0.33}$	0.49
Au_5Ca	0.38	$Cr_{0.8-0.6}Rh_{0.2-0.4}$	0.5
RhV_3	0.38	$K_{0.27-0.031}O_3W$	0.5
AlOs	0.39	BMo	0.5 (extrapolated)
$B_{12}Sc$	0.39	$La_{0.28}Pt_{0.72}$	0.54
C_8K	0.39	$Ba_xO_3Sr_{1-x}Ti$	0.55
$Cr_{0.6}Ir_{0.4}$	0.4	$Ca_xO_3Sr_{1-x}Ti$	0.55
GePt	0.4	Al_2Pt	0.55
Ge_xTe_{1-x}	0.41	C_8K (gold)	0.55

Critical Temperature Data for Type II
Superconducting Compounds (Continued)

Compound	T_c, K	Compound	T_c, K
CK (excess K)	0.55	P_3Pd_7 (low temp.)	0.7
$AsRh_{1.4-1.6}$	0.56	CoTi	0.71
$CoHf_2$	0.56	$AlZr_3$	0.73
$Na_{0.28-0.35}O_3W$	0.56	AuV_3	0.74
Rh5Y	0.56	B_6Th	0.74
OTi	0.58	$Cr_{0.0175}U_{0.9825}$	0.75
$AlMo_3$	0.58	Al_3Th	0.75
AsRh	0.58	$Cr_{0.7}Ir_{0.3}$	0.76
$AuSb_2$	0.58	$GaMo_3$	0.76
$PtTi_3$	0.58	$PbZr_3$	0.76
$Cr_{0.65}Ir_{0.35}$	0.59	$IrSn_2$	0.78
PtTe	0.59	SbV_3	0.8
AuIn	0.6	Pt_3Y_7	0.82
$AsPd_2$ (low-T phase)	0.6	$Cr_{0.72}Ir_{0.28}$	0.83
$As_{0.5}InTe_{0.5}$	0.62	Al_3Mg_2	0.84
Pd_3Sn_2	0.64	$CeCO_2$	0.84
$Ag_{0.438}Hg_{0.562}$	0.64	$Ag_xAl_yAn_{1-x-y}$	0.845
RhY_3	0.65	Ag_xZn_{1-x}	0.845
$Pd_{6-7}Se$	0.66	Al_xZn_{1-x}	0.845
Al (1 to 21 katm)	0.687	Cu_xZn_{1-x}	0.845
$PPd_{3.0-3.2}$	0.7	$Au_xZn(1-x)$	0.845
$AlAu_4$	0.7	Ag_4Ge	0.85
Au_5Ba	0.7	$PdTh_2$	0.85
InPd	0.7	$LuRu_2$	0.86

Critical Temperature Data for Type II Superconducting Compounds (Continued)

Compound	T_c, K	Compound	T_c, K
$Pt_{0.02}U_{0.98}$	0.87	Rh_5Th	1.07
PtSi	0.88	Rh_3Y	1.07
Bi_3Zn	0.9	RuTi	1.07
Pt_2Y_3	0.9	Ir_2Y	1.09
$Ag_7F_{0.25}N_{0.75}O_{10.25}$	0.90	Au_5Sn	1.1
$Rh_{0.24}Sc_{0.76}$	0.92	$Cs_{0.32}O_3W$	1.12
$AuZr_3$	0.92	$As_{0.33}InTe_{0.67}$	1.15
PdSi	0.93	BiPtTe	1.15
$Rh_{0.02}U_{0.98}$	0.96	$IrTe_3$	1.18
Pt_3Th_7	0.98	As_3Sn_4	1.19
Ir_xOs_{1-x}	0.98(max.)	AgLa (9.5 kbar)	1.2
BiPdSe	1	AuGa	1.2
CW	1	$AuNb_3$	1.2
$Fe_{0.1}Zr_{0.9}$	1	BiPdTe	1.2
P_3Pd_7 (high temperature)	1	PdSbTe	1.2
PdSbSe	1	Hg_2K	1.2
IrMo	1	$AuZn_3$	1.21
$Bi_{4-9}Mg$	1.0	BiPt	1.21
Bi_3Te	1.0	PRh	1.22
$Ag_{0.2}In_{0.8}Te$	1.00	$Ti_{0.5}Zr_{0.5}$ (annealed)	1.23
Ag_7NO_{11}	1.04	In_3Te_4	1.25
LaZn	1.04	AuSn	1.25
$LaMg_2$	1.05	NbO	1.25
Nb_3Os	1.05	$PdSb_2$	1.25

Critical Temperature Data for Type II
Superconducting Compounds (Continued)

Compound	T_c, K	Compound	T_c, K
$PtV_{3.5}$	1.26	$Re_{0.64}Ta_{0.36}$	1.46
$Lu_{0.275}Rh_{0.725}$	1.27	Rh_2Y_3	1.48
Al_2NNb3	1.3	$K_{0.40-0.57}O_3W$	1.5
		Nb_3Si	1.5
$C_{0.44}Mo_{0.56}$	1.3	$Ni_{0.1}Zr_{0.9}$	1.5
Ge_2Sc	1.3	$PdSb$	1.5
PRh_2	1.3	$Hf_{0.99-0.96}Rh_{0.01-0.04}$	1.51
$Ta_{0.225}Ti_{0.975}$	1.3	$Bi_{0.3}Zr_{0.7}$	1.51
Mo_3Si	1.3	Os_3Th_7	1.51
$BiCu$	1.33	$RhTe_2$	1.51
$MoSi_{0.7}$	1.34	Rh_xTe_{1-x}	1.51max.
LaN	1.35	Ir_3Th_7	1.52
$PtV_{2.5}$	1.36	$NiZr_2$	1.52
Ga_3Zr	1.38	Ru_2Y	1.52
$As_{0.50}Ni_{0.06}Pd_{0.44}$	1.39	CuS_2	1.53
IrV_2	1.39	Ce_xPt_{1-x}	1.55
$CoSi_2$	1.4	AlN	1.55
$Co_xFe_{1-x}Si_2$	1.4max.	$Nb_{0.6}Pd_{0.4}$	1.6
$Co_xNi_{1-x}Si_2$	1.4max.	$Sn_{0.02-0.057}V_{0.98-0.943}$	1.6
$SbSn$	1.42	Ir_2Y_3	1.61
$GeMo_3$	1.43	Au_3Te_5	1.62
$BiPtSe$	1.45	CuS	1.62
$Pt_{0.5}W_{0.5}$	1.45	Hg_2Na	1.62
$Al_{0.131}Cr_{0.088}V_{0.781}$	1.46	$LaRh_5$	1.62

Critical Temperature Data for Type II
Superconducting Compounds (Continued)

Compound	T_c, K	Compound	T_c, K
$Au_{0.40-0.92}Ge_{0.60-0.08}$	1.63	$PdTe_{2.3}$	1.85
$LaRu_2$	1.63	Fe_3Th_7	1.86
$Pd_{2.2}S$ (quenched)	1.63	$PdTe_{1.02-1.03}$	1.88
$Cr_{0.1-0.5}Ru_{0.9-0.5}$	1.65	$Ag_{0.1}In_{0.9}Te$	1.89
$Pd_{0.9}Pt_{0.1}Te_2$	1.65	$PdTe_{2.1}$	1.89
$Mo_{0.2}Ru_{0.8}$	1.66	$Nb_{0.6}Os_{0.4}$	1.89
Ru_2Sc	1.67	$BaO_{0.13}O_3W$	1.9
$PdTe_2$	1.69	$GeNb_2$	1.9
Pt_2Y	1.7	$IrNb_3$	1.9
CoU	1.7	$Co_{0.02-0.10}Nb_3Rh_{0.098-0.90}$	1.90
Hg_3Li	1.7	$Cd_{0.3-0.5}Hg_{0.7-0.5}$	1.92
$AsPd_2$ (high-temp phase)	1.7	$AuTl$	1.92
Bi_2Pd	1.7	$OsTa$	1.95
$GaPt$	1.74	Nb_xU_{1-x}	1.95max.
$AlGe_2$	1.75	$MoRh$	1.97
$Th_{0-0.55}Y_{1-0.45}$	1.8	$Ni_{0.3}Th_{0.7}$	1.98
$Mo_{0.25}Ti_{0.95}$	1.8	$O_3Rb_{0.27-0.29}W$	1.98
Au_2Bi	1.8	$Ir_{0.69}Y_{0.31}$	1.98
Ge_3Te_4	1.80	Bi_3Ca	2
$Al_{0.108}V_{0.897}$	1.82	$Mo_{0.04}Ti_{0.96}$	2
CeI_5	1.82	$OsReY$	2
Co_3Th_7	1.83	$Ti_{0.5}Zr_{0.5}$ (quenched)	2
Re_2Y	1.83	$CuSSe$	2.0
Ru_2Zr	1.84	$CuSeTe$	2.0

Critical Temperature Data for Type II
Superconducting Compounds (Continued)

Compound	T_c, K	Compound	T_c, K
Cr_2Ru	2.02	BiNa	2.25
BiRh	2.06	Bi_2Y_3	2.25
Ir_2Sc	2.07	Rh_xTi_{1-x}	2.25
Mo_xU_{1-x}	2.1	$AgTh_2$	2.26
$AlMo_6Pd$	2.1	PW_3	2.26
InSb	2.1	CoU_6	2.29
Mo_3Sb_4	2.1	$LaSi_2$	2.3
PtSb	2.1	$Pd_{2.3}Se$	2.3
$CTi_{0.5-0.7}W_{0.5-0.3}$	2.1	$Ir_{0.265}O_{0.035}Ti_{0.65}$	2.3
Ge_3Rh_5	2.12	Bi_2Ir	2.3
Ir_5La	2.13	Mn_xTi_{1-x}	2.3max.
$O_3Tl_{0.30}W$	2.14	$Nb_3Os_{0.02-0.10}Rh_{0.98-0.90}$	2.30
CdHg	2.15	Al (film)	2.31
$Cr_{0.40}Re_{0.060}$	2.15	$Ag_{0.94}Tl_{0.06}$	2.32
Rh_3Th_7	2.15	Ir_3La	2.32
$Ir_{0.70}Y_{0.30}$	2.16	$Ir_{0.37}Nb_{0.63}$	2.32
W_2Zr	2.16	MnU_6	2.32
Ir_2Y	2.18	$Rh_{0.4}Ta_{0.6}$	2.35
$BaO_{14}O_3W$	2.2	SbSn	2.37
Ge_2La	2.2	$Ir_{0.05}Nb_3Rh_{0.95}$	2.38
$BiRe_2$	2.2	$AgIn_2$	2.4
Bi_xCu_{1-x} (electrodeposited)	2.2	$Ge_{0.62}Y$	2.4
$La_{0.55}LU_{0.45}$	2.2	$Ir_{1.5}Os_{0.5}$	2.4
Ir_3La_7	2.24	$O_3Sr_{0.08}W$	2.4

Critical Temperature Data for Type II
Superconducting Compounds (Continued)

Compound	T_c, K	Compound	T_c, K
Si_2Th	2.4	$Ta_{0.05-0.75}V_{0.95-0.25}$	2.65
$CuSe_2$	2.43	Pb_2Rh	2.66
$Ir_{0.2}Nb_3Rh_{0.98}$	2.43	$Ag_{0.03}Tl_{0.97}$	2.67
$Nb_3Rh_{0.98-0.90}Ru_{0.02-0.10}$	2.44	$HfOs_2$	2.69
$Ir_{2.5}SC$	2.46	Pt_xW_{1-x}	2.7
$BiLi$	2.47	Bi_4Rh	2.7
Ir_2Lu	2.47	CW_2	2.74
$Co_{0.5}Rh_{0.5}Si_2$	2.5	$Mg_{\sim0.047}Tl_{\sim0.53}$	2.75
$Cr_{0.80}Os_{0.20}$	2.5	$In_{0.62}Tl_{0.38}$	2.76
$Ru_{0.05}Ti_{0.95}$	2.5	$AgBi_2$	2.78
Pd_xSe_{1-x}	2.5max.	$In_{0.11}O_3W$	2.8
$Be_{22}Mo$	2.51	Au_xZr_y	2.8
$NbOs_2$	2.52	$BiZr_3$	2.8
$Cd_{0.18}Tl_{0.82}$	2.54	$SnTaV_2$	2.8
$Nb_3Pd_{0.02-0.10}Rh_{0.98-0.90}$	2.55	Pb_4Pt	2.8
$B_{0.3}Ru_{0.7}$	2.58	Co_xTi_{1-x}	2.8max.
La_7Rh_3	2.58	$H_{0.12}Nb_{0.88}$	2.81
$AgTe_3$	2.6	Si_2W_3	2.84
$IrOsY$	2.6	Ir_3Lu	2.89
$LaRh_3$	2.6	$Ga_{0.07}Pt_{0.03}$	2.9
$NbSn_2$	2.6	$SiV_{2.7}Ru_{0.3}$	2.9
$Sb_{0.01-0.03}V_{0.99-0.97}$	2.63	$Ta_{0.5}Ti_{0.95}$	2.9
Nb_3Rh	2.64	$Mo_{0.913}Ti_{0.087}$	2.95
$AuBe$	2.64	Pb_2Pd	2.95

Critical Temperature Data for Type II
Superconducting Compounds (Continued)

Compound	T_c, K	Compound	T_c, K
$Cd_{0.97}Tl_{0.03}$	3	Cr_3Ru (annealed)	3.3
OsW_3	3	$In_{0.78-0.69}Tl_{0.22-0.31}$	3.32
$Ir_{0.5}Te_{0.5}$	3	$CeIr_3$	3.34
Os_2Zr	3	$PdSn_2$	3.34
$PtZr$	3	Al_5Re_{24}	3.35
Mn_xTi_{1-x}	3.0	OW_3 (film)	3.35
$BiNb$ (high P and T)	3.05	$Cd_{0.0075-0.05}In_{1-x}$	3.36
Hg_4Na	3.05	$In_{1-0.86}Mg_{0-0.14}$	3.363
$AuTh_2$	3.08	In_xTl_{1-x}	3.374
BHf	3.1	$Ca_{0.1}O_3W$	3.4
BW_2	3.1	BZr	3.4
Pt_5Th	3.13	$CHf_{0.5}Mo_{0.5}$	3.4
$AuPb_2$	3.15	$La_{0.8}Lu_{0.2}$	3.4
Hg_3K	3.18	RhW	3.4
Bi_3Rh	3.2	Hg_8K	3.42
Si_2Th	3.2	$CoTi_2$	3.44
Fe_xTi_{1-x}	3.2max.	$In_{0.82-1}Te$	3.45
PtV_3	3.20	$CuTh_2$	3.49
$In_{0.8}Tl_{0.2}$	3.223	$LuOs_2$	3.49
Al_2La	3.23	$Ru_{0.1}Ti_{0.9}$	3.5
CTa_2	3.26	$Mo_{0.5}Pd_{0.05}$	3.52
$H_{0.08}Nb_{0.12}$	3.26	Ga_5V_2	3.55
Hg_4K	3.27	Ru_2Th	3.56
$In_{0.69-0.62}Tl_{0.31-0.38}$	3.3	Bi_2K	3.58

Critical Temperature Data for Type II Superconducting Compounds (Continued)

Compound	T_c, K	Compound	T_c, K
BiK	3.6	Bi_3Sn	3.8
$As_{\sim 2}Sn_{\sim 3}$	3.6	BiSn	3.8
Cr_xTi_{1-x}	3.6max.	SnV_3	3.8
$Gd_xRu_2Th_{1-x}$	3.6max.	Ge_2Y	3.8
$Al_xTi_yV_{1-y}$	3.62	Co_xTi_{1-x}	3.8max.
$H_{0.04}Nb_{0.96}$	3.62	HgIn	3.81
CdSn	3.65	IrW_3	3.82
AsSn	3.65	PdTe	3.85
$Co_xRh_{1-x}Si_2$	3.65max.	$Ga_{0-1}Sn_{1-0}$ (annealed)	3.85
$Pb_{1-0.26}Tl_{0-0.74}$	3.68	CTc (excess C)	3.85
Ir_xY_{1-x}	3.7	$Bi_{0.019}In_{0.981}$	3.86
Ag_xSn_{1-x}	3.7	FeU_6	3.86
Cu_xSn_{1-x} (film)	3.7	Hg_5Tl_2	3.86
Bi_3Mo	3.7	$Fe_{0.05}Ni_{0.05}Zr_{0.9}$	3.9
Cu_xSn_{1-x}	3.7	$Co_{0.1}Zr_{0.9}$	3.9
$In_{1.000}Te_{1.002}$	3.7	Ir5Th	3.93
BiPd	3.7	Bi_2Ir (quenched)	3.96
$SnTa_2V$	3.7	$Bi_{0.4}Pd_{0.6}$	4
Fe_xTi_{1-x}	3.7max.	BTa	4
$NbsPt_3$	3.73	$BiPd_2$	4
$(InSb)_{0-0.7}Sn_{1-0.93}$	3.74	Nb_3SiSnV_3	4
$Cu_{0.15}In_{0.85}$ (film)	3.75	$Ru_{0.45}V_{0.55}$	4
Ag_xSn_{1-x} (film)	3.8	Bi_3Ni	4.06
$Au_xSn_{(1-x)}$ (film)	3.8	$Pd_{1.1}Te$	4.07

Critical Temperature Data for Type II
Superconducting Compounds (Continued)

Compound	T_c, K	Compound	T_c, K
Os_xW_{1-x}	4.1	BiNi	4.25
$CGaMo_2$	4.1	Bi_2Pd	4.25
$PbTl_2$	4.1	Bi_2Rb	4.25
$Bi_{0.34-0.48}In_{0.66-0.52}$	4.1	$CoLa_3$	4.28
$Be_{23}W$	4.1	$AuPb_2$ (film)	4.3
Bi_3In_5	4.1	$Ta_{0.8-1}W_{0.2-0}$	4.4
AsSn	4.1	$Bi_{0.26}Tl_{0.74}$	4.4
Ir_2Zr	4.1	$Cu_{0.04-0.09}In_{1-x}$	4.4
Hg_xTl_{1-x}	4.109	$AuPb_3$	4.4
$Be_{22}W$	4.12	$Ir_{0.28}W_{0.72}$	4.49
$Bi_{0.26}Tl_{0.74}$	4.15	$CHf_{0.6}Nb_{0.4}$	4.5
$Ga_{0.1}Sn_{1-0}$ (quenched)	4.18	$Re_{0.76}V_{0.24}$	4.52
Bi_xSn_y	4.18	Hg_xIn_{1-x}	4.55
$In_{0.98-0.91}Pb_{0.02-0.09}$	4.2	BRe_2	4.6
$Cd_{0.97}Pb_{0.03}$	4.2	$Ce_{0.005}La_{0.995}$	4.6
HgSn	4.2	Os_2Sc	4.6
$Hf_{0.75}Nb_{0.25}$	4.2	Pb_3Zr_5	4.6
$Sn_{0.174-0.104}Ta_{0.826-0.896}$	4.2	$Bi_{0.05}In_{0.95}$	4.65
Cr_xTi_{1-x}	4.2max.	$Gd_xOs_2Y_{1-x}$	4.7
$Nb_{0.62}Pt_{0.38}$	4.21	$B_{12}Y$	4.7
$Nb_{0.60}Rh_{0.40}$	4.21	GeIr	4.7
GaSb	4.24	Os_2Y	4.7
$Mo_{0.16}Ti_{0.84}$	4.25	Ir_3Th	4.71
$AuPb_3$ (film)	4.25	BMo_2	4.74

Critical Temperature Data for Type II
Superconducting Compounds (Continued)

Compound	T_c, K	Compound	T_c, K
Mo_2Zr	4.75	BeTc	5.21
Bi_2Cs	4.75	PbTe (plus 0.1 *w/o* Tl)[b]	5.27
Nb_xRu_{1-x}	4.8	Mo_xV_{1-x}	5.3
InSb	4.8	$Ce_xPr_{1-x}Ru_2$	5.3
$CHf_{0.5}Nb_{0.5}$	4.8	$Cu_{0-0.27}V$	5.3
$CTa_{0.4}Ti_{0.6}$	4.8	Mo_3P	5.31
$HfRe_2$	4.8	La_xY_{1-x}	5.4
NTa (film)	4.84	$Bi_{0.15-0.30}In_{0.85-0.70}$	5.4
$N_{0.34}Re$	5	$IrTi_3$	5.4
$CRe_{0.06}W$	5	Gd_xLa_{1-x}	5.5
Mo_2N	5	NbS_2	5.5
$CRe_{0.01-0.08}W$	5.0	In_3Sn	5.5
$In_{0.95-0.85}Pb_{0.05-0.15}$	5.05	$CHf_{0.3}Mo_{0.7}$	5.5
$Bi_{0.010}In_{0.90}$	5.05	$Ir_{0.287}O_{0.14}Ti_{0.573}$	5.5
CTa (film)	5.09	$Ir_{0.1}Zr_{0.9}$	5.5
$(InSb)_{0.95-0.10}Sn_{0.05-0.90}$	5.1	$NbSnV_2$	5.5
$Re_{0.5}W_{0.5}$	5.12	$AlLa_3$	5.57
PbTe (plus 0.1 *w/o* Ph)[b]	5.19	Os_xZr_{1-x}	5.6
Cr_xRe_{1-x}	5.2	$N_xO_yTi_z$	5.6
Sn_xTl_{1-x}	5.2	$CHf_{0.4}Nb_{0.6}$	5.6
$Ce_xGd_{1-x}Ru_2$	5.2	$Cr_{0.1}Ti_{0.3}V_{0.6}$	5.6
$Be_8Nb_5Zr_2$	5.2	Mg_2Nb	5.6
CW_2	5.2	$NbSe_2$	5.62
Sb_2Tl_7	5.2	Bi_3Sr	5.62

Critical Temperature Data for Type II
Superconducting Compounds (Continued)

Compound	T_c, K	Compound	T_c, K
$BiIn_2$	5.65	$Re_{0.6}W_{0.4}$	6
$Mo_{0.62}Os_{0.38}$	5.65	$CHf_{0.25}Mo_{0.75}$	6
$BaBi_3$	5.69	Rh_xSe_{1-x}	6.0max.
B_6La	5.7	GeV_3	6.01
$Nb_{0.88}V_{0.12}$	5.7	$CHf_{0.7}Nb_{0.3}$	6.1
$Ru_{0.1}Zr_{0.9}$	5.7	$Ti_{0.70}V_{0.30}$	6.14
Ir_2Sr	5.7	$CaIr_2$	6.15
$N_{0.6-0.987}Ti$	5.8	NbS_2	6.163
Al_2CaSi	5.8	$NbSnTaV$	6.2
$Rh_{17}S_{15}$	5.8	Rh_2Sr	6.2
$Rh_{0.005}Zr$ (annealed)	5.8	$SnTa_3$	6.2
$SbTi_3$	5.8	Ce_xLa_{1-x}	6.3
$B_{12}Zr$	5.82	Ho_xLa_{1-x}	6.3
$GaLa_3$	5.84	Er_xLa_{1-x}	6.3
$CuLa$	5.85	$CoZr_2$	6.3
GaN (black)	5.85	$Mo_{0.42}Re_{0.58}$	6.35
$Hf_{0.14}Re_{0.86}$	5.86	Bi_5Tl_3	6.4
$GaSb$ (unannealed)	5.9	$In_{0.5}Nb_3Zr_{0.5}$	6.4
$V_{0.26}Zr_{0.74}$	5.9	$CaRh_2$	6.4
Al_3Os	5.9	$PbTl_{0.27}$	6.43
Re_2Zr	5.9	$Hf_{0-0.55}Ta_{1-0.45}$	6.5
$Ge_{0.29}Nb_{0.71}$	6	$LaOs_2$	6.5
$BaRh_2$	6	La_3S_4	6.5
$CeRu_2$	6	Ir_2Th	6.5

Critical Temperature Data for Type II
Superconducting Compounds (Continued)

Compound	T_c, K	Compound	T_c, K
$HfN_{0.989}$	6.6	$BCMo_2$	7.2
$PbSb$	6.6	$LiPb$	7.2
$Re_{24}Ti_5$	6.6	Mo_3Os	7.2
Re_xTi_{1-x}	6.6max.	$Na_{0.28}Pb_{0.72}$	7.2
$Ru_xTi_{0.6}V_y$	6.6max.	Ag_xPb_{1-x}	7.2max.
$InPb$	6.65	Hg_xPb_{1-x}	7.26
$PbTl_{0.17}$	6.73	$H_{0.33}Nb_{0.67}$	7.28
$IrMo_3$	6.8	In_xSn_{1-x}	7.3
$Re_{0.92}V_{0.08}$	6.8	$Ga_{0.5}Ge_{0.5}Nb_3$	7.3
$Fe_xTi_{0.6}V_{1-x}$	6.8max.	$H_{0.1}Nb_{0.9}$	7.38
$PbTl_{0.12}$	6.88	Re_6Zr	7.4
NNb (film)	6.9	Ti_xV_{1-x}	7.5
$PbTl_{0.075}$	6.98	$CMo_{0.2}Ta_{0.8}$	7.5
Au_2Pb	7	$Pd_{0.1}Zr_{0.9}$	7.5
Cu_xSn_{1-x} (film)	7	RuW	7.5
$CHf_{0.25}Nb_{0.75}$	7	OV_3Zr_3	7.5
$CaPb$	7	$Tc_{0.50}W_{0.50}$	7.52
$AlPb$ (films)	7	$C_{0.6}Mo_{48}Si_3$	7.6
$Nb_{1-0.05}Se_2$	7.0	Cu_xPb_{1-x}	7.7
$BCMo_2$	7.0	$CMo_{0.5}Ta_{0.5}$	7.7
$PbTl_{0.04}$	7.06	$CHf_{0.2}Nb_{0.8}$	7.8
B_6Y	7.1	PPb	7.8
$Tc_{0.1-0.4}W_{0.9-0.6}$	7.18	$H_{0.05}Nb_{0.85}$	7.83
$Mo_{0.61}Ru_{0.39}$	7.18	$Tc_{0.60}W_{0.40}$	7.88

Critical Temperature Data for Type II
Superconducting Compounds (Continued)

Compound	T_c, K	Compound	T_c, K
$N_{0.82-0.99}V$	7.9	AsBiPb	9
IrNb	7.9	AsBiPbSb	9
$Ag_{0.8-0.3}Ga_{0.2-0.7}$	8	$Rh_{0.1}Zr_{0.9}$	9
Mo_xRh_{1-x}	8.2	$CHf_{0.9-0.1}Ta_{1-0.9}$	9.0
$N_xO_yV_z$	8.2	CNb_2	9.1
$Bi_{0.5}Cd_{0.13}Pb_{0.25}Sn_{0.12}$	8.2	$Bi_{1-0}Pb_{0-1}$	9.14
BNb	8.25	$GaV_{4.5}$	9.15
$CTa_{0.2-0.9}Zr_{0.8-0.1}$	8.3	Mo_xNb_{1-x}	9.2
$SnTa_3$	8.35	$InNb_3$ (high P and temp.)	9.2
$CNb_{0.6-0.9}Zr_{0.4-0.1}$	8.4	Nb_xTa_{1-x}	9.2
$Bi_{0.05-0.49}Pb_{0.95-0.60}$	8.4	$Cr_{0-0.1}Nb_{1-0.9}$	9.2
AsPb	8.4	CMo	9.26
$Bi_{0.5}Pb_{0.31}Sn_{0.19}$ (*w/o*)	8.5	CMo_xV_{1-x}	9.3
$Bi_{0.5}Pb_{0.25}Sn_{0.25}$	8.5	$N_{0.906-0.984}Zr$	9.5
$CMo_{0.75}Ta_{0.25}$	8.5	CMo_xZr_{1-x}	9.5
La_3Se_4	8.6	$Hf_{0-0.5}Nb_{1-0.5}$	9.5
$Bi_{1-0}Pb_{0-1}$ (film)	8.67	Ga_2Mo	9.5
$CMo_{0.8}Ta_{0.2}$	8.7	$Nb_3Pt_{0.02-0.98}Rh_{0.98-0.02}$	9.6
CNb_xTi_{1-x}	8.8	HfV_2	9.6
$IrMo_3$	8.8	$Be_{0.957}Re_{0.043}$	9.62
V_2Zr	8.8	$C_{0.848-0.987}Ta$	9.7
BiPbSb	8.9	$C_{0.987}Ta$	9.7
$CMo_{0.85}Ta_{0.15}$	8.9	Tc_6Zr	9.7
$Pd_{0.05}Ru_{0.05}Zr_{0.9}$	9	$Nb_{0.38-0.18}Re_{0.62-0.82}$	9.70

Critical Temperature Data for Type II
Superconducting Compounds (Continued)

Compound	T_c, K	Compound	T_c, K
$Be_{0.98-0.92}Re_{0.02-0.08}$	9.75	$NbSnTa_2$	10.8
Nb_xTi_{1-x}	9.8	$Nb_{0.75}Zr_{0.25}$	10.8
$C_{0.984}Nb$	9.8	$Nb_{0.66}Zr_{0.33}$	10.8
Ga_4Mo	9.8	$RhZr_2$	10.8
$Ir_{0.4}Nb_{0.6}$	9.8	$Au_{0.2-0.98}Nb_3Rh_{0.98-0.2}$	10.9
NZr	9.8	Nb_3Pt	10.9
Nb_2SnV	9.8	$SiV_{2.7}Ti_{0.3}$	10.9
$Nh_{0.6}Ti_{0.4}$	9.8	NNb_xO_y	11
$MoRe_3$	9.89	$C_{0.7-1.0}Nb_{0.3-0}$	11
$Fe_{0-0.5}Mo_{0.8}Re_{0.2}$	10	CTa	11(extrapolated)
$Co_{0-0.01}Mo_{0.8}RC_{0.02}$	10	$Au_{0.03}Nb_{1-0.7}$	11.0
Al_2CMo_3	10	$AuNb_{3(1-x)}V_{3x}$	11.0
Mo_3Re	10	$Mo_{0.52}Re_{0.48}$	11.1
Al_2CMo_3	10.2	$C_{0.-0.38}N_{1-0.62}Ta$	11.3
$CMo_{0.83}Ti_{0.17}$	10.2	$Ge_{0.5}Nb_{35}Sn_{0.5}$	11.3
CMo_xTi_{1-x}	10.2 max.	$SiV_{2.7}Cr_{0.3}$	11.3
$InLa_3$	10.4	$AuNb_3$	11.5
CNb (whiskers)	10.5	$CNb_{0.6-0.9}W_{0.4-0.1}$	11.6
$CTa_{1-0.4}W_{0-0.6}$	10.5	$SiV_{2.7}Mo_{0.3}$	11.7
$MoRu$	10.5	MoN	12
$NbTc_3$	10.5	Mo_xRe_{1-x}	12.2
$InLa_3$ (0–35 kbar)	10.55	CMo_2	12.2
$Mo_{\sim0.60}Re_{0.395}$	10.6	$Nb_2SnTa_{0.5}V_{0.5}$	12.2
$Rh_{0-0.45}Zr_{1-0.55}$	10.8	$C_{0.5}Mo_xNb_{1-x}$	12.5

Critical Temperature Data for Type II Superconducting Compounds (Continued)

Compound	T_c, K	Compound	T_c, K
$B_{0.03}C_{0.51}Mo_{0.47}$	12.5	Mo_xSiV_{3-x}	16.0
$Al_{0.5}Ge_{0.5}Nb$	12.6	NNb (diffusion wires)	16.1
$SiV_{2.7}Nb_{0.3}$	12.8	$N_{100-75w/o}Nb_{0-25w/o}Zr$[b]	16.35
$N_{0.70-0.795}Nb$	12.9	Nb_2SnTa	16.4
$Co_{0.40-0.44}Mo_{0.60-0.56}$	13	$Si_{0.9}V_3C_{0.1}$	16.4
$SiV_{2.7}Zr_{0.3}$	13.2	Nb_3Sn_2	16.6
Al_xNb_{1-x}	13.5	$N_{100-42w/o}Nb_{0-58w/o}Ti$[b]	16.8
$AlNb_xV_{1-x}$	13.5	GaV_3	16.8
NNb_xZr_{1-x}	13.8	$GeNb_3$ (quenched)	17
$N_{0.93}Nb_{0.85}Zr_{0.15}$	13.8	NNb_xO_y	17.0
CNb_xTa_{1-x}	13.9	SiV_3	17.1
NTa (extrapolated value)	14	Al_xNb_{1-x}	17.5
CNb (extrapolated)	14	$Al_{0.27}Nb_{0.73-0.48}V_{0-0.25}$	17.5
$Mo_{0.57}Re_{0.43}$	14	$Nb_{2.5}SnTa_{0.5}$	17.6
$Si_{0.9}V_3Ge_{0.1}$	14	$Nb_{2.75}SnTa_{0.25}$	17.8
$Si_{0.9}V_3Al_{0.1}$	14.05	$C_{0.1-0.9}N_{0.9-0.1}Nb$	17.9
$Nb_{2.5}SnV_{0.5}$	14.2	$Nb_3Sb_{0-0.7}Sn_{1-0.3}$	18
$GaV_{2.1-3.5}$	14.45	$AlNo_3$	18
NNb (whiskers)	14.5	$Nb_{3x}SnTa_{3(1-x)}$	18.0
$GaNb_3$	14.5	$Ge_xNb_3Sn_{1-x}$	18.0
$N_{0.988}Nb$	14.9	Nb_3Sn	18.05
$N_{0.824-0.988}Nb$	15.3	$In_{0-0.3}Nb_3Sn_{1-0.07}$	18.19
Mo_xTc_{1-x}	15.8	$Ga_xNb_3Sn_{1-x}$	18.37
$Si_{0.9}V_3B_{0.1}$	15.8	Nb_xSn_{1-x} (film)	18.5

Critical Temperature Data for Type II
Superconducting Compounds (Continued)

Compound	T_c, K
$Nb_{0.8}Sn_{0.2}$	18.5
$Al_{\sim0.8}Ge_{\sim0.2}Nb_3$	20.7

Data Compiled by J.S. Park from: CRC Handbook of Materials Science, Vol. 3, Charles T. Lynch, Ed., CRC Press, Cleveland, (1974)

SELECTED SUPERCONDUCTIVE COMPOUNDS AND ALLOYS: CRITICAL FIELD DATA

Substance	H_0 oersteds	Substance	H_0 oersteds
Ag_2F	2.5	$In_{0.8}Tl_{0.2}$	252
Ag_7NO_{11}	57	$Mg_{\sim0.47}Tl_{0.53}$	220
Al_2CMo_3	1,700	$Mo_{0.6}Ti_{0.84}$	<985
$BaBi_3$	740	$NbSn_2$	620
Bi_2Pt	10		
		$PbTl_{0.27}$	756
Bi_3Sr	530	$PbTl_{0.17}$	796
Bi_5Tl_3	>400	$PbTl_{0.12}$	849
$CdSn$	>266	$PbTl_{0.75}$	880
$CoSi_2$	105		
		$PbTl_{0.04}$	864
$CrO_{0.1}Ti_{0.3}V_{0.6}$	1,360		
$In_{1-0.86}Mg_{0-0.14}$	272.4-259.2		
$InSb$	1,100		
In_xTl_{1-x}	252-284		

From Roberts, B. W., *Properties of Selected Superconductive Materials,* National Bureau of Standards Technical Notes 482 and 724, U.S. Government Printing Office, Washington, D.C., 1969, 1972.

T_c DATA FOR HIGH TEMPERATURE SUPERCONDUCTING COMPOUNDS

Compound	T_c, K	Compound	T_c, K
$Nd_2Ce_{0.5}Sr_{0.5}Cu_{1.2}O_y$	~20	$Bi_{2-x}Pb_xSr_2Ca_{1-x}Y_xCu_2O_8$	~85
$Bi_2Sr_2CuO_6$	22	$Tl_{0.5}Pb_{0.5}Sr_2CaCu_2O_7$	~85
$Nd_{2-x}Ce_xCuO_y$	~24	$Y_{0.5}La_{0.5}Ba_2Cu_3O_{7-x}$	87
$(La_{0.925}Ba_{0.075})_2CuO_4$	~30	$Y(Ba_{0.75}Sr_{0.25})_2Cu_3O_{7-x}$	87
$Tl_2Ba_2CuO_6$	~30	$SmBa_2Cu_3O_{7-x}$	88.6
$Nd_{2-x-z}Ce_xSr_zCuO_{4-y}$	~30	$YbBa_2Cu_3O_{7-x}$	89.3
$Ba_xLa_{5-x}Cu_5O_{5(3-y)}$	35	$Y_{0.5}Sc_{0.5}Ba_2Cu_3O_{7-x}$	90
$(La_{0.9}Sr_{0.1})_2CuO_4$	36	$TmBa_2Cu_3O_{7-x}$	90.5
$(La_{1-x}Sr_x)_2CuO_4$	37	$ErBa_2Cu_3O_{7-x}$	90.7
$(La_{0.925}Sr_{0.075})_2CuO_4$	38.5	$GdBa_2Cu_3O_{7-x}$	90.9
$TlSr_2CaCu_2O_7$	~50	$Y_{0.75}Sc_{0.25}Ba_2Cu_3O_{7-x}$	91
$(La_{0.9}Ba_{0.1})_2CuO_{4-y}$, at 1Gpa	52.5	$EuBa_2Cu_3O_{7-x}$	91.1
$LaBa_2Cu_3O_{7-x}$	59.2	$HoBa_2Cu_3O_{7-x}$	91.1
$TlBa_2CaCu_2O_7$	~60	$DyBa_2Cu_3O_{7-x}$	91.8
$LaBa_2Cu_3O_{7-x}$	60	$YBa_2Cu_3O_{7-x}$	93
$LuBa_2Cu_3O_{7-x}$	72.6	$Eu_{0.75}Sc_{0.25}Ba_2Cu_3O_{7-x}$	93
$NdBa_2Cu_3O_{7-x}$	78.3	$Y_{0.1}Eu_{0.9}Ba_2Cu_3O_{7-x}$	94.5
$La(Ba_{0.5}Ca_{0.5})_2Cu_3O_{7-x}$	79	$Y_{0.25}Eu_{0.75}Ba_2Cu_3O_{7-x}$	95
$Pr_{0.1}Eu_{0.9}Ba_2Cu_3O_{7-x}$	82	$Tl_2Ba_2CaCu_2O_8$	~105
$Bi_2Sr_2CaCuO_8$	~85	$TlBa_2Ca_3Cu_4O_{11}$	~108

T_c DATA FOR HIGH TEMPERATURE SUPERCONDUCTING COMPOUNDS

Compound	T_c, K	Compound	T_c, K
$Bi_{2-x}Pb_xSr_2Ca_2Cu_3O_{10}$	~110	$Tl_2Ba_2Ca_2Cu_3O_{10}$	~125
$Tl_2Ba_2Ca_3Cu_4O_{12}$	~115		
$TlBa_2Ca_2Cu_3O_9$	~120		
$Tl_{0.5}Pb_{0.5}Sr_2Ca_2Cu_3O_9$	~120		

Data Compiled by J.S. Park from:

[a] *Source:* data from *Copper Oxide Superconductors*, edited by Charles P. Poole, Jr. *et al*, John Wiley & Sons, New York, 1988.

[b] *Source:* data from "Structural Problems and Non Stoichiometry in Thallium, Bismuth, and Lead Oxides," B. Raveau, et al in *High Temperature Superconductors: Relationships Between Properties, Structure, and Solid State Chemistry* edited by James D. Jorgensen *et al*, Materials Research Society Symposium Proceedings, Vol. 156, Pittsburgh, PA, 1989.

[c] *Source:* data from "A New Class of Oxide Superconductor (Nd, Ce, Sr)$_2$CuO$_4$" E. Takayama-Muromachi, in *High Temperature Superconductors: Relationships Between Properties, Structure, and Solid State Chemistry* edited by James D. Jorgensen *et al*, Materials Research Society Symposium Proceedings, Vol. 156, Pittsburgh, PA, 1989.

Bonding, Thermodynamic, and Kinetic Data

BOND STRENGTHS IN DIATOMIC MOLECULES
Listed by Molecules

Molecule	kcal · mol^{-1}		Molecule	kcal · mol^{-1}	
H-H	104.207	± 0.001	H-Br	87.4	± 0.5
H-D	105.030	± 0.001	H-Rb	40	± 5
D-D	106.010	± 0.001	H-Sr	39	± 2
H-Li	56.91	± 0.01	H-Ag	59	± 1
H-Be	54		H-Cd	16.5	± 0.1
H-B	79	± 1	H-In	59	± 2
H-C	80.9		H-Sn	63	± 1
H-N	75	± 4	H-Te	64	± 1
H-O	102.34	± 0.30	H-I	71.4	± 0.2
H-F	135.9	± 0.3	H-Cs	42.6	± 0.9
H-Na	48	± 5	H-Ba	42	± 4
H-Mg	47	± 12	H-Yb	38	± 1
H-Al	68	± 2	H-Pt	84	± 9
H-Si	71.4	± 1.2	H-Au	75	± 3
H-P	82	± 7	H-Hg	9.5	
H-S	82.3	± 2.9	H-Ti	45	± 2
H-Cl	103.1		H-Pb	42	± 5
H-K	43.8	± 3.5	H-Bi	59	± 7
H-Ca	40.1		Li-Li	24.55	± 0.14
H-Cr	67	± 12	Li-O	78	± 6
H-Mn	56	± 7	Li- F	137.5	± 1
H-Ni	61	± 7	Li-Cl	111.9	± 2
H -Cu	67	± 2	Li-Br	100.2	± 2
H-Zn	20.5	± 0.5	Li-I	84.6	± 2
H -Ga	68	± 5	Be-Be	17	
H-Ge	76.8	± 0.2	Be-0	98	± 7
H-As	65	± 3	Be-F	136	± 2
H-Se	73	± 1	Be-S	89	± 14

Bond Strengths in Diatomic Molecules (Continued)
Listed by Molecules

Molecule	kcal • mol^{-1}		Molecule	kcal • mol^{-1}	
Be-Cl	92.8	± 2.2	C-V	133	
Be-Au	~ 67		C-Ge	110	± 5
B-B	~ 67	± 5	C-Se	139	± 23
B-N	93	± 12	C-Br	67	± 5
B-0	192.7	± 1.2	C-Ru	152	± 3
B-F	180	± 3	C-Rh	139	± 2
B-S	138.8	± 2.2	C-I	50	± 5
B-Cl	119		C-Ce	109	± 7
B-Se	110	± 4	C-Ir	149	± 3
B-Br	101	± 5	C-Pt	146	± 2
B-Ru	107	± 5	C-U	111	± 7
B-Rh	114	± 5	N-N	226.8	± 1.5
B-Pd	79	± 5	N-O	150.8	± 0.2
B-Te	85	± 5	N-F	62.6	± 0.8
B-Ce	~ 100		N-Al	71	± 23
B-Ir	123	± 4	N-Si	105	± 9
B-Pt	114	± 4	N-P	148	± 5
B-Au	82	± 4	N-S	~ 120	± 6
B-Th	71		N-Cl	93	± 12
C-C	144	± 5	N-Ti	111	
C-N	184	± 1	N-As	116	± 23
C-0	257.26	± 0.77	N-Se	105	± 23
C-F	128	± 5	N-Br	67	± 5
C-Si	104	± 5	N-Sb	72	± 12
C-P	139	± 23	N-I	~.38	
C-S	175	± 7	N-Xe	55	
C-Cl	93		N-Th	138	± 1
C-Ti	~128		N-U	127	± 1

Bond Strengths in Diatomic Molecules (Continued)
Listed by Molecules

Molecule	kcal • mol^{-1}		Molecule	kcal • mol^{-1}	
O-O	118.86	± 0.04	O-Y	162	± 5
O-F	56	± 9	O-Zr	181	± 10
O-Na	61	± 4	O-Nb	189	± 10
O-Mg	79	± 7	O-Mo	115	± 12
O- Al	116	± 5	O-Ru	115	± 15
O-Si	184	± 3	O-Rh	90	± 15
O-P	119.6	± 3	O-Pd	56	± 7
O-S	124.69	± 0.03	O-Ag	51	± 20
O-Cl	64.29	± 0.03	O-Cd	≤ 67	
O-K	57	± 8	O-In	≤ 77	
O-Ca	84	± 7	O-Sn	127	± 2
O-Sc	155	± 5	O-Sb	89	± 20
O-Ti	158	± 8	O-Fe	93.4	± 2
O-V	154	± 5	O-I	47	± 7
O-Cr	110	± 10	O-Xe	9	± 5
O-Mn	96	± 8	O-Cs	67	± 8
O-Fe	96	± 5	O-Ba	131	± 6
O-Co	88	± 5	O-La	188	± 5
O-Ni	89	± 5	O-Ce	188	± 6
O-Cu	82	± 15	O-Pr	183.7	
O-Zn	≤ 66		O-Nd	168	± 8
O-Ga	68	± 15	O-Sm	134	± 8
O-Ge	158.2	± 3	O-Eu	130	± 10
O-As	115	± 3	O-Gd	162	± 6
O-Se	101		O-Tb	165	± 8
O-Br	56.2	± 0.6	O-Dy	146	± 10
O-Rb	(61)	± 20	O-Ho	149	± 10
O-Sr	93	± 6	O-Er	147	± 10

Bond Strengths in Diatomic Molecules (Continued)
Listed by Molecules

Molecule	kcal • mol^{-1}		Molecule	kcal • mol^{-1}	
O-Tm	122	± 15	F-Mn	101.2	± 3.5
O-Yb	98	± 15	F-Ni	89	± 4
O-Lu	159	± 8	F-Cu	88	± 9
O-Hf	185	± 10	F-Ga	138	± 4
O-Ta	183	± 15	F-Ge	116	± 5
O-W	156	± 6	F-Br	55.9	
O-Os	< 142		F-Rb	116.1	± 1
O-Ir	≤ 94		F-Sr	129.5	± 1.6
O-Pt	83	± 8	F-Y	144	± 5
O-Pb	90.3	± 1.0	Mg-I	~.68	
O-Bi	81.9	± 1.5	Mg-Au	59	± 23
O-Th	192	± 10	Al-Al	44	
O-U	182	± 8	Al-P	52	± 3
O-Np	172	± 7	Al-S	79	
O-Pu	163	± 15	Al-Cl	119.0	± 1
O-Cm	≤ 134		Al-Br	103.1	
F-F	37.5	± 2.3	Al-I	88	
F-Na	114	± I	Al-Au	65	
F-Mg	110	± 1	Al-U	78	± 7
F-Al	159:	± 3	Si-Si	76	± 5
F-Si	116	± 12	Si-S	148	± 3
F-P	105	± 23	Si-Cl	105	± 12
F-Cl	59.9	± 0.1	Si-Fe	71	± 6
F-K	118.9	± 0.6	Si-Co	66	± 4
F-Ca	125	± 5	Si-Ni	76	± 4
F-Sc	141	± 3	Si-Ge	72	± 5
F-Ti	136	± 8	Si-Se	127	± 4
F-Cr	104.5	± 4.7	Si-Br	82	± 12

Bond Strengths in Diatomic Molecules (Continued)
Listed by Molecules

Molecule	kcal • mol^{-1}		Molecule	kcal • mol^{-1}	
Si-Ru	95	± 5	Na-K	15.2	± 0.7
Si-Rh	95	± 5	Na-Br	86.7	± 1
Si-Pd	75	± 4	Na-Rb	14	± 1
Si-Te	121	± 9	Na-I	72.7	± 1
Si-Ir	110	± 5	Mg-Mg	8?	
Si-Pt	120	± 5	Mg-S	56?	
Si-Au	75	± 3	Mg-Cl	76	± 3
P-P	117	± 3	Mg-Br	75	± 23
F-Ag	84.7	± 3.9	P-S	70	
F-Cd	73	± 5	P-Ga	56	
F-In	121	± 4	P-W	73	± 1
F-Sn	111.5	± 3	P-Th	90	
F-Sb	105	± 23	S-S	101.9	± 2.5
F-I	67?		S-Ca	75	± 5
F-Xe	11		S-Sc	114	± 3
F-Cs	119.6	± 1	S-Mn	72	± 4
F-Ba	140.3	± 1.6	S-Fe	78	
F-Nd	130	± 3	S-Cu	72	± 12
F-Sm	126.9	± 4.4	S-Zn	49	± 3
F-Eu	126.1	± 4.4	S-Ge	131.7	± 0.6
F-Gd	141.	± 46.5	S-Se	911	± 5
F-Hg	31	± 9	S-Sr	75	± 5
F-Ti	106.4	± 4.6	S-Y	127	± 3
F-Pb	85	± 2	S-Cd	48	
F-Bi	62		S-In	69	± 4
F-Pu	129	± 7	S-Sn	111	± 1
Na-Na	18.4		S-Te	81	± 5
Na-Cl	97.5	± 0.5	S-Ba	96	± 5

Bond Strengths in Diatomic Molecules (Continued)
Listed by Molecules

Molecule	kcal • mol^{-1}		Molecule	kcal • mol^{-1}	
S-La	137	± 3	Cl-Sr	97	± 3
S-Ce	137	± 3	Cl-Y	82	± 23
S-Pr	122.7		Cl-Ag	75	± 9
S- Nd	113	± 4	Cl-Cd	49.9	
S-Eu	87	± 4	Cl-In	103.3	
S-Gd	126	± 4	CI-Sn	75?	
S-Ho	102	± 4	Cl-Sb	86	± 12
S-Lu	121	± 4	Cl-I	50.5	± 0.1
S-Au	100	± 6	Cl-Cs	106.2	± 1
S-Hg	51		Cl-Ba	106	± 3
S-Pb	82.7	± 0.4	Cl-Au	82	± 2
S-Bi	75.4	± 1.1	Cl-Hg	24	± 2
S-U	135	± 2	Cl-Ti	89.0	± 0.5
Cl-Cl	58.066	± 0.001	Cl-Pb	72	± 7
Cl-K	101.3	± 0.5	Cl-Bi	72	± 1
Cl-Ca	95	± 3	Cl-Ra	82	± 18
Cl-Sc	79		Ar-Ar	0.2	
Cl-Ti	26	± 2	K-K	12.8	
Cl-Cr	87.5	± 5.8	K-Br	90.9	± 0.5
Cl-Mn	86.2	± 2.3	K-I	76.8	± 0.5
Cl-Fe	84?		Ca-I	70	± 23
Cl-Ni	89	± 5	Ca-Au	18	
Cl-Cu	84	± 6	Sc-Sc	25.9	± 5
Cl-Zn	54.7	± 4.7	Ti-Ti	34	± 5
Cl-Ga	114.5		V-V	58	± 5
Cl-Ge	82?		Cr-Cr	<37	
Cl-Br	52.3	± 0.2	Cr -Cu	37	± 5
Cl-Rb	100.7	± 1	Cr-Ge	41	± 7

Bond Strengths in Diatomic Molecules (Continued)
Listed by Molecules

Molecule	kcal • mol^{-1}		Molecule	kcal • mol^{-1}	
Cr-Br	78.4	± 5	Cu-Te	42	± 9
Cr-I	68.6	± 5.8	Cu-I	47?	
Cr-Au	51.3	± 3.5	Cu-Au	55.4	± 2.2
Mn-Mn	4	± 3	Zn-Zn	7	
Mn-Se	48	± 3	Zn-Se	33	± 3
Mn-Br	75.1	± 23	Zn-Te	49?	
Mn-I	67.6	± 2.3	Zn-I	33	± 7
Mn-Au	44	± 3	Ga-Ga	3	± 3
Fe-Fe	24	± 5	Ga-As	50.1	± 0.3
Fe-Ge	50	± 7	Ga-Br	101	± 4
Fe-Br	59	± 23	Ga-Ag	4	± 3
Fe-Au	45	± 4	Ga-Te	60	± 6
Co-Co	40	± 6	Ga-I	81	± 2
Co-Cu	39	± 5	Ga-Au	51	± 23
Co-Ge	57	± 6	Ge-Ge	65.8	± 3
Co-Au	51	± 3	Ge-Se	114	±
Ni-Ni	55.5	± 5	Ge-Br	61	± 7
Ni-Cu	48	± 5	Ge-Te	93	± 2
Ni-Ge	67.3	± 4	Ge-Au	70	± 23
Ni-Br	86	± 3	As-As	91.7	
Ni-I	70	± 5	As-Se	23	
Ni-Au	59	± 5	Se-Se	79.5	± 0.1
Cu-Cu	46.6	± 2.2	Se-Cd	~75	
Cu-Ge	49	± 5	Se-in	59	± 4
Cu-Se	70	± 9	Se-Sn	95.9	± 1.4
Cu-Br	79	± 6 5	Se-Te	64	± 2
Cu-Ag	41.6	± 2.2	Se-La	114	± 4
Cu-Sn	42.3	± 4	Se-Nd	92	± 4

Bond Strengths in Diatomic Molecules (Continued)
Listed by Molecules

Molecule	kcal • mol^{-1}		Molecule	kcal • mol^{-1}	
Se-Eu	72	± 4	Ag-Te	70	± 23
Se-Gd	103	± 4	Ag-I	56	± 7
Se-Ho	80	± 4	Ag-Au	48.5	± 2.2
Se-Lu	100	± 4	Cd-Cd	2.7	± 0.2
Se-Pb	72.4	± 1	Cd-I	33	± 5
Se-Bi	67.0	± 1.5	In-In	23.3	± 2.5
Bi-Br	46.336	± 0.001	In-Sb	36.3	± 2.5
Br-Rb	90.4	± 1	In-Te	52	± 4
Br-Ag	70	± 7	In-I	80	
Br-Cd	~38		Sn-Sn	46.7	± 4
Br-In	93		Sn-Te	76	± 1
Bi-Sn	47	± 23	Sn-Au	58.4	± 4
Br-Sb	75	± 14	Sb-Sb	71.5	± 1.5
Br-I	42.8	± 0.1	Sb-Te	61	± 4
Br-Cs	96.5	± 1	Sb-Bi	60	± 1
Br-Hg	17.3		Te-Te	63.2	± 0.2
Br-Ti	79.8	± 0.4	Te-La	91	± 4
Br-Pb	59	± 9	Te-Nd	73	± 4
Br-Bi	63.9	± 1	Te-Eu	58	± 4
Rb-Rb	12.2		Te-Gd	82	± 4
Rb-I	76.7	± 1	Te-Ho	62	± 4
Sr-Au	63	± 23	Te-Lu	78	± 4
Y-Y	38.3		Te-Au	59	± 16
Y-La	48.3		Te-Pb	60	± 3
Pd-Pd	33?		Te-Bi	56	± 3
Pd-Au	34.2	± 5	I-I	36.460	± 0.002
Ag-Ag	41	± 2	I-Cs	82.4	± 1
Ag-Sn	32.5	± 5	I-Hg	9	

Bond Strengths in Diatomic Molecules (Continued)
Listed by Molecules

Molecule	kcal $\cdot$ mol^{-1}		Molecule	kcal $\cdot$ mol^{-1}	
I-Ti	65	± 2	Au-Au	52.4	± 2.2
I-Pb	47	± 9	Au-Pb	31	± 23
I-Bi	52	± 1	Au-U	76	± 7
Xe-Xe	~ 0.7		Hg-Hg	4.1	± 0.5
Cs-Cs	11.3		Hg-Tl	1	
Ba-Au	38	± 14	Tl-Tl	15?	
La-Ld	58.6		Pb-Pb	24	± 5
La-Au	80	± 5	Pb-Bi	32	± 5
Ce-Ce	66	± 1	Bi-Bi	45	± 2
Ce-Au	76	± 4	Po-Po	44.4	± 2.3
Pr-Au	74	± 5	At-At	19	
Nd-Au	70	± 6	Th-Th	<69	

To convert kcal to KJ, multiply by 4.184.

The strength of a chemical bond, D(R - X), often known as the bond dissociation energy, is defined as the heat of the reaction: RX -> R + X.

It is given by: D(R - X) = DHf°(R) + DHf°(X) - DHf°(RX).

Some authors list bond strengths for 0°K, but here the values for 298°K are given because more thermodynamic data are available for this temperature. Bond strengths, or bond dissociation energies, are not equal to, and may differ considerable from, mean bond energies derived solely from thermochemical data on molecules and atoms.

The values in this table have usually been measured spectroscopically or by mass spectrometric analysis of hot gases effusing from a Knudsen cell.

From Kerr, J. A., Parsonage, M. J., and Trotman-Dickenson, A. F., in *Handbook of Chemistry and Physics*, 55th ed., Weast, R. C., Ed., CRC Press, Cleveland, 1974, F-204.

BOND STRENGTHS IN DIATOMIC MOLECULES
Listed by Values

Molecule	kcal • mol^{-1}		Molecule	kcal • mol^{-1}	
Ar-Ar	0.2		At-At	19	
N-I	~.38		H-Zn	20.5	± 0.5
Mg-I	~.68		As-Se	23	
Xe-Xe	~ 0.7		In-In	23.3	± 2.5
Hg-Tl	1		Cl-Hg	24	± 2
Cd-Cd	2.7	± 0.2	Fe-Fe	24	± 5
Ga-Ga	3	± 3	Pb-Pb	24	± 5
Mn-Mn	4	± 3	Li-Li	24.55	± 0.14
Ga-Ag	4	± 3	Sc-Sc	25.9	± 5
Hg-Hg	4.1	± 0.5	Cl-Ti	26	± 2
			F-Hg	31	± 9
Zn-Zn	7		Au-Pb	31	± 23
Mg-Mg	8?		Pb-Bi	32	± 5
O-Xe	9	± 5	Ag-Sn	32.5	± 5
I-Hg	9		Zn-Se	33	± 3
H-Hg	9.5		Zn-I	33	± 7
F-Xe	11		Cd-I	33	± 5
Cs-Cs	11.3		Pd-Pd	33?	
Rb-Rb	12.2		Ti-Ti	34	± 5
K-K	12.8		Pd-Au	34.2	± 5
Na-Rb	14	± 1	In-Sb	36.3	± 2.5
Tl-Tl	15?		I-I	36.460	± 0.002
Na-K	15.2	± 0.7	Cr -Cu	37	± 5
H-Cd	16.5	± 0.1	Cr-Cr	<37	
Be-Be	17		F-F	37.5	± 2.3
Br-Hg	17.3		H-Yb	38	± 1
Ca-Au	18		Br-Cd	~38	
Na-Na	18.4		Ba-Au	38	± 14

Bond Strengths in Diatomic Molecules (Continued)
Listed by Values

Molecule	kcal • mol^{-1}		Molecule	kcal • mol^{-1}	
Y-Y	38.3		I-Pb	47	± 9
H-Sr	39	± 2	Cu-I	47?	
Co-Cu	39	± 5	H-Na	48	± 5
H-Rb	40	± 5	S-Cd	48	
Co-Co	40	± 6	Mn-Se	48	± 3
H-Ca	40.1		Ni-Cu	48	± 5
Cr-Ge	41	± 7	Y-La	48.3	
Ag-Ag	41	± 2	Ag-Au	48.5	± 2.2
Cu-Ag	41.6	± 2.2	S-Zn	49	± 3
H-Ba	42	± 4	Cu-Ge	49	± 5
H-Pb	42	± 5	Zn-Te	49?	
Cu-Te	42	± 9	Cl-Cd	49.9	
Cu-Sn	42.3	± 4	C-I	50	± 5
H-Cs	42.6	± 0.9	Fe-Ge	50	± 7
Br-I	42.8	± 0.1	Ga-As	50.1	± 0.3
H-K	43.8	± 3.5	Cl-I	50.5	± 0.1
Al-Al	44		O-Ag	51	± 20
Mn-Au	44	± 3	S-Hg	51	
Po-Po	44.4	± 2.3	Co-Au	51	± 3
H-Ti	45	± 2	Ga-Au	51	± 23
Fe-Au	45	± 4	Cr-Au	51.3	± 3.5
Bi-Bi	45	± 2	Al-P	52	± 3
Bi-Br	46.336	±0.001	In-Te	52	± 4
Cu-Cu	46.6	± 2.2	I-Bi	52	± 1
Sn-Sn	46.7	± 4	Cl-Br	52.3	± 0.2
H-Mg	47	± 12	Au-Au	52.4	± 2.2
O-I	47	± 7	H-Be	54	
Bi-Sn	47	± 23	Cl-Zn	54.7	± 4.7

Bond Strengths in Diatomic Molecules (Continued)
Listed by Values

Molecule	kcal • mol^{-1}		Molecule	kcal • mol^{-1}	
N-Xe	55		Te-Au	59	± 16
Cu-Au	55.4	± 2.2	F-Cl	59.9	± 0.1
Ni-Ni	55.5	± 5	Ga-Te	60	± 6
F-Br	55.9		Sb-Bi	60	± 1
H-Mn	56	± 7	Te-Pb	60	± 3
O-F	56	± 9	O-Rb	(61)	± 20
O-Pd	56	± 7	H-Ni	61	± 7
P-Ga	56		O-Na	61	± 4
Ag-I	56	± 7	Ge-Br	61	± 7
Te-Bi	56	± 3	Sb-Te	61	± 4
Mg-S	56?		F-Bi	62	
O-Br	56.2	± 0.6	Te-Ho	62	± 4
H-Li	56.91	± 0.01	N-F	62.6	± 0.8
O-K	57	± 8	H-Sn	63	± 1
Co-Ge	57	± 6	Sr-Au	63	± 23
V-V	58	± 5	Te-Te	63.2	± 0.2
Te-Eu	58	± 4	Br-Bi	63.9	± 1
Cl-Cl	58.066	±0.001	H-Te	64	± 1
Sn-Au	58.4	± 4	Se-Te	64	± 2
La-Ld	58.6		O-Cl	64.29	± 0.03
H-Ag	59	± 1	H-As	65	± 3
H-In	59	± 2	Al-Au	65	
H-Bi	59	± 7	I-Ti	65	± 2
Mg-Au	59	± 23	Ge-Ge	65.8	± 3
Fe-Br	59	± 23	Si-Co	66	± 4
Ni-Au	59	± 5	Ce-Ce	66	± 1
Se-in	59	± 4	O-Zn	≤ 66	
Br-Pb	59	± 9	B-B	~ 67	± 5

Bond Strengths in Diatomic Molecules (Continued)
Listed by Values

Molecule	kcal • mol^{-1}		Molecule	kcal • mol^{-1}	
H-Cr	67	± 12	H-I	71.4	± 0.2
H -Cu	67	± 2	Sb-Sb	71.5	± 1.5
C-Br	67	± 5	N-Sb	72	± 12
N-Br	67	± 5	Si-Ge	72	± 5
O-Cs	67	± 8	S-Mn	72	± 4
O-Cd	≤ 67		S-Cu	72	± 12
Se-Bi	67.0	± 1.5	Cl-Pb	72	± 7
F-I	67?		Cl-Bi	72	± 1
Ni-Ge	67.3	± 4	Se-Eu	72	± 4
Mn-I	67.6	± 2.3	Se-Pb	72.4	± 1
H-Al	68	± 2	Na-I	72.7	± 1
H -Ga	68	± 5	H-Se	73	± 1
O-Ga	68	± 15	F-Cd	73	± 5
Cr-I	68.6	± 5.8	P-W	73	± 1
S-In	69	± 4	Te-Nd	73	± 4
Th-Th	<69		Pr-Au	74	± 5
P-S	70		H-N	75	± 4
Ca-I	70	± 23	H-Au	75	± 3
Ni-I	70	± 5	Si-Pd	75	± 4
Cu-Se	70	± 9	Si-Au	75	± 3
Ge-Au	70	± 23	Mg-Br	75	± 23
Br-Ag	70	± 7	S-Ca	75	± 5
Ag-Te	70	± 23	S-Sr	75	± 5
Nd-Au	70	± 6	Cl-Ag	75	± 9
B-Th	71		Se-Cd	~75	
N-Al	71	± 23	Br-Sb	75	± 14
Si-Fe	71	± 6	CI-Sn	75?	
H-Si	71.4	± 1.2	Mn-Br	75.1	± 23

Bond Strengths in Diatomic Molecules (Continued)
Listed by Values

Molecule	kcal · mol^{-1}		Molecule	kcal · mol^{-1}	
S-Bi	75.4	± 1.1	S-Te	81	± 5
Si-Si	76	± 5	Ga-I	81	± 2
Si-Ni	76	± 4	O-Bi	81.9	± 1.5
Mg-Cl	76	± 3	H-P	82	± 7
Sn-Te	76	± 1	B-Au	82	± 4
Ce-Au	76	± 4	O-Cu	82	± 15
Au-U	76	± 7	Si-Br	82	± 12
Rb-I	76.7	± 1	Cl-Y	82	± 23
H-Ge	76.8	± 0.2	Cl-Au	82	± 2
K-I	76.8	± 0.5	Cl-Ra	82	± 18
O-In	≤ 77		Te-Gd	82	± 4
Li-O	78	± 6	Cl-Ge	82?	
Al-U	78	± 7	H-S	82.3	± 2.9
S-Fe	78		I-Cs	82.4	± 1
Te-Lu	78	± 4	S-Pb	82.7	± 0.4
Cr-Br	78.4	± 5	O-Pt	83	± 8
H-B	79	± 1	H-Pt	84	± 9
B-Pd	79	± 5	O-Ca	84	± 7
O-Mg	79	± 7	Cl-Cu	84	± 6
Al-S	79		Cl-Fe	84?	
Cl-Sc	79		Li-I	84.6	± 2
Cu-Br	79	± 6 5	F-Ag	84.7	± 3.9
Se-Se	79.5	± 0.1	B-Te	85	± 5
Br-Ti	79.8	± 0.4	F-Pb	85	± 2
Se-Ho	80	± 4	Cl-Sb	86	± 12
In-I	80		Ni-Br	86	± 3
La-Au	80	± 5	Cl-Mn	86.2	± 2.3
H-C	80.9		Na-Br	86.7	± 1

Bond Strengths in Diatomic Molecules (Continued)
Listed by Values

Molecule	kcal $\cdot$ mol^{-1}		Molecule	kcal $\cdot$ mol^{-1}	
S-Eu	87	± 4	O-Fe	93.4	± 2
H-Br	87.4	± 0.5	O-Ir	≤ 94	
Cl-Cr	87.5	± 5.8	Si-Ru	95	± 5
O-Co	88	± 5	Si-Rh	95	± 5
F-Cu	88	± 9	Cl-Ca	95	± 3
Al-I	88		Se-Sn	95.9	± 1.4
Be-S	89	± 14	O-Mn	96	± 8
O-Ni	89	± 5	O-Fe	96	± 5
O-Sb	89	± 20	S-Ba	96	± 5
F-Ni	89	± 4	Br-Cs	96.5	± 1
Cl-Ni	89	± 5	Cl-Sr	97	± 3
Cl-Ti	89.0	± 0.5	Na-Cl	97.5	± 0.5
O-Rh	90	± 15	Be-0	98	± 7
P-Th	90		O-Yb	98	± 15
O-Pb	90.3	± 1.0	S-Au	100	± 6
Br-Rb	90.4	± 1	Se-Lu	100	± 4
K-Br	90.9	± 0.5	B-Ce	~ 100	
S-Se	91	± 5	Li-Br	100.2	± 2
Te-La	91	± 4	Cl-Rb	100.7	± 1
As-As	91.7		B-Br	101	± 5
Se-Nd	92	± 4	O-Se	101	
Be-Cl	92.8	± 2.2	Ga-Br	101	± 4
B-N	93	± 12	F-Mn	101.2	± 3.5
C-Cl	93		Cl-K	101.3	± 0.5
N-Cl	93	± 12	S-S	101.9	± 2.5
O-Sr	93	± 6	S-Ho	102	± 4
Ge-Te	93	± 2	H-O	102.34	± 0.30
Br-In	93		Se-Gd	103	± 4

Bond Strengths in Diatomic Molecules (Continued)
Listed by Values

Molecule	kcal $\cdot$ mol^{-1}		Molecule	kcal $\cdot$ mol^{-1}	
H-Cl	103.1		S- Nd	113	± 4
Al-Br	103.1		B-Rh	114	± 5
Cl-In	103.3		B-Pt	114	± 4
C-Si	104	± 5	F-Na	114	$\pm I$
H-H	104.207	± 0.001	S-Sc	114	± 3
F-Cr	104.5	± 4.7	Ge-Se	114	$\pm$
N-Si	105	± 9	Se-La	114	± 4
N-Se	105	± 23	Cl-Ga	114.5	
F-P	105	± 23	O-As	115	± 3
Si-Cl	105	± 12	O-Mo	115	± 12
F-Sb	105	± 23	O-Ru	115	± 15
H-D	105.030	± 0.001	N-As	116	± 23
Cl-Ba	106	± 3	O- Al	116	± 5
D-D	106.010	± 0.001	F-Si	116	± 12
Cl-Cs	106.2	± 1	F-Ge	116	± 5
F-Ti	106.4	± 4.6	F-Rb	116.1	± 1
B-Ru	107	± 5	P-P	117	± 3
C-Ce	109	± 7	O-O	118.86	± 0.04
B-Se	110	± 4	F-K	118.9	± 0.6
C-Ge	110	± 5	B-Cl	119	
O-Cr	110	± 10	Al-Cl	119.0	± 1
F-Mg	110	± 1	O-P	119.6	± 3
Si-Ir	110	± 5	F-Cs	119.6	± 1
C-U	111	± 7	N-S	~ 120	± 6
N-Ti	111		Si-Pt	120	± 5
S-Sn	111	± 1	Si-Te	121	± 9
F-Sn	111.5	± 3	F-In	121	± 4
Li-Cl	111.9	± 2	S-Lu	121	± 4

Bond Strengths in Diatomic Molecules (Continued)
Listed by Values

Molecule	kcal • mol⁻¹		Molecule	kcal • mol⁻¹	
O-Tm	122	± 15	S-Ce	137	± 3
S-Pr	122.7		Li- F	137.5	± 1
B-Ir	123	± 4	N-Th	138	± 1
O-S	124.69	± 0.03	F-Ga	138	± 4
F-Ca	125	± 5	B-S	138.8	± 2.2
S-Gd	126	± 4	C-P	139	± 23
F-Eu	126.1	± 4.4	C-Se	139	± 23
F-Sm	126.9	± 4.4	C-Rh	139	± 2
N-U	127	± 1	F-Ba	140.3	± 1.6
O-Sn	127	± 2	F-Sc	141	± 3
Si-Se	127	± 4	F-Gd	141.	± 46.5
S-Y	127	± 3	O-Os	< 142	
C-F	128	± 5	C-C	144	± 5
C-Ti	~128		F-Y	144	± 5
F-Pu	129	± 7	C-Pt	146	± 2
F-Sr	129.5	± 1.6	O-Dy	146	± 10
O-Eu	130	± 10	O-Er	147	± 10
F-Nd	130	± 3	N-P	148	± 5
O-Ba	131	± 6	Si-S	148	± 3
S-Ge	131.7	± 0.6	C-Ir	149	± 3
C-V	133		O-Ho	149	± 10
O-Sm	134	± 8	N-O	150.8	± 0.2
O-Cm	≤ 134		C-Ru	152	± 3
S-U	135	± 2	O-V	154	± 5
H-F	135.9	± 0.3	O-Sc	155	± 5
Be-F	136	± 2	O-W	156	± 6
F-Ti	136	± 8	O-Ti	158	± 8
S-La	137	± 3	O-Ge	158.2	± 3

Bond Strengths in Diatomic Molecules (Continued)
Listed by Values

Molecule	kcal • mol^{-1}		Molecule	kcal • mol^{-1}	
O-Lu	159	± 8	O-Ta	183	± 15
F-Al	159	± 3	O-Pr	183.7	
O-Y	162	± 5	C-N	184	± 1
O-Gd	162	± 6	O-Si	184	± 3
O-Pu	163	± 15	O-Hf	185	± 10
O-Tb	165	± 8	O-La	188	± 5
O-Nd	168	± 8	O-Ce	188	± 6
O-Np	172	± 7	O-Nb	189	± 10
C-S	175	± 7	O-Th	192	± 10
B-F	180	± 3	B-0	192.7	± 1.2
O-Zr	181	± 10	N-N	226.8	± 1.5
O-U	182	± 8	C-0	257.26	± 0.77

To convert kcal to KJ, multiply by 4.184.

The strength of a chemical bond, D(R - X), often known as the bond dissociation energy, is defined as the heat of the reaction: RX -> R + X. It is given by: D(R - X) = DHf°(R) + DHf°(X) - DHf°(RX). Some authors list bond strengths for 0°K, but here the values for 298°K are given because more thermodynamic data are available for this temperature. Bond strengths, or bond dissociation energies, are not equal to, and may differ considerable from, mean bond energies derived solely from thermochemical data on molecules and atoms. The values in this table have usually been measured spectroscopically or by mass spectrometric analysis of hot gases effusing from a Knudsen cell.

From Kerr, J. A., Parsonage, M. J., and Trotman-Dickenson, A. F., in *Handbook of Chemistry and Physics*, 55th ed., Weast, R. C., Ed., CRC Press, Cleveland, 1974, F-204.

BOND STRENGTHS OF POLYATOMIC MOLECULES
Listed by Molecule

Bond	Kcal mol^{-1}	Bond	Kcal mol^{-1}
H-CH	102 ± 2	H-ONO$_2$	101.2 ± 0.5
H-CH$_2$	110 ± 2	H-SH	90 ± 2
H-CH$_3$	104 ± 1	H-SCH?	≥ 88
H-ethynyl	128 ± 5	H-SiH$_3$	94 ± 3
H-vinyl	≥ 108 ± 2	H-Si(CH$_3$)$_3$	90 ± 3
H-C$_2$H$_5$	98 ± 1	BH$_3$-BH$_3$	35
H-propargyl	93.9 ± 1.2	HC=CH	230 ± 2
H-allyl	89 ± 1	H$_2$C=CH$_2$	172 ± 2
H-cyclopropyl	100.7 ± 1	H$_3$C-CH$_3$	88 ± 2
H-n-C$_3$H$_7$	98 ± 1	CH$_3$-C(CH$_3$)$_2$CH:CH$_2$	69.4
H-i-C$_3$H$_7$	95 ± 1	C$_6$H$_5$CH$_2$-C$_2$H$_5$	69 ± 2
H-cyclobutyl	96.5 ± 1	C$_6$H$_5$CH(CH$_3$) - CH$_3$	71
H-cyclopropycarbinyl	97.4 ± 1.6	C$_6$H$_5$CH$_2$-n -C$_3$H$_7$	67 ± 2
H-methdllyl	83 ± 1	CH$_3$-CH$_2$CN	72.7 ± 2
H-s-C$_4$H$_9$	95 ± 1	CH$_3$-C(CH$_3$)$_2$CN	70.2 ± 2
H-t-C$_4$H$_9$	92 ± 1.2	C$_6$H$_5$C(CH$_3$)(CN) - CH$_3$	59.9
H-cyclopentadien-1,3-yl-5	81.2 ± 1.2	NC-CN	128 ± 1
H-pentadien-1,4-yl-3	80 ± 1	C$_6$H$_5$CH$_2$CO - CH$_2$C$_6$H$_5$	65.4
H-OH	119 ± 1	C$_6$H$_5$CO - CF$_3$	73.8
H-OCH$_3$	103.6 ± 1	CH$_3$CO - COCH$_3$	67.4 ± 2.3
H-OC$_2$H$_5$	103.9 ± 1	C$_6$H$_5$CH$_2$ - COOH	68.1
H-OC(CH$_3$)$_3$	104.7 ± 1	C$_6$H$_5$CH$_2$ - O$_2$CCH$_3$	67
H-OC$_6$H$_5$	88 ± 5	C$_6$H$_5$CO - COC$_6$H$_5$	66.4
H-O$_2$H	90 ± 2	C$_6$H$_5$CH$_2$ - O$_2$CC$_6$H$_5$	69
H-O$_2$CCH$_3$	112 ± 4	(C$_6$H$_5$CH$_2$)$_2$CH-COOH	59.4
H-O$_2$CC$_2$H$_3$	110 ± 4	CH$_2$F - CH$_2$F	88 ± 2
H-O$_2$Cn-C$_3$H$_7$	103 ± 4	CF$_2$ = CF$_2$	76.3 ± 3
H-ONO	78.3 ± 0.5	CF$_3$ - CF$_3$	96.9 ± 2

Bond Strengths of Polyatomic Molecules (Continued)
Listed by Molecule

Bond	Kcal mol^{-1}	Bond	Kcal mol^{-1}
$C_6H_5CH_2 - NH_2$	71.9 ± 1	$Cl - C_2F_5$	82.7 ± 1.7
$C_6H_5NH-CH_3$	67.7	$Br - CH_3$	70.0 ± 1.2
$C_6H_5CH_2 - NHCH_3$	68.7 ± 1	$Br - CN$	83 ± 1
$C_6H_5N(CH_2) - CH_3$	65.2	$Br - COC_6H_5$	64.2
$C_6H_5CH_2 - N(CH_3)_2$	60.9 ± 1	$Br - CF_3$	70.6 ± 1.0
$CF_3 - NF_2$	65 ± 2.5	$Br - CBr_3$	56.2 ± 1.8
$CH_2 = N_2$	$\leq 41.7 \pm 1$	$Br - C_2F_5$	68.7 ± 1.5
$CH_3N:N - CH_3$	52.5	$Br -n -C_3F$	66.5 ± 2.5
$C_2H_5N:N-C_2H_5$	50.0	$I - CH_3$	56.3 ± 1
$i -C_3H_7N:N-i -C_3H_7$	47.5	1-norbornyl	62.5 ± 2.5
$n -C_4H_9N:N-n -C_4H_9$	50.0	$I - CN$	73 ± 1
$i -C_4H_9N:N-i -C_4H_9$	49.0	$I - CF_3$	53.5 ± 2
$s -C_4H_9N:N-s -C_4H_9$	46.7	$CH_3 - Ga(CH_3)_2$	59.5
$t -C_4H_9N:N-t -C_4H_9$	43.5	$CH_3 - CdCH_3$	54.4
$C_6H_5CH_2N:N-C_6H_5CH_2$	37.6	$CH_3 - HgCH_3$	57.5
$CF_3N:N - CF_3$	55.2	$C_2H_5 - HgC_2H_5$	43.7 ± 1
$C_2H_5 - NO_2$	62	$n -C_3H_7 - Hg \ n -C_3H_7$	47.1
$O=CO$	127.2 ± 0.1	$i -C_3H_7 - Hg \ i -C_3H_7$	40.7
$CH_3 - O_2SCH_3$	66.8	$C_6H_5 - HgC_6H_5$	68
Allyl-O_2SCH_3	49.6	$CH_3 -TlCH_3)_2$	36.4 ± 0.6
$C_6H_5CH_2 - O_2SCH_3$	52.9	$CH_3 - Pb(CH_3)_3$	49.4 ± 1
$C_6H_5S - CH_3$	60	$NH_2 - NH_2$	70.8 ± 2
$C_6H_5CH_2 - SCH_3$	53.8	$NH_2 - NHCH_3$	64.8
$F - CH_3$	103 ± 3	$NH_2 - N(CH_3)_2$	62.7
$Cl - CN$	97 ± 1	$NH_2 - NHC_6H_5$	51.1
$Cl - COC_6H_5$	74 ± 3	$NO - NO_2$	9.5 ± 0.5
$Cl - CF_3$	86.1 ± 0.8	$NO_2 - NO_2$	12.9 ± 0.5
$Cl - CCl_2F$	73 ± 2	$NF_2 - NF_2$	21 ± 1

Bond Strengths of Polyatomic Molecules (Continued)
Listed by Molecule

Bond	Kcal mol^{-1}	Bond	Kcal mol^{-1}
$O - N_2$	40	n -$C_3H_7CO_2$ - O_2Cn -C_3H_7	30.4 ± 2
$O - NO$	73	$O - SO$	132 ± 2
$HO - N{:}CHCH_3$	49.7	$F - OCF_3$	43.5 ± 0.5
$Cl - NF_2$	≈ 32	$Cl - OH$	60 ± 3
$HO - OH$	51 ± 1	$O - ClO$	59 ± 3
$CH_3O - OCH_3$	36.9 ± 1	$Br - OH$	56 ± 3
$HO - OC(CH_3)_3$	42.5	$I - OH$	56 ± 3
$C_2H_5O - OC_2H_5$	37.3 ± 1.2	$ClO3 - ClO_4$	58.4
n -C_3H_7O - O n -C_3H_7	37.2 ± 1	$O = PF_3$	130 ± 5
i -C_3H_7O - O i -C_3H_7	37.0 ± 1	$O = PCl_3$	122 ± 5
s -C_4H_9O - O s -C_4H_9	36.4 ± 1	$O = PBr_3$	119 ± 5
t -C_4H_9O - O t -C_4H_9	37.4 ± 1	$SiH_3 - SiH_3$	81 ± 4
$(CH_3)_3CCH_2O$ - $OCH_2C(CH_3)_3$	36.4 ± 1	$(CH_3)_3Si - Si(CH_3)_3$	80.5
$O - O_2ClF$	58.4		
CH_3CO_2 - O_2CCH_3	30.4 ± 2		
$C_2H_5CO_2$ - $O_2CC_2H_5$	30.4 ± 2		

To convert kcal to KJ, multiply by 4.184.

The values refer to a temperature of 298 K and have mostly been determined by kinetic methods. Some have been calculated from formation of the species involved according to equations:

$$D(R\text{-}X) = \Delta H_f^\circ (R^\bullet) + \Delta H_f^\circ(X^\bullet) - \Delta H_f^\circ (RX)$$

$$\text{or} \quad D(R\text{-}X) = 2\Delta H_f^\circ (R^\bullet) - \Delta H_f^\circ (RR)$$

Data from: Kerr, J. A., Parsonage, M. J., and Trotman-Dickenson, A. F., in *Handbook of Chemistry and Physics, 55th ed.*, Weast, R. C., Ed., CRC Press, Cleveland, 1974, F-213.

BOND STRENGTHS OF POLYATOMIC MOLECULES
Listed by Values

Bond	Kcal mol^{-1}	Bond	Kcal mol^{-1}
$NO - NO_2$	9.5 ± 0.5	$CH_3 - Pb(CH_3)_3$	49.4 ± 1
$NO_2 - NO_2$	12.9 ± 0.5	Allyl-O_2SCH_3	49.6
$NF_2 - NF_2$	21 ± 1	$HO - N{:}CHCH_3$	49.7
$CH_3CO_2 - O_2CCH_3$	30.4 ± 2	$C_2H_5N{:}N-C_2H_5$	50.0
$C_2H_5CO_2 - O_2CC_2H_5$	30.4 ± 2	$n\text{-}C_4H_9N{:}N\text{-}n\text{-}C_4H_9$	50.0
$n\text{-}C_3H_7CO_2 - O_2Cn\text{-}C_3H_7$	30.4 ± 2	$HO - OH$	51 ± 1
$Cl - NF_2$	≈ 32	$NH_2 - NHC_6H_5$	51.1
BH_3-BH_3	35	$CH_3N{:}N - CH_3$	52.5
$CH_3 - Tl(CH_3)_2$	36.4 ± 0.6	$C_6H_5CH_2 - O_2SCH_3$	52.9
$s\text{-}C_4H_9O - O\ s\text{-}C_4H_9$	36.4 ± 1	$I - CF_3$	53.5 ± 2
$(CH_3)_3CCH_2O - OCH_2C(CH_3)_3$	36.4 ± 1	$C_6H_5CH_2 - SCH_3$	53.8
$CH_3O - OCH_3$	36.9 ± 1	$CH_3 - CdCH_3$	54.4
$i\text{-}C_3H_7O - O\ i\text{-}C_3H_7$	37.0 ± 1	$CF_3N{:}N - CF_3$	55.2
$n\text{-}C_3H_7O - O\ n\text{-}C_3H_7$	37.2 ± 1	$Br - OH$	56 ± 3
$C_2H_5O - OC_2H_5$	37.3 ± 1.2	$I - OH$	56 ± 3
$t\text{-}C_4H_9O - O\ t\text{-}C_4H_9$	37.4 ± 1	$Br - CBr_3$	56.2 ± 1.8
$C_6H_5CH_2N{:}N-C_6H_5CH_2$	37.6	$I - CH_3$	56.3 ± 1
$O - N_2$	40	$CH_3 - HgCH_3$	57.5
$i\text{-}C_3H_7 - Hg\ i\text{-}C_3H_7$	40.7	$O - O_2ClF$	58.4
$CH_2 = N_2$	≤ 41.7 ± 1	$ClO3 - ClO_4$	58.4
$HO - OC(CH_3)_3$	42.5	$O - ClO$	59 ± 3
$t\text{-}C_4H_9N{:}N\text{-}t\text{-}C_4H_9$	43.5	$(C_6H_5CH_2)_2CH-COOH$	59.4
$F - OCF_3$	43.5 ± 0.5	$CH_3 - Ga(CH_3)_2$	59.5
$C_2H_5 - HgC_2H_5$	43.7 ± 1	$C_6H_5C(CH_3)(CN) - CH_3$	59.9
$s\text{-}C_4H_9N{:}N\text{-}s\text{-}C_4H_9$	46.7	$C_6H_5S - CH_3$	60
$n\text{-}C_3H_7 - Hg\ n\text{-}C_3H_7$	47.1	$Cl - OH$	60 ± 3
$i\text{-}C_3H_7N{:}N\text{-}i\text{-}C_3H_7$	47.5	$C_6H_5CH_2 - N(CH_3)_2$	60.9 ± 1
$i\text{-}C_4H_9N{:}N\text{-}i\text{-}C_4H_9$	49.0	$C_2H_5 - NO_2$	62

Bond Strengths of Polyatomic Molecules (Continued)
Listed by Values

Bond	Kcal mol^{-1}		Bond	Kcal mol^{-1}	
1-norbornyl	62.5	± 2.5	Cl - CCl$_2$F	73	± 2
NH$_2$ - N(CH$_3$)$_2$	62.7		I - CN	73	± 1
Br - COC$_6$H$_5$	64.2		O - NO	73	
NH$_2$ - NHCH$_3$	64.8		C$_6$H$_5$CO - CF$_3$	73.8	
CF$_3$ - NF$_2$	65	± 2.5	Cl - COC$_6$H$_5$	74	± 3
C$_6$H$_5$N(CH$_2$) - CH$_3$	65.2		CF$_2$ = CF$_2$	76.3	± 3
C$_6$H$_5$CH$_2$CO - CH$_2$C$_6$H$_5$	65.4		H-ONO	78.3	± 0.5
C$_6$H$_5$CO - COC$_6$H$_5$	66.4		H-pentadien-1,4-yi-3	80	± 1
Br -n -C$_3$F	66.5	± 2.5	(CH$_3$)$_3$Si - Si(CH$_3$)$_3$	80.5	
CH$_3$ - O$_2$SCH$_3$	66.8		SiH$_3$ - SiH$_3$	81	± 4
C$_6$H$_5$CH$_2$-n -C$_3$H$_7$	67	± 2	H-cyclopentadien-1,3-yl-5	81.2	± 1.2
C$_6$H$_5$CH$_2$ - O$_2$CCH$_3$	67		Cl - C$_2$F5	82.7	± 1.7
CH$_3$CO - COCH$_3$	67.4	± 2.3	H-methdllyl	83	± 1
C$_6$H$_5$NH-CH$_3$	67.7		Br - CN	83	± 1
C$_6$H$_5$ - HgC$_6$H$_5$	68		Cl - CF$_3$	86.1	± 0.8
C$_6$H$_5$CH$_2$ - COOH	68.1		H-OC$_6$H$_5$	88	± 5
C$_6$H$_5$CH$_2$ - NHCH$_3$	68.7	± 1	H$_3$C-CH$_3$	88	± 2
Br - C$_2$F$_5$	68.7	± 1.5	CH$_2$F - CH$_2$F	88	± 2
C$_6$H$_5$CH$_2$-C$_2$H$_5$	69	± 2	H-SCH?	≥ 88	
C$_6$H$_5$CH$_2$ - O$_2$CC$_6$H$_5$	69		H-allyl	89	± 1
CH$_3$-C(CH$_3$)$_2$CH:CH$_2$	69.4		H-O$_2$H	90	± 2
Br - CH$_3$	70.0	± 1.2	H-SH	90	± 2
CH$_3$-C(CH$_3$)$_2$CN	70.2	± 2	H-Si(CH$_3$)$_3$	90	± 3
Br - CF$_3$	70.6	± 1.0	H-t-C$_4$H$_9$	92	± 1.2
NH$_2$ - NH$_2$	70.8	± 2	H-propargyl	93.9	± 1.2
C$_6$H$_5$CH(CH$_3$) - CH$_3$	71		H-SiH$_3$	94	± 3
C$_6$H$_5$CH$_2$ -NH$_2$	71.9	± 1	H-i-C$_3$H$_7$	95	± 1
CH$_3$-CH$_2$CN	72.7	± 2	H-s-C$_4$H$_9$	95	± 1

Bond Strengths of Polyatomic Molecules (Continued)
Listed by Values

Bond	Kcal mol^{-1}	Bond	Kcal mol^{-1}	
H-cyclobutyl	96.5 ± 1	H-CH$_2$	110	± 2
CF$_3$ - CF$_3$	96.9 ± 2	H-O$_2$CC$_2$H$_3$	110	± 4
Cl - CN	97 ± 1	H-O$_2$CCH$_3$	112	± 4
H-cyclopropycarbinyl	97.4 ± 1.6	H-OH	119	± 1
H-C$_2$H$_5$	98 ± 1	O = PBr$_3$	119	± 5
H-n-C$_3$H$_7$	98 ± 1	O = PCl$_3$	122	± 5
H-cyclopropyl	100.7 ± 1	O=CO	127.2	± 0.1
H-ONO$_2$	101.2 ± 0.5	H-ethynyl	128	± 5
H-CH	102 ± 2	NC-CN	128	± 1
H-O$_2$Cn-C$_3$H$_7$	103 ± 4	O = PF$_3$	130	± 5
F - CH$_3$	103 ± 3	O - SO	132	± 2
H-OCH$_3$	103.6 ± 1	H$_2$C=CH$_2$	172	± 2
H-OC$_2$H$_5$	103.9 ± 1	HC=CH	230	± 2
H-CH$_3$	104 ± 1			
H-OC(CH$_3$)$_3$	104.7 ± 1			
H-vinyl	≥ 108 ± 2			

To convert kcal to KJ, multiply by 4.184.

The values refer to a temperature of 298 K and have mostly been determined by kinetic methods. Some have been calculated from formation of the species involved according to equations:

$$D(R\text{-}X) = \Delta H_f^{\circ}(R^{\bullet}) + \Delta H_f^{\circ}(X^{\bullet}) - \Delta H_f^{\circ}(RX) \quad \text{or} \quad D(R\text{-}X) = 2\Delta H_f^{\circ}(R^{\bullet}) - \Delta H_f^{\circ}(RR)$$

From Kerr, J. A., Parsonage, M. J., and Trotman-Dickenson, A. F., in *Handbook of Chemistry and Physics, 55th ed.*, Weast, R. C., Ed., CRC Press, Cleveland, 1974, F-213.

CARBON BOND LENGTHS

Bonds Between Carbon and Other Elements

Bond Lengths of Carbon with other Elements (Å)
Range of Values

1 IA	2 IIA	3 IIIB	4 IVB	5 VB	6 VIB	7 VIIB	8 VIII	9 VIII	10 VIII	11 IB	12 IIB	13 IIIA	14 IVA	15 VA	16 VIA	17 VIIA	18 VIIIA
H 1.056 -1.115	Be 1.93											B 1.56	C 1.54 -1.20	N 1.47 -1.1	O 1.15 -1.43	F 1.30 -1.831	
					Cr 1.92		Fe 1.94	Co 1.93	Ni 1.82	------		Al 2.24	Si 1.84 -1.88	P 1.87	S 1.55 -1.81	Cl 1.70 -1.79	
					Mo 2.08				Pd 2.27				Ge 1.98	As 1.98	Se 1.71 -1.98	Br 1.79 -1.937	
					W 2.06						Hg 2.07	In 2.16	Sn 2.143 -2.18	Sb 2.202	Te 2.05	I 2.05 -2.13	
													Pb 2.29	Bi 2.30			

CARBON BOND LENGTHS
Bonds Between Carbon and Other Elements

Group No.	Element	At. No.	Sym	Bond Length (Å)		Bond Type
1	Hydrogen	1	H	1.056	± 1.115	
2	Beryllium	4	Be	1.93		
6	Chromium	24	Cr	1.92	± 0.04	
	Molybdenum	42	Mo	2.08	± 0.04	
	Tungsten	74	W	2.06	± 0.01	
8	Iron	26	Fe	1.94	± 0.02	
9	Cobalt	27	Co	1.93	± 0.02	
10	Nickel	28	Ni	1.82	± 0.03	
	Palladium	46	Pd	2.27	± 0.04	
12	Mercury	80	Hg	2.07	± 0.01	
13	Aluminum	13	Al	2.24	± 0.04	
	Boron	5	B	1.56	± 0.01	
	Indium	49	In	2.16	± 0.04	
14	Carbon	6	C	1.20	± 1.54	Alkyls (CH_3XH_3)
	Germanium	32	Ge	1.98	± 0.03	Alkyls (CH_3XH_3)
	Lead	82	Pb	2.29	± 0.05	Alkyls (CH_3XH_3)
	Silicon	14	Si	1.865	± 0.008	Alkyls (CH_3XH_3)
				1.84	± 0.01	Aryls (C_6H5XH_3)
				1.88	± 0.01	Neg. Subst. (CH_3XCl_3)
	Tin	50	Sn	2.143	± 0.008	Alkyls (CH_3XH_3)
				2.18	± 0.02	Neg. Subst. (CH_3XCl_3)
15	Arsenic	33	As	1.98	± 0.02	Paraffinic ($(CH_3)_3X$)
	Bismuth	83	Bi	2.30		Paraffinic ($(CH_3)_3X$)
	Nitrogen	7	N	1.47	± 1.1	
	Phosphorus	15	P	1.87	± 0.02	Paraffinic ($(CH_3)_3X$)
	Antimony	51	Sb	2.202	± 0.016	Paraffinic ($(CH_3)_3X$)
16	Oxygen	8	O	1.43	± 1.15	
	Sulfur	16	S	1.81	± 1.55	
	Selenium	34	Se	1.98	± 1.71	

Carbon Bond Lengths (Continued)
Bonds Between Carbon and Other Elements

Group No.	Element	At. No.	Sym	Bond Length (Å)		Bond Type
	Tellurium	52	Te	2.05	± 0.14	
17	Bromine	35	Br	1.937	± 0.003	Paraffinic (mono. substituted) (CH_3X)
				1.937	± 0.003	Paraffinic (disubstituted) (CH_2X_2)
				1.89	± 0.01	Olfinic$(CH_2:CHX)$
				1.85	± 0.01	Aromatic (C_6H_3X)
				1.79	± 0.01	Acetylenic $(HC:CX)$
	Chlorine	17	Cl	1.767	± 0.002	Paraffinic (mono. substituted) (CH_3X)
				1.767	± 0.002	Paraffinic (disubstituted) (CH_2X_2)
				1.72	± 0.01	Olfinic $(CH_2:CHX)$
				1.70	± 0.01	Aromatic (C_6H_3X)
				1.79	± 0.01	Acetylenic $(HC:CX)$
	Fluorine	9	F	1.831	± 0.005	Paraffinic (mono. substituted) (CH_3X)
				1.334	± 0.004	Paraffinic (disubstituted) (CH_2X_2)
				1.325	± 0.1	Olfinic$(CH_2:CHX)$

CARBON BOND LENGTHS
Bonds Between Carbon and Other Elements

Group No.	Element	At. No.	Sym	Bond Length (Å)		Bond Type
				1.30	± 0.01	Aromatic (C_6H_3X)
				1.635	± 0.004	Acetylenic (HC:CX)
	Iodine	53	I	2.13	± 0.1	Paraffinic (mono. substituted) (CH_3X)
				2.13	± 0.1	Paraffinic (disubstituted) (CH_2X_2)
				2.092	± 0.005	Olfinic (CH_2:CHX)
				2.05	± 0.01	Aromatic (C_6H_3X)
				1.99	± 0.02	Acetylenic (HC:CX)

Data from: David R., *CRC Handbook of Chemistry and Physics*, CRC Press, Boca Raton, (1990); and *"Tables of interatomic distances"* Chem. Soc. of London, 1958.

   The tables are based on bond distance determinations, by experimental methods, mainly x-ray and electron diffraction, and include, values published up to January 1, 1956. In the present tables, for the sake of completeness individual values of bond distances of lower accuracy are quoted with limits of error indicated where possible. Values for tungsten and bismuth should be treated with particular caution.

   According to the statistical theory of errors if an average quantity μ and a standard deviation s can be evaluated there is a 95% probability that the true value lies within the interval m ± 2a. Too much reliance should, however, not be placed on a values in bond distance determinations since the derivation of these certain sources of error may have been neglected. Values of the bond lengths and the limits of error are each given in Angstrom units.

   Reproduced by permission from International Tables for X-ray Crystallography.

BOND LENGTH VALUES BETWEEN ELEMENTS
Listed by bonds

Elements	Compound	Bond length (Å)		
B–B	B_2H_6	1.770	±	0.013
B–Br	BBF	1.88		
	BBr_3	1.87	±	0.02
B–Cl	BCl	1.715		
	BCl_3	1.72	±	0.01
B–F	BF	1.262		
	BF_3	1.29	±	0.01
B–H	Hydrides	1.21	±	.02
B–H bridge	Hydrides	1.39	±	.02
B–N	$(BClNH)_3$	1.42	±	.01
B–0	BO	1.2049		
	$B(OH)_3$	1.362	±	0.005 (av)
N–Cl	NO_2Cl	1.79	±	0.02
N–F	NF_3	1.36	±	0.02
N–H	$[NH_4]^+$	1.034	±	0.003
	NH	1.038		
	ND	1.041		
	HNCS	1.013	±	0.005
N–N	N_3H	1.02	±	0.01
	N_2O	1.126	±	0.002
	$[N_2]^+$	1.116		
N–O	NO_2Cl	1.24	±	0.01
	NO_2	1.188	±	0.005
N=O	N_2O	1.186	±	0.002
	$[NO]^+$	1.0619		
N–Si	SiN	1.572		

Bond Lengths Between Elements (Continued)
Listed by bonds

Elements	Compound	Bond length (Å)		
O–H	$[OH]^+$	1.0289		
	OD	0.9699		
	H_2O_2	0.960	±	0.005
O–O	H_2O_2	1.48	±	0.01
	$[O_2]^+$	1.227		
	$[O_2]^-$	1.26	±	0.2
	$[O_2]^{--}$	1.49	±	0.02
P–D	PD	1.429		
P–H	$[PH_4]^+$	1.42	±	0.02
P–N	PN	1.4910		
P–S	$PSBr_3$ (Cl_3,F_3)	1.86	±	0.02
S–Br	$SOBr_2$	2.27	±	0.02
S–F	SOF_2	1.585	±	0.005
S–D	SD	1.3473		
	SD_2	1.345		
S–O	SO_2	1.4321		
	$SOCl_2$	1.45	±	0.02
S–S	S_2Cl_2	2.04	±	0.01
Si–Br	$SiBr_4$	2.17	±	1.01
Si–Cl	$SiCl_4$	2.03	±	1.01 (av)
Si–F	SiF_4	1.561	±	0.003 (av)
Si–H	SiH_4	1.480	±	0.005
Si–O	$[SiO]^+$	1.504		
Si–Si	$[Si_2Cl_2]$	2.30	±	0.02

To convert Å to nm, multiply by 10^{-1}

Data from: Kennard, O., in *Handbook of Chemistry and Physics*, 69th ed., Weast, R. C., Ed., CRC Press, Boca Raton, Fla., 1988, F-167.

BOND LENGTH VALUES BETWEEN ELEMENTS
Listed by Bond Length Values

Elements	Compound	Bond length (Å)		
O–H	H_2O_2	0.960	±	0.005
O–H	OD	0.9699		
N–H	HNCS	1.013	±	0.005
N–N	N_3H	1.02	±	0.01
O–H	$[OH]^+$	1.0289		
N–H	$[NH4]^+$	1.034	±	0.003
N–H	NH	1.038		
N–H	ND	1.041		
N=O	$[NO]+$	1.0619		
N–N	$[N_2]^+$	1.116		
N–N	N2O	1.126	±	0.002
N=O	N_2O	1.186	±	0.002
N–O	NO_2	1.188	±	0.005
B–0	BO	1.2049		
B–H	Hydrides	1.21	±	.02
O–O	$[O2]+$	1.227		
N–O	NO2Cl	1.24	±	0.01
O–O	$[O_2]^-$	1.26	±	0.2
B–F	BF	1.262		
B–F	BF_3	1.29	±	0.01
S–D	SD2	1.345		
S–D	SD	1.3473		
N–F	NF3	1.36	±	0.02
B–0	B(OH)3	1.362	±	0.005 (av)
B–H bridge	Hydrides	1.39	±	.02
B–N	$(BClNH)_3$	1.42	±	.01
P–H	[PH4]+	1.42	±	0.02
P–D	PD	1.429		

Bond Length Values Between Elements (Continued)
Listed by Bond Length Values

Elements	Compound	Bond length (Å)		
S–O	SO2	1.4321		
S–O	$SOCl_2$	1.45	±	0.02
O–O	H_2O_2	1.48	±	0.01
Si–H	SiH_4	1.480	±	0.005
O–O	$[O2]^{--}$	1.49	±	0.02
P–N	PN	1.4910		
Si–O	[SiO]+	1.504		
Si–F	SiF_4	1.561	±	0.003 (av)
N–Si	SiN	1.572		
S–F	SOF2	1.585	±	0.005
B–Cl	BCl	1.715		
B–Cl	BCl_3	1.72	±	0.01
B–B	B2H6	1.770	±	0.013
N–Cl	NO2Cl	1.79	±	0.02
P–S	$PSBr_3$ (Cl_3,F_3)	1.86	±	0.02
B–Br	BBr3	1.87	±	0.02
B–Br	BBF	1.88		
Si–Cl	$SiCl_4$	2.03	±	1.01 (av)
S–S	S_2Cl_2	2.04	±	0.01
Si–Br	SiBr4	2.17	±	1.01
S–Br	$SOBr_2$	2.27	±	0.02
Si–Si	Si_2Cl_2	2.30	±	0.02

To convert Å to nm, multiply by 10^{-1}

From Kennard, O., in *Handbook of Chemistry and Physics*, 69th ed., Weast, R. C., Ed., CRC Press, Boca Raton, Fla., 1988, F–167.

BOND ANGLE VALUES BETWEEN ELEMENTS
Listed by Bonds

Element	Bond	Compound	Bond angle (°)		
B	H-B-H	B_2H_6	121.5	±	7.5
B	Br-B-Br	BBr_3	120	±	6
B	Cl- B-Cl	BCl_3	120	±	3
B	F-B-F	BF_3	120		
B	O-B-O	$B(OH)_3$	119.7		
N	B-N-B	$(BClNH)_3$	121		
N	F-N-F	NF_3	102.5	±	1.5
N	H-N-C	HNCS	130.25	±	0.25
N	H-N-N'	N_3H	112.65	±	0.5
N	O-N-O	NO_2Cl	126	±	2
N	O-N-O	NO_2	134.1	±	0.25
O	O-O-H	H_2O_2	100	±	2
S	Br-S-Br	$SOBr_2$	96	±	2
S	F-S-F	SOF_2	92.8	±	1
S	O-S-O	SO_2	119.54		

To convert Å to nm, multiply by 10^{-1}

Data from: Kennard, O., in *Handbook of Chemistry and Physics*, 69th ed., Weast, R. C., Ed., CRC Press, Boca Raton, Fla., 1988, F-167.

BOND ANGLE VALUES BETWEEN ELEMENTS
Listed by Bond Angle Value

Bond	Compound	Bond angle (°)		
F-S-F	SOF_2	92.8	±	1
Br-S-Br	$SOBr_2$	96	±	2
O-O-H	H_2O_2	100	±	2
F-N-F	NF_3	102.5	±	1.5
H-N-N'	N_3H	112.65	±	0.5
O-S-O	SO_2	119.54		
O-B-O	$B(OH)_3$	119.7		
Br-B-Br	BBr_3	120	±	6
Cl- B-Cl	BCl_3	120	±	3
F-B-F	BF_3	120		
B-N-B	$(BClNH)_3$	121		
H-B-H	B_2H_6	121.5	±	7.5
O-N-O	NO_2Cl	126	±	2
H-N-C	HNCS	130.25	±	0.25
O-N-O	NO_2	134.1	±	0.25

To convert Å to nm, multiply by 10^{-1}

Data from: Kennard, O., in *Handbook of Chemistry and Physics*, 69th ed., Weast, R. C., Ed., CRC Press, Boca Raton, Fla., 1988, F-167.

HEAT OF FORMATION OF SELECTED INORGANIC OXIDES

Reaction	Temperature range of validity	ΔH_0	2.303a	b	c	I
2 Ac(c) + 3/2 O_2(g) = Ac_2O_3(c)	298.16–1,000K	–446,090	–16.12	–	–	+109.89
2 Al(c) + 1/2 O_2(g) = Al_2O(g)	298.16–931.7K	–31,660	+14.97	–	–	–72.74
2 Al(l) + 1/2 O_2(g) = Al_2O(g)	931.7–2,000K	–38,670	+10.36	–	–	–51.53
Al(c) + 1/2 O_2(g) = AlO(g)	298.16–931.7K	+10,740	+5.76	–	–	–37.61
Al(l) + 1/2 O_2(g) = AlO(g)	931.7–2,000K	+8,170	+5.76	–	–	–34.85
2 Al(c) + 3/2 O_2(g) = Al_2O_3 (corundum)	298.16–931.7K	–404,080	–15.68	+2.18	+3.935	+123.64
2 Al(l) + 3/2 O_2(g) = Al_2O_3 (corundum)	931.7–2,000K	–407,950	–6.19	–0.78	+3.935	+102.37
2 Sb(c) + 3/2 O_2(g) = Sb_2O_3 (cubic)	298.16–842K	–169,450	+6.12	–6.01	–0.30	+52.21
2 Sb(c) + 3/2 O_2(g) = Sb_2O_3 (orthorhombic)	298.16–903K	–168,060	+6.12	–6.01	–0.30	+50.56
2 As(c) + 3/2 O_2(g) = As_2O_3 (orthorhombic)	298.16–542K	–154,870	+29.54	–21.33	–0.30	–8.83
2 As(c) + 3/2 O_2(g) = As_2O_3 (monoclinic)	298.16–586K	–150,760	+29.54	–21.33	–0.30	–16.95
2 As(c) + 5/2 O_2(g) = As_2O_5(c)	298.16–883K	–217,080	+12.32	–4.65	–0.50	+80.50
Ba(α) + 1/2 O_2(g) = BaO(c)	298.16–648K	–134,590	–7.60	+0.87	+0.42	+45.76
Ba(β) + 1/2 O_2(g) = BaO(c)	648–977K	–134,140	–3.34	–0.56	+0.42	+34.01
Be(c) + 1/2 O_2(g) = BeO(c)	298.16–1,556K	–144,220	–1.91	–0.46	+1.24	+30.64
Bi(c) + 1/2 O_2(g) = BiO(c)	298.16–544K	–50,450	–4.61	–	–	+35.51

Heat of Formation of Selected Inorganic Oxides (Continued)

Reaction	Temperature range of validity	ΔH_0	2.303a	b	c	I
Bi(l) + 1/2 O$_2$(g) = BiO(c)	544–1,600K	−52,920	−4.61	—	—	+40.05
2 Bi(c) + 3/2 O$_2$(g) = Bi$_2$O$_3$(c)	298.16–544K	−139,000	−11.56	+2.15	−0.30	+96.52
2 Bi(l) + 3/2 O$_2$(g) = Bi$_2$O$_3$(c)	544–1,090K	−142,270	+2.30	−3.25	−0.30	+67.55
2 B(c) + 3/2 O$_2$(g) = B$_2$O(c)	298.16–723K	−304,690	+11.72	−7.55	+0.355	+34.25
2 B(c) + 3/2 O$_2$(g) = B$_2$O$_3$(gl)	298.16–723K	−298,670	+26.57	−15.90	−0.30	−10.40
Cd(c) + 1/2 O$_2$(g) = CdO(c)	298.16–594K	−62,330	−2.05	+0.71	−0.10	+29.17
Cd(l) + 1/2 O$_2$(g) = CdO(c)	594–1,038K	−63,240	+2.07	−0.76	−0.10	+20.14
Ca(α) + 1/2 O$_2$(g) = CaO(c)	298.16–673K	−151,850	−6.56	+1.46	+0.68	+43.93
Ca(β) + 1/2 O$_2$(g) = CaO(c)	673–1,124K	−151,730	−4.14	+0.41	+0.68	+37.63
C(graphite) + 1/2 O$_2$(g) = CO(g)	298.16–2,000K	−25,400	+2.05	+0.27	−1.095	−28.79
C(graphite) + O$_2$(g) = CO$_2$(g)	298.16–2,000K	−93,690	+1.63	−0.7	−0.23	−5.64
2 Ce(c) + 3/2 O$_2$(g) = Ce$_2$O$_3$(c)	298.16–1,048K	−435,600	−4.60	—	—	+92.84
2 Ce(l) + 3/2 O$_2$(g) = Ce$_2$O$_3$(c)	1,048–1,900K	−440,400	−4.60	—	—	+97.42
Ce(c) + O$_2$(g) = CeO$_2$(c)	298.16–1,048K	−245,490	−6.42	+2.34	−0.20	+67.79
Ce(l) + O$_2$(g) = CeO$_2$(c)	1,048–2,000K	−247,930	+0.71	−0.66	−0.20	+51.73
2 Cs(c) + 1/2 O$_2$(g) = Cs$_2$O(c)	298.16–301.5K	−75,900	—	—	—	+36.60

Heat of Formation of Selected Inorganic Oxides (Continued)

Reaction	Temperature range of validity	ΔH_0	2.303a	b	c	I
$2\ Cs(l) + 1/2\ O_2(g) = Cs_2O(c)$	301.5–763K	−76,900	–	–	–	+39.92
$2\ Cs(l) + 1/2\ O_2(g) = Cs_2O(l)$	763–963K	−75,370	−9.21	–	–	+64.47
$2\ Cs(g) + 1/2\ O_2(g) = Cs_2O(l)$	963–1,500K	−113,790	−23.03	–	–	+145.60
$2\ Cs(c) + 3/2\ O_2(g) = Cs_2O_3(c)$	298.16–301.5K	−112,690	−11.51	–	–	+110.10
$2\ Cs(l) + 3/2\ O_2(g) = Cs_2O_3(c)$	301.5–775K	−113,840	−12.66	–	–	+116.77
$2\ Cs(l) + 3/2\ O_2(g) = Cs_2O_3(l)$	775–963K	−110,740	−26.48	–	–	+152.70
$2\ Cs(g) + 3/2\ O_2(g) = Cs_2O_3(l)$	963–1,500K	−148,680	−39.14	–	–	+229.87
$Cl_2(g) + 1/2\ O_2(g) = Cl_2O(g)$	298.16–2,000K	+17,770	−0.71	−0.12	+0.49	+16.81
$1/2\ Cl_2(g) + 1/2\ O_2(g) = ClO(g)$	298.16–1,000K	+33,000	–	–	–	0.24
$2\ Cl_2(g) + 3/2\ O_2(g) = ClO(g)$	298.16–500K	+37,740	+5.76	–	–	+21.42
$2\ Cr(c) + 3/2\ O_2(g) = Cr_2O_3(\beta)$	298.16–1,823K	−274,670	−14.07	+2.01	+0.69	+105.65
$2\ Cr(l) + 3/2\ O_2(g) = Cr_2O_3(\beta)$	1,823–2,000K	−278,030	+2.33	−0.35	+1.57	+58.29
$Cr(c) + O_2(g) = CrO_2\ (c)$	298.16–1,000K	−142,500	–	–	–	+42.00
$Cr(c) + 3/2\ O_2(g) = CrO_3(c)$	298.16–471K	−141,590	−13.82	–	–	+103.90
$Cr(c) + 3/2\ O_2(g) = Cr_2O_3(l)$	471–600K	−141,580	−32.24	–	–	+153.14
$Co(\alpha,\beta) + 1/2\ O_2(g) = CoO(c)$	298.16–1,400K	−56,910	+0.69	–	–	+16.03

Heat of Formation of Selected Inorganic Oxides (Continued)

Reaction	Temperature range of validity	ΔH_0	2.303a	b	c	I
$Co(\gamma) + 1/2\ O_2(g) = CoO(c)$	1,400–1,763K	−58,160	−1.15	–	–	+22.71
$2\ Cu(c) + 1/2\ O_2(g) = Cu_2O(c)$	298.16–1,357K	+10,550	−1.15	−1.10	−0.10	+21.92
$2\ Cu(l) + 1/2\ O_2(g) = Cu_2O(c)$	1,357–1,502K	−43,880	+8.47	−2.60	−0.10	−3.72
$2\ Cu(l) + 1/2\ O_2(g) = Cu_2O(l)$	1,502–2,000K	−37,710	−12.48	+0.25	−0.10	+54.44
$Cu(c) + 1/2\ O_2(g) = CuO(c)$	298.16–1,357K	−37,740	−0.64	−1.40	−0.10	+24.87
$Cu(l) + 1/2\ O_2(g) = CuO(c)$	1,357–1,720K	−39,410	+4.17	−2.15	−0.10	+12.05
$Cu(l) + 1/2\ O_2(g) = CuO(l)$	1,720–2,000K	−41,060	−11.35	+0.25	−0.10	+59.09
$2\ Au(c) + 3/2\ O_2(g) = Au_2O_3(c)$	298.16–500K	−2,160	−10.36	–	–	+95.14
$Hf(c) + O_2(g) = HfO_2$ (monoclinic)	298.16–2,000K	−268,380	−9.74	−0.28	+1.54	+78.16
$H_2(g) + 1/2\ O_2(g) = H_2O(l)$	298.16–373.16K	−70,600	−18.26	+0.64	−0.04	+91.67
$H_2(g) + 1/2\ O_2(g) = H_2O(g)$	298.16–2,000K	−56,930	+6.75	−0.64	−0.08	−8.74
$D_2(g) + 1/2\ O_2(g) = D_2O(l)$	298.16–374.5K	−72,760	−18.10	–	–	+93.59
$D_2(g) + 1/2\ O_2(g) = D_2O(g)$	298.16–2,000K	−58,970	+5.50	−0.75	+0.085	−3.74
$0.947\ Fe(\alpha) + 1/2\ O_2(g) = Fe_{0.9470}(c)$	298.16–1,033K	−65,320	−11.26	+2.61	+0.44	+48.60
$0.947\ Fe(\alpha) + 1/2\ O_2(g) = Fe_{0.9470}(c)$	1,033–1,179K	−62,380	+4.08	−0.75	+0.235	+3.00
$0.947\ Fc(\beta) + 1/2\ O_2(g) = Fe_{0.9470}(c)$	1,179–1,650K	−66,750	−8.04	+0.67	−0.10	+42.28

Heat of Formation of Selected Inorganic Oxides (Continued)

Reaction	Temperature range of validity	ΔH_0	2.303a	b	c	I
$0.947\ Fe(\gamma) + 1/2\ O_2(g) = Fe_{0.9470}(l)$	1,650–1,674K	−64,200	−18.72	+1.67	−0.10	+73.45
$0.947\ Fe(\gamma) + 1/2\ O_2(g) = Fe_{0.9470}(l)$	1,647–1,803K	−59,650	−6.84	+0.25	−0.10	+34.81
$0.947\ Fe(\delta) + 1/2\ O_2(g) = Fe_{0.9470}(l)$	1,803–2,000K	−63,660	−7.48	+0.25	−0.10	+39.12
$3\ Fe(\alpha) + 2\ O_2(g) = Fe_3O_4(magnetite)$	298.16–900K	−268,310	+5.87	−12.45	+0.245	+73.11
$3\ Fe(\alpha) + 2\ O_2(g) = Fe_3O_4(\beta)$	900–1,033K	−272,300	−54.27	+11.65	+0.245	+233.52
$3\ Fe(\beta) + 2\ O_2(g) = Fe_3O_4(\beta)$	1,033–1,179K	−262,990	−5.71	+1.00	−0.40	+89.19
$3\ Fe(\gamma) + 2\ O_2(g) = Fe_3O_4(\beta)$	1,179–1,674K	−276,990	~4.05	+5.50	−0.40	+213.52
$2\ Fe(\alpha) + 3/2\ O_2(g) = Fe_2O_3(hematite)$	298.16–950K	−200,000	−13.84	−1.45	+1.905	+108.26
$2\ Fe(\alpha) + 3/2\ O_2(g) = Fe_2O_3(\beta)$	950–1,033K	−202,960	−42.64	+7.85	+0.13	+188.48
$2\ Fe(\beta) + 3/2\ O_2(g) = Fe_2O_3(\beta)$	1,033–1,050K	−196,740	−10.27	+0.75	−0.30	+92.26
$2\ Fe(\beta) + 3/2\ O_2(g) = Fe_2O_3(\gamma)$	1,050–1,179K	−193,200	−0.39	−0.13	−0.30	+59.96
$2\ Fe(\gamma) + 3/2\ O_2(g) = Fe_2O_3(\gamma)$	1,179–1,674K	−202,540	−25.95	+2.87	−0.30	+142.85
$2\ Fe(\alpha) + 3/2\ O_2(g) = Fe_2O_3(\gamma)$	1,674–1,800K	−192,920	−0.85	−0.13	−0.30	+61.21
$Pb(c) + 1/2\ O_2(g) = PbO\ (red)$	298.16–600.5K	−52,800	−2.76	−0.80	−0.10	+32.49
$Pb(l) + 1/2\ O_2(g) = PbO\ (red)$	600.5–762K	−53,780	−0.51	−1.75	−0.10	+28.44
$Pb(c) + 1/2\ O_2(g) = PbO\ (yellow)$	298.16–600.5K	−52,040	+0.81	−2.00	−0.10	+22.13

Heat of Formation of Selected Inorganic Oxides (Continued)

Reaction	Temperature range of validity	ΔH_0	2.303a	b	c	I
$Pb(l) + 1/2\ O_2(g) = PbO$ (yellow)	600.5–1,159K	−53,020	+3.06	−2.95	−0.10	+18.08
$I_2(c) + 5/2\ O_2(g) = I_2O_5(c)$	298.16–386.8K	−42,040	+2.30	—	—	+113.71
$I_2(l) + 5/2\ O_2(g) = I_2O_5(c)$	386.8–456K	−43,490	+16.12	—	—	+81.70
$I_2(g) + 5/2\ O_2(g) = I_2O_5(c)$	456–500K	−58,020	−6.91	—	—	+174.79
$Ir(c) + O_2(g) = IrO_2(c)$	298.16–1,300K	−39,480	+8.17	−6.39	−0.20	+20.33
$3\ Pb(c) + 2\ O_2(g) = Pb_3O_4(c)$	298.16–600.5K	−174,920	+8.82	−8.20	−0.40	+72.78
$Pb(c) + O_2(g) = PbO_2(c)$	298.16–600.5K	−66,120	+0.64	−2.45	−0.20	+45.58
$2\ Li(c) + 1/2\ O_2(g) = Li_2O(c)$	298.16–452K	−142,220	−3.06	+5.77	−0.10	+34.19
$Mg(c) + 1/2\ O_2(g) = MgO$ (periclase)	298.16–923K	−144,090	−1.06	+0.13	+0.25	+29.16
$Mg(l) + 1/2\ O_2(g) = MgO$ (periclase)	923–1,393K	−145,810	+1.84	−0.62	+0.64	+23.07
$Mg(g) + 1/2\ O_2(g) = MgO$ (periclase)	1,393–2,000K	−180,700	−3.75	−0.62	+0.64	+65.69
$Mn(\alpha) + 1/2\ O_2(g) = MnO(c)$	298.16–1,000K	−92,600	−4.21	+0.97	+0.155	+29.66
$Mn(\beta) + 1/2\ O_2(g) = MnO(c)$	1,000–1,374K	−91,900	+1.84	−0.39	+0.34	+12.15
$Mn(\gamma) + 1/2\ O_2(g) = Mno(c)$	1,374–1,410K	−89,810	+7.30	−0.72	+0.34	−6.05
$Mn(\delta) + 1/2\ O_2(g) = MnO(c)$	1,410–1,517K	−89,390	+8.68	−0.72	+0.34	−10.70
$Mn(l) + 1/2\ O_2(g) = MnO(c)$	1,517–2,000K	−93,350	+7.99	−0.72	+0.34	−5.90

Heat of Formation of Selected Inorganic Oxides (Continued)

Reaction	Temperature range of validity	ΔH_0	2.303a	b	c	I
3 Mn(α) + 2 O$_2$(g) = Mn$_3$O$_4$(α)	298.16–1,000K	–332,400	–7.41	+0.66	+0.145	+106.62
2 Mn(α) + 3/2 O$_2$(g) = Mn$_2$O$_3$(c)	298.16–1,000K	–230,610	–5.96	–0.06	+0.945	+80.74
Mn(α) + O$_2$(g) = MnO$_2$(c)	298.16–1,000K	–126,400	–8.61	+0.97	+1.555	+70.14
2 Hg(l) + 1/2 O$_2$(g) = Hg$_2$O(c)	298.16–629.88K	–22,400	–4.61	–	–	+43.29
Hg(l) + 1/2 O$_2$(g) = HgO (red)	298.16–629.88K	–21,760	+0.85	–2.47	–0.10	+24.81
Mo(c) + O$_2$(g) = MoO$_2$(c)	298.16–2,000K	–132,910	–3.91	–	–	+47.42
Mo(c) + 3/2 O$_2$(g) = MoO$_3$(c)	298.16–1,068K	–182,650	–8.86	–1.55	+1.54	+90.07
Ni(α) + 1/2 O$_2$(g) = NiO(c)	298.16–633K	–57,640	–4.61	+2.16	–0.10	+34.41
Ni(β) + 1/2 O$_2$(g) = NiO(c)	633–1,725K	–57,460	–0.14	–0.46	–0.10	+23.27
2 Nb(c) + 2 O$_2$(g) = Nb$_2$O$_4$(c)	298.16–2,000K	–382,050	–9.67	–	–	+116.23
2 Nb(c) + 5/2 O$_2$(g) = Nb$_2$O$_5$(c)	298.16–1,785K	–458,640	–16.14	–0.56	+1.94	+157.66
2 Nb(c) + 5/2 O$_2$(g) = Nb$_2$O$_5$(l)	1,785–2,000K	–463,630	–66.04	+2.21	–0.50	+317.84
N$_2$(g) + 1/2 O$_2$(g) = N$_2$O(g)	298.16–2,000K	+0,019	–1.57	–0.27	+0.92	+23.47
3/2 O$_2$(g) = O$_3$(g)	298.16–2,000K	+33,980	+2.03	–0.48	+0.36	+11.45
P (white) + 1/2 O$_2$(g) = PO(g)	298.16–317.4K	–9,370	+2.53	–	–	–25.40
P(l) + 1/2 O$_2$(g) = PO(g)	317.4–553K	–9,390	+3.45	–	–	–27.63

Heat of Formation of Selected Inorganic Oxides (Continued)

Reaction	Temperature range of validity	ΔH_0	2.303a	b	c	I
4 P (white) + 5 O_2(g) = P_4H_{10} (hexagonal)	298.16–317.4K	−711,520	+95.67	−51.50	−1.00	−28.24
2 K(c) + 1/2 O_2(g) = K_2O(c)	298.16–336.4K	−86,400	–	–	–	+33.90
2 K(l) + 1/2 O_2(g) = K_2O(c)	336.4–1,049K	−87,380	+1.15	–	–	+33.90
2 K(g) + 1/2 O_2(g) = K_2O(c)	1,049–1,500K	−133,090	−16.12	–	–	+129.64
Ra(c) + 1/2 O_2(g) = RaO(c)	298.16–1,000K	−130,000	–	–	–	+23.50
Re(c) + 3/2 O_2(g) = ReO_3(c)	298.16–433K	−149,090	−16.12	–	–	+110.49
Re(c) + 3/2 O_2(g) = ReO_3(l)	433–1,000K	−146,750	−31.32	–	–	+145.16
2Re(c) + 7/2 O_7(g) = Re_2O_7(c)	298.16–569K	−301,470	−34.64	–	–	+250.57
2 Re(c) + 7/2 O_7(g) = Re_2O_7(l)	569–635.5K	−295,810	−73.68	–	–	+348.45
2 Re(c) + 4 O_2(g) = Re_2O_8(l)	420–600K		−87.50	–	–	+425.32
2 Rb(c) + 1/2 O_2(g) = Rb_2O(c)	298.16–312.2K	−78,900	–	–	–	+32.20
2 Rb(l) + 1/2 O_2(g) = Rb_2O(c)	312.2–750K	−79,950	–	–	–	+35.56
Se(c) + 1/2 O_2(g) = SeO(g)	298.16–490K	+9,280	−3.04	+4.40	+0.30	−14.78
Se(l) + 1/2 O_2(g) = SeO(g)	490–1,027K	+9,420	+8.70	–	+0.30	−44.50
1/2 Se_2(g) + 1/2 O_2(g) = SeO(g)	1,027–2,000K	−7,400	−0.37	–	+0.19	−0.80
Si(c) + 1/2 O_2(g) = SiO(g)	298.16–1,683K	−21,090	+3.84	−0.16	~0.295	−33.14

Heat of Formation of Selected Inorganic Oxides (Continued)

Reaction	Temperature range of validity	ΔH_0	2.303a	b	c	I
$Si(l) + 1/2\ O_2(g) = SiO(g)$	1,683–2,000K	–30,170	–7.78	–0.12	+0.25	–40.01
$Si(c) + O_2(g) = SiO_2(\alpha$–quartz$)$	298.16–848K	–210,070	+3.98	–3.32	+0.605	+34.59
$Si(c) + O_2(g) = SiO_2(\beta$–quartz$)$	848–1,683K	–209,920	–3.36	–0.19	–0.745	+53.44
$Si(l) + O_2(g) = SiO_2(l)$	1,883–2,000K	–228,590	–15.66	–	–	+103.97
$Si(c) + O_2(g) = SiO_2(\alpha$–cristobalite$)$	298.16–523K	–207,330	+19.96	–9.75	–0.745	–9.78
$Si(c) + O_2(g) = SiO_2(\beta$–cristobalite$)$	523–1,683K	–209,820	–3.34	–0.24	–0.745	+53.35
$Si(c) + O_2(g) = SiO_2(\alpha$–tridymite$)$	298.16–390K	–207,030	+22.29	–11.62	–0.745	–15.64
$Si(c) + O_2(g) = SiO_2(\beta$–tridymite$)$	390–1,683K	–209,350	–1.59	–0.54	–0.745	+47.86
$2\ Ag(c) + 1/2\ O_2(g) = Ag_2O_2(c)$	298.16–1,000K	–7,740	–4.14	–	–	+27.84
$2\ Ag(c) + O_2(g) = Ag_2O_2(c)$	298.16–500K	–6,620	–3.22	–	–	+52.17
$2\ Na(c) + 1/2\ O_2(g) = Na_2O(c)$	298.16–371K	–99,820	–7.51	+5.47	–0.10	+50.43
$2\ Na(l) + 1/2\ O_2(g) = Na_2O(c)$	371–1,187K	–100,150	+4.97	–2.45	–0.10	+22.19
$2\ Na(c) + O_2(g) = Na_2O_2(c)$	298.16–371K	–122,500	–2.30	–	–	+57.51
$Sr(c) + 1/2\ O_2(g) = SrO(c)$	298.16–1,043K	–142,410	–6.79	+0.305	+0.675	+44.33
$S(rhombohedral) + 1/2\ O_2(g) = SO(g)$	298.16–368.6K	+19,250	–1.24	+2.95	+0.225	–18.84
$S(monoclinic) + 1/2\ O_2(g) = SO(g)$	368.6–392K	+19,200	–1.29	+3.31	+0.225	–18 72

Heat of Formation of Selected Inorganic Oxides (Continued)

Reaction	Temperature range of validity	ΔH_0	2.303a	b	c	I
$S(\lambda,\mu) + 1/2\ O_2(g) = SO(g)$	392–718K	+20,320	+10.22	−0.17	+0.225	−50 05
$1/2\ S_2(g) + 1/2\ O_2(g) = SO(g)$	298.16–2,000K	+3,890	+0 07	−	−	−1.50
$S(rhombohedral) + O_2(g) = SO_2(g)$	298.16–368.6K	−70,980	+0 83	+2.35	+0.51	−5 85
$S(monoclinic) + O_2(g) = SO_2(g)$	368.6–392K	−71,020	+0.78	+2.71	+0_51	−574
$S(\lambda,\mu) + O_2(g) = SO_2(g)$	392–718K	−69,900	+12.30	−0.77	+0.51	−37 10
$1/2\ S_2(g) + O_2\ (g) = SO_2(g)$	298.16–2,000K	−86,330	+2.42	−0.70	+0.31	+10 71
$S(rhombohedral) + 3/2\ O_2(g) = SO_3(c-I)$	298.16–335.4K	−111,370	−6.45	−	−	+88.32
$S(rhombohedral) + 3/2\ O_2(g) = SO_3(c-II)$	298.16–305.7K	−108,680	−11.97	−	−	+94 95
$S(rhombohedral) + 3/2\ O_2(g) = SO_3(l)$	298.16–335.4K	−107,430	−21.18	−	−	+113 76
$S(rhombohedral) + 3/2\ O_2(g) = SO_3(g)$	298.16–368.6K	−95,070	+1.43	+0.66	+1.26	+16.81
$S(monoclinic) + 3/2\ O_2(g) = SO_3(g)$	368.6–392K	−95,120	+1.38	+1.02	+1.26	+16.93
$S(\lambda,\mu) + 3/2\ O_2(g) = SO_3(g)$	392–718K	−94,010	+12.89	−2.46	+126	−14 40
$1/2\ S_2(g) + 3/2\ O_2(g) = SO_3(g)$	298.16–1,500K	−110,420	+3.02	−2.39	+106	+33 41
$2\ Ta(c) + 5/2\ O_2(g) = Ta_2O_5(c)$	298.16–2,000K	−492,790	−17.18	−1.25	+2.46	+161.68
$Te(c) + 1/2\ O_2(g) = TeO(g)$	298.16–723K	+43,110	+1.91	+0.84	+0.315	−27.22
$Te(l) + 1/2\ O_2(g) = TeO(g)$	723–1,360K	+39,750	+6.08	+0.09	+0.315	−33.94

Heat of Formation of Selected Inorganic Oxides (Continued)

Reaction	Temperature range of validity	ΔH_0	2.303a	b	c	I
$2\,Tl(\alpha) + O_2(g) = Tl_{2r}O(c)$	298.16–505.5K	−44,110	−6.91	—	—	+42.30
$2\,Tl(\beta) + O_2(g) = Tl_{2r}O(c)$	505.5–573K	−44,260	−6.91	—	—	+42.60
$2\,Tl(\alpha) + 3/2\,O_2(g) = Tl_{2r}O_3(c)$	298.16–505.5K	−99,410	−16.12	—	—	+119.09
$Th(c) + O_2(g) = ThO_2(c)$	298.16–2,000K	−294,350	−5.25	+0.59	+0.775	+62.81
$Sn(c) + 1/2\,O_2(g) = SnO(c)$	298.16–505K	−68,600	−3.57	+1.65	−0.10	+32.59
$Sn(l) + 1/2\,O_2(g) = SnO(c)$	505–1,300K	−69,670	+3.06	−1.50	−0.10	+18.39
$Sn(c) + O_2(g) = SnO_2(c)$	298.16–505K	−0,142	−14.00	+2.45	+2.38	+90.74
$Ti(\alpha) + 1/2\,O_2(g) = TiO(\alpha)$	298.16–1,150K	−125,010	−4.01	−0.29	+0.83	+36.28
$Ti(\alpha) + 1/2\,O_2(g) = TiO(\alpha)$	1,150–1,264K	−125,040	+1.17	−1.55	+0.83	+21.90
$2\,Ti(\alpha) + 3/2\,O_2(g) = Ti_2O_3(\alpha)$	298.16–473K	−360,660	+32.08	−23.49	−0.30	−10.66
$2\,Ti(\alpha) + 3/2\,O_2(g) = Ti_2O_3(\beta)$	473–1,150K	−369,710	−30.95	+2.62	+4.80	+162.79
$Ti(\alpha) + O_2(g) = TiO_2$ (rutile)	298.16–1,150K	−228,360	−12.80	+1.62	+1.975	+82.81
$Ti(\alpha) + O_2(g) = TiO_2$ (rutile)	1,150–2,000K	−228,380	−7.62	+0.36	+1.975	+68.43
$W(c) + O_2(g) = WO_2(c)$	298.16–1,500K	−137,180	−1.38	—	—	+45.56
$4W(c) + 11/2\,O_2(g) = W_4O_{11}(c)$	298.16–1,700K	−745,730	−32.70	—	—	+321.84
$W(c) + 3/2\,O_2(g) = WO_3(c)$	298.16–1,743K	−201,180	−2.92	−1.81	−0.30	+70.89

Heat of Formation of Selected Inorganic Oxides (Continued)

Reaction	Temperature range of validity	ΔH_0	2.303a	b	c	I
$W(c) + 3/2\ O_2(g) = WO_3(l)$	1,743–2,000K	–203,140	–35.74	+1.13	–0.30	+173.27
$U(\alpha) + O_2(g) = UO_2(c)$	298.16–935K	–262,880	–19.92	+3.70	+2.13	+100.54
$U(\beta) + O_2(g) = UO_2(c)$	935–1,045K	–260,660	–4.28	–0.31	+1.78	+55.50
$U(\gamma) + O_2(g) = UO_2(c)$	1,045–1,405K	–262,830	–6.54	–0.31	+1.78	+64.41
$U(l) + O_2(g) = UO_2(l)$	1,405–1,500K	–264,790	–5.92	–	–	+63.50
$3\ U(\alpha) + 4\ O_2(g) = U_3O_8(c)$	298.16–935K	–863,370	–56.57	+10.68	+5.20	+330.19
$3\ U(\beta) + 4\ O_2(g) = U_3O_8(c)$	935–1,045K	–856,720	–9.67	–1.35	+4.15	+195.12
$3\ U(\gamma) + 4\ O_2(g) = U_3O_8(c)$	1,045–1,405K	–863,230	–16.44	–1.35	+4.15	+221.79
$3\ U(l) + 4\ O_2(g) = U_3O_8(c)$	1,405–1,500K	–869,460	–10.91	–1.35	+4.15	+208.82
$U(\alpha) + 3/2\ O_2(g) = UO_3$ (hexagonal)	298.16–935K	–294,090	–18.33	+3.49	+1.535	+114.94
$U(\beta) + 3/2\ O_2(g) = UO_3$ (hexagonal)	935–1,045K	–291,870	–2.69	–0.52	+1.185	+69.90
$U(\gamma) + 3/2\ O_2(g) = UO_3$ (hexagonal)	1,045–1,400K	–294,040	–4.95	–0.52	+1.185	+78.80
$V(c) + 1/2\ O_2(g) = VO(c)$	298.16–2,000K	–101,090	–5.39	–0.36	+0.53	+38.69
$V(c) + 1/2\ O_2(g) = VO(g)$	298.16–2,000K	+52,090	+1.80	+1.04	+0.35	–28.42
$2\ V(c) + 3/2\ O_2(g) = V_2O_3(c)$	298.16–2,000K	–299,910	–17.98	+0.37	+2.41	+118.83
$2\ V(c) + 2\ O_2(g) = V_2O_4(\alpha)$	209.16–345K	–342,890	–11.03	+3.00	–0.40	+117.38

Heat of Formation of Selected Inorganic Oxides (Continued)

Reaction	Temperature range of validity	ΔH_0	2.303a	b	c	I
$2\,V(c) + 2\,O_2(g) = V_2O_4(\beta)$	345–1,818K	–345,330	–24.36	+1.30	+3.545	+155.55
$6\,V(c) + 13/2\,O_2(g) = V_6O_{13}(c)$	298.16–1,000K	–1,076,340	–95.33	–	–	+557.61
$2\,V(c) + 5/2\,O_2(g) = V_2O_5(c)$	298.16–943K	–381,960	–41.08	+5.20	+6.11	+228.50
$2\,Y(c) + 3/2\,O_2(g) = Y_2O_3(c)$	298.16–1,773K	–419,600	+2.76	–1.73	–0.30	+66.36
$Zn(c) + 1/2\,O_2(g) = ZnO(c)$	298.16–692.7K	–84,670	–6.40	+0.84	+0.99	+43.25
$Zr(\alpha) + O_2(g) = ZrO_2(\alpha)$	298.16–1,135K	–262,980	–6.10	+0.16	+1.045	+65.00
$Zr(\beta) + O_2(g) = ZrO_2(\alpha)$	1,135–1,478K	–264,190	–5.09	–0.40	+1.48	+63.58
$Zr(\beta) + O_2(g) = ZrO_2(\beta)$	1,478–2,000K	–262,290	–7.76	+0.50	–0.20	+69.50

The ΔH_0 values are given in gram calories per mole. The a, b, and I values listed here make it possible for one to calculate the $°\Delta F$ and $°\Delta S$ values by use of the following equations:

$$°\Delta F_t = °\Delta H + 2.303aT\log T + b \times 10^{-3}T^2 + c \times 10^{-1}T + IT$$

$$°\Delta F_t = °\Delta\log T + b \times 10^{-3}T^2 + c \times 10^{-1}T + IT$$

$$°\Delta S_t = -a - 2.303a\log T - 2b \times 10^{-3}T + c \times 10^{-5}T^{-2} - I$$

Data from: CRC Handbook of Materials Science, Vol II, Charles T. Lynch, Ed., CRC Press, Cleveland, (1974).

HEATS OF SUBLIMATION (AT 25°C) OF SELECTED METALS AND THEIR OXIDES

Metal	kcal/mole	kJ/mole
Al	78	326
Cu	81	338
Fe	100	416
Mg	113	473

Metal Oxide		
FeO	122	509
MgO	145	605
α-TiO	143	597
TiO_2 (rutile)	153	639

Data from: JANAF *Thermochemical Tables*, 2nd ed., National Standard Reference Data Series, Natl. Bur. Std. (U.S.), *37* (1971) and Supplement in *J. Phys. Chem. Ref. Data* 4(1), 1-175 (1975).

MELTING POINTS OF SELECTED ELEMENTS AND INORGANIC COMPOUNDS
Listed by Compound

Compound	Formula	Melting Point °C
Actinium[227]	Ac	1050±50
Aluminum	Al	658.5
Aluminum bromide	Al_2Br_6	87.4
Aluminum chloride	Al_2Cl_6	192.4
Aluminum iodide	Al_2I_6	190.9
Aluminum oxide	Al_2O_3	2045.0
Antimony	Sb	630
Antimony pentachloride	$SbCl_5$	4.0
Antimony tribromide	$SbBr_3$	96.8
Antimony trichloride	$SbCl_3$	73.3
Antimony trioxide	Sb_4O_6	655.0
Antimony trisulfide	Sb_4S_6	546.0
Argon	Ar	190.2
Arsenic	As	816.8
Arsenic pentafluoride	AsF_5	80.8
Arsenic tribromide	$AsBr_3$	30.0
Arsenic trichloride	$AsCl_3$	−16.0
Arsenic trifluoride	AsF_3	−6.0
Arsenic trioxide	As_4O_6	312.8
Barium	Ba	725
Barium bromide	$BaBr_2$	846.8
Barium chloride	$BaCl_2$	959.8
Barium fluoride	BaF_2	1286.8
Barium iodide	BaI_2	710.8
Barium nitrate	$Ba(NO_3)_2$	594.8
Barium oxide	BaO	1922.8
Barium phosphate	$Ba_3(PO_4)_2$	1727
Barium sulfate	$BaSO_4$	1350

Melting Points of Selected Elements and
Inorganic Compounds (Continued)
Listed by Compound

Compound	Formula	Melting Point °C
Beryllium	Be	1278
Beryllium bromide	$BeBr_2$	487.8
Beryllium chloride	$BeCl2$	404.8
Beryllium oxide	BeO	2550.0
Bismuth	Bi	271
Bismuth trichloride	$BiCl_3$	223.8
Bismuth trifluoride	BiF_3	726.0
Bismuth trioxide	Bi_2O_3	815.8
Boron	B	2300
Boron tribromide	BBr_3	−48.8
Boron trichloride	BCl_3	−107.8
Boron trifluoride	BF_3	−128.0
Boron trioxide	B_2O_3	448.8
Bromine	Br_2	−7.2
Bromine pentafluoride	BrF_5	−61.4
Cadmium	Cd	320.8
Cadmium bromide	$CdBr_2$	567.8
Cadmium chloride	$CdCl_2$	567.8
Cadmium fluoride	CdF_2	1110
Cadmium iodide	CdI_2	386.8
Cadmium sulfate	$CdSO_4$	1000
Calcium	Ca	851
Calcium bromide	$CaBr_2$	729.8
Calcium carbonate	$CaCO_3$	1282
Calcium chloride	$CaCl_2$	782
Calcium fluoride	CaF_2	1382
Calcium metasilicate	$CaSiO_3$	1512
Calcium nitrate	$Ca(NO_3)_2$	560.8

Melting Points of Selected Elements and
Inorganic Compounds (Continued)
Listed by Compound

Compound	Formula	Melting Point °C
Calcium oxide	CaO	2707
Calcium sulfate	$CaSO_4$	1297
Carbon dioxide	CO_2	−57.6
Carbon monoxide	CO	−205
Cyanogen	C_2N_2	−27.2
Cyanogen chloride	CNCl	−5.2
Cerium	Ce	775
Cesium	Cs	28.3
Cesium chloride	CsCl	38.5
Cesium nitrate	$CsNO_3$	406.8
Chlorine	Cl_2	−103±5
Chromium	Cr	1890
Chromium (II) chloride	$CrCl_2$	814
Chromium (III) sequioxide	Cr_2O_3	2279
Chromium trioxide	CrO_3	197
Cobalt	Co	1490
Cobalt (II) chloride	$CoCl_2$	727
Copper	Cu	1083
Copper (II) chloride	$CuCl_2$	430
Copper (I) chloride	CuCl	429
Copper(l) cyanide	$Cu_2(CN)_2$	473
Copper (I) iodide	CuI	587
Copper (II) oxide	CuO	1446
Copper (I) oxide	Cu_2O	1230
Copper (I) sulfide	Cu_2S	1129
Dysprosium	Dy	1407
Erbium	Er	1496
Europium	Eu	826

Melting Points of Selected Elements and Inorganic Compounds (Continued)
Listed by Compound

Compound	Formula	Melting Point °C
Europium trichloride	$EuCl_3$	622
Fluorine	F_2	−219.6
Gadolinium	Gd	1312
Gallium	Ga	29
Germanium	Ge	959
Gold	Au	1063
Hafnium	Hf	2214
Holmium	Ho	1461
Hydrogen	H_2	−259.25
Hydrogen bromide	HBr	−86.96
Hydrogen chloride	HCl	−114.3
Hydrogen fluoride	HF	83.11
Hydrogen iodide	HI	−50.91
Hydrogen nitrate	HNO_3	−47.2
Hydrogen oxide (water)	H_2O	0
Deuterium oxide	D_2O	3.78
Hydrogen peroxide	H_2O_2	−0.7
Hydrogen selenate	H_2SeO_4	57.8
Hydrogen sulfate	H_2SO_4	10.4
Hydrogen sulfide	H_2S	−85.6
Hydrogen sulfide, di–	H_2S_2	−89.7
Hydrogen telluride	H_2Te	−49.0
Indium	In	156.3
Iodine	I_2	112.9
Iodine chloride (α)	ICl	17.1
Iodine chloride (β)	ICl	13.8
Iron	Fe	1530.0
Iron carbide	Fe_3C	1226.8

Melting Points of Selected Elements and Inorganic Compounds (Continued)
Listed by Compound

Compound	Formula	Melting Point °C
Iron (III) chloride	Fe_2Cl_6	303.8
Iron (II) chloride	$FeCl_2$	677
Iron (II) oxide	FeO	1380
Iron oxide	Fe_3O_4	1596
Iron pentacarbonyl	$Fe(CO)_5$	−21.2
Iron (II) sulfide	FeS	1195
Lanthanum	La	920
Lead	Pb	327.3
Leadbromide	$PbBr_2$	487.8
Lead chloride	$PbCl_2$	497.8
Lead fluoride	PbF_2	823
Lead iodide	PbI_2	412
Lead molybdate	$PbMoO_4$	1065
Lead oxide	PbO	890
Lead sulfate	$PbSO_4$	1087
Lead sulfide	PbS	1114
Lithium	Li	178.8
Lithium bromide	$LiBr$	552
Lithium chloride	$LiCl$	614
Lithium fluoride	LiF	896
Lithium hydroxide	$LiOH$	462
Lithium iodide	LiI	440
Lithium metasilicate	Li_2SiO_3	1177
Lithium molybdate	Li_2MoO_4	705
Lithium nitrate	$LiNO_3$	250
Lithium orthosilicate	Li_4SiO_4	1249
Lithium sulfate	Li_2SO_4	857
Lithium tungstate	Li_2WO_4	742

Melting Points of Selected Elements and Inorganic Compounds (Continued)
Listed by Compound

Compound	Formula	Melting Point °C
Lutetium	Lu	1651
Magnesium	Mg	650
Magnesium bromide	$MgBr_2$	711
Magnesium chloride	$MgCl_2$	712
Magnesium fluoride	MgF_2	1221
Magnesium oxide	MgO	2642
Magnesium silicate	$MgSiO_3$	1524
Magnesium sulfate	$MgSO_4$	1327
Manganese	Mn	1220
Manganese dichloride	$MnCl_2$	650
Manganese metasilicate	$MnSiO_3$	1274
Manganese (II) oxide	MnO	1784
Manganese oxide	Mn_3O_4	1590
Mercury	Hg	−39
Mercury bromide	$HgBr_2$	241
Mercury chloride	$HgCl_2$	276.8
Mercury iodide	HgI_2	250
Mercury sulfate	$HgSO_4$	850
Molybdenum	Mo	2622
Molybdenum dichloride	$MoCl_2$	726.8
Molybdenum hexafluoride	MoF_6	17
Molybdenum trioxide	MoO_3	795
Neodymium	Nd	1020
Neon	Ne	− 248.6
Nickel	Ni	1452
Nickel chloride	$NiCl_2$	1030
Nickel subsulfide	Ni_3S_2	790
Niobium	Nb	2496

Melting Points of Selected Elements and
Inorganic Compounds (Continued)
Listed by Compound

Compound	Formula	Melting Point °C
Niobium pentachloride	$NbCl_5$	211
Niobium pentoxide	Nb_2O_5	1511
Nitric oxide	NO	−163.7
Nitrogen	N_2	−210
Nitrogen tetroxide	N_2O_4	−13.2
Nitrous oxide	N_2O	−90.9
Osmium	Os	2700
Osmium tetroxide (white)	OsO_4	41.8
Osmium tetroxide (yellow)	OsO_4	55.8
Oxygen	O_2	−218.8
Palladium	Pd	1555
Phosphoric acid	H_3PO_4	42.3
Phosphoric acid. hypo–	$H_4P_2O_6$	54.8
Phosphorus acid, hypo–	H_3PO_2	17.3
Phosphorus acid, ortho–	H_3PO_3	73.8
Phosphorus oxychloride	$POCl_3$	1.0
Phosphorus pentoxide	P_4O_{10}	569.0
Phosphorus trioxide	P_4O_6	23.7
Phosphorus, yellow	P_4	44.1
Platinum	Pt	1770
Potassium	K	63.4
Potassium borate, meta–	KBO_2	947
Potassium bromide	KBr	742
Potassium carbonate	K_2CO_3	897
Potassium chloride	KCl	770
Potassium chromate	K_2CrO_4	984
Potassium cyanide	KCN	623
Potassium dichromate	$K_2Cr_2O_7$	398

Melting Points of Selected Elements and Inorganic Compounds (Continued)
Listed by Compound

Compound	Formula	Melting Point °C
Potassium fluoride	KF	875
Potassium hydroxide	KOH	360
Potassium iodide	Kl	682
Potassium nitrate	KNO_3	338
Potassium peroxide	K_2O_2	490
Potassium phosphate	K_3PO4	1340
Potassium pyro– phosphate	$K_4P_2O_7$	1092
Potassium sulfate	K_2SO_4	1074
Potassium thiocyanate	KSCN	179
Praseodymium	Pr	931
Rhenium	Re	3167±60
Rhenium heptoxide	Re_2O_7	296
Rhenium hexafluoride	ReF_6	19.0
Rubidium	Rb	38 .9
Rubidium bromide	RbBr	677
Rubidium chloride	RbCl	717
Rubidium fluoride	RbF	833
Rubidium iodide	Rbl	638
Rubidium nitrate	$RbNO_3$	305
Samarium	Sm	1072
Scandium	Sc	1538
Selenium	Se	217
Seleniumoxychloride	$SeOCl_3$	9.8
Silane, hexaHuoro–	Si_2F_6	−28.6
Silicon	Si	1427
Silicon dioxide (Cristobalite)	SiO_2	2100
Silicon dioxide (Quartz)	SiO_2	1470
Silicon tetrachloride	$SiCl_4$	−67.7

Melting Points of Selected Elements and
Inorganic Compounds (Continued)
Listed by Compound

Compound	Formula	Melting Point °C
Silver	Ag	961
Silver bromide	AgBr	430
Silver chloride	AgCl	455
Silver cyanide	AgCN	350
Silver iodide	AgI	557
Silver nitrate	$AgNO_3$	209
Silver sulfate	Ag_2SO_4	657
Silver sulfide	Ag_2S	841
Sodium	Na	97.8
Sodium borate, meta–	$NaBO_2$	966
Sodium bromide	NaBr	747
Sodium carbonate	Na_2CO_3	854
Sodium chlorate	$NaClO_3$	255
Sodium chloride	NaCl	800
Sodium cyanide	NaCN	562
Sodium fluoride	NaF	992
Sodium hydroxide	NaOH	322
Sodium iodide	NaI	662
Sodium molybdate	Na_2MoO_4	687
Sodium nitrate	$NaNO_3$	310
Sodium peroxide	Na_2O_2	460
Sodium phosphate, meta–	$NaPO_3$	988
Sodium pyrophosphate	$Na_4P_2O_7$	970
Sodiumsilicate,aluminum–	$NaAlSi_3O_8$	1107
Sodium silicate, di–	$Na_2Si_2O_5$	884
Sodium silicate, meta–	Na_2SiO_3	1087
Sodium sulfate	Na_2SO_4	884
Sodium sulfide	Na_2S	920

Melting Points of Selected Elements and Inorganic Compounds (Continued)
Listed by Compound

Compound	Formula	Melting Point °C
Sodium thiocyanate	NaSCN	323
Sodium tungstate	Na_2WO_4	702
Strontium	Sr	757
Strontium bromide	$SrBr_2$	643
Strontium chloride	$SrCl_2$	872
Strontium fluoride	SrF_2	1400
Strontium oxide	SrO	2430
Sulfur (monatomic)	S	119,
Sulfur dioxide	SO_2	− 73.2
Sulfur trioxide (α)	SO_3	16.8
Sulfur trioxide (β)	SO_3	32.3
Sulfur trioxide (γ)	SO_3	62.1
Tantalum	Ta	2996 ± 50
Tantalum pentachloride	$TaCl_5$	206.8
Tantalum pentoxide	Ta_2O_5	1877
Tellurium	Te	453
Terbium	Tb	1356
Thallium	Tl	302.4
Thallium bromide, mono–	TlBr	460
Thallium carbonate	Tl_2CO_3	273
Thallium chloride, mono–	TlCl	427
Thallium iodide, mono–	TlI	440
Thallium nitrate	$TlNO_3$	207
Thallium sulfate	Tl_2SO_4	632
Thallium sulfide	Tl_2S	449
Thorium	Th	1845
Thorium chloride	$ThCl_4$	765
Thorium dioxide	ThO_2	2952

Melting Points of Selected Elements and
Inorganic Compounds (Continued)
Listed by Compound

Compound	Formula	Melting Point °C
Thulium	Tm	1545
Tin	Sn	231.7
Tin bromide, di–	$SnBr_2$	231.8
Tin bromide, tetra–	$SnBr_4$	29.8
Tin chloride, di–	$SnCl_2$	247
Tinchloride,tetra–	$SnCl_4$	–33.3
Tin iodide, tetra–	SnI_4	143.4
Tin oxide	SnO	1042
Titanium	Ti	1800
Titanium bromide, tetra–	$TiBr_4$	38
Titanium chloride, tetra–	$TiCl_4$	–23.2
Titanium dioxide	TiO_2	1825
Titanium oxide	TiO	991
Tungsten	W	3387
Tungsten dioxide	WO_2	1270
Tungsten hexafluoride	WF_6	–0.5
Tungsten tetrachloride	WCl_4	327
Tungsten trioxide	WO_3	1470
Uranium[235]	U	~1133
Uranium tetrachloride	UCl_4	590
Vanadium	V	1917
Vanadium dichloride	VCl_2	1027
Vanadium oxide	VO	2077
Vanadium pentoxide	V_2O_5	670
Xenon	Xe	–111.6
Ytterbium	Yb	823
Yttrium	Y	1504
Yttrium oxide	Y_2O_3	2227

Melting Points of Selected Elements and
Inorganic Compounds (Continued)
Listed by Compound

Compound	Formula	Melting Point °C
Zinc	Zn	419.4
Zincchloride	$ZnCl_2$	283
Zinc oxide	ZnO	1975
Zinc sulfide	ZnS	1745
Zirconium	Zr	1857
Zirconium dichloride	$ZrCl_2$	727
Zirconium oxide	ZrO_2	2715

Data from: Weast, R C., Ed., *Handbook of Chemistry and Physics, 55th ed.*, CRC Press, Cleveland, (1974); and Bolz, R. E. and Tuve, G. L., Eds., *Handbook of Tables for Applied Engineering Science,* 2nd ed., CRC Press, Cleveland, (1973), p.479 .

MELTING POINTS OF SELECTED ELEMENTS AND INORGANIC COMPOUNDS
Listed by Melting Point

Compound	Formula	Melting Point °C
Hydrogen	H_2	−259.25
Neon	Ne	− 248.6
Fluorine	F_2	−219.6
Oxygen	O_2	−218.8
Nitrogen	N_2	−210
Carbon monoxide	CO	−205
Nitric oxide	NO	−163.7
Boron trifluoride	BF_3	−128.0
Hydrogen chloride	HCl	−114.3
Xenon	Xe	−111.6
Boron trichloride	BCl_3	−107.8
Chlorine	Cl_2	−103±5
Nitrous oxide	N_2O	−90.9
Hydrogen sulfide, di–	H_2S_2	−89.7
Hydrogen bromide	HBr	−86.96
Hydrogen sulfide	H_2S	−85.6
Sulfur dioxide	SO_2	− 73.2
Silicon tetrachloride	$SiCl_4$	−67.7
Bromine pentafluoride	BrF_5	−61.4
Carbon dioxide	CO_2	−57.6
Hydrogen iodide	HI	−50.91
Hydrogen telluride	H_2Te	−49.0
Boron tribromide	BBr_3	−48.8
Hydrogen nitrate	HNO_3	−47.2
Mercury	Hg	−39
Tinchloride,tetra–	$SnCl_4$	−33.3
Silane, hexaHuoro–	Si_2F_6	−28.6
Cyanogen	C_2N_2	−27.2

Melting Points of Selected Elements and Inorganic Compounds (Continued)
Listed by Melting Point

Compound	Formula	Melting Point °C
Titanium chloride, tetra–	$TiCl_4$	–23.2
Iron pentacarbonyl	$Fe(CO)_5$	–21.2
Arsenic trichloride	$AsCl_3$	–16.0
Nitrogen tetroxide	N_2O_4	–13.2
Bromine	Br_2	–7.2
Arsenic trifluoride	AsF_3	–6.0
Cyanogen chloride	$CNCl$	–5.2
Hydrogen peroxide	H_2O_2	–0.7
Tungsten hexafluoride	WF_6	–0.5
Hydrogen oxide (water)	H_2O	0
Phosphorus oxychloride	$POCl_3$	1.0
Deuterium oxide	D_2O	3.78
Antimony pentachloride	$SbCl_5$	4.0
Seleniumoxychloride	$SeOCl_3$	9.8
Hydrogen sulfate	H_2SO_4	10.4
Iodine chloride (β)	ICl	13.8
Sulfur trioxide (α)	SO_3	16.8
Molybdenum hexafluoride	MoF_6	17
Iodine chloride (α)	ICl	17.1
Phosphorus acid, hypo–	H_3PO_2	17.3
Rhenium hexafluoride	ReF_6	19.0
Niobium pentachloride	$NbCl_5$	21 1
Phosphorus trioxide	P_4O_6	23.7
Cesium	Cs	28.3
Gallium	Ga	29
Tin bromide, tetra–	$SnBr_4$	29.8
Arsenic tribromide	$AsBr_3$	30.0
Sulfur trioxide (β)	SO_3	32.3

Melting Points of Selected Elements and
Inorganic Compounds (Continued)
Listed by Melting Point

Compound	Formula	Melting Point °C
Titanium bromide, tetra–	$TiBr_4$	38
Cesium chloride	$CsCl$	38.5
Rubidium	Rb	38 .9
Osmium tetroxide (white)	OsO_4	41.8
Phosphoric acid	H_3PO_4	42.3
Phosphorus, yellow	P_4	44.1
Phosphoric acid. hypo–	$H_4P_2O_6$	54.8
Osmium tetroxide (yellow)	OsO_4	55.8
Hydrogen selenate	H_2SeO_4	57.8
Sulfur trioxide (γ)	SO_3	62.1
Potassium	K	63.4
Antimony trichloride	$SbCl_3$	73.3
Phosphorus acid, ortho–	H_3PO_3	73.8
Arsenic pentafluoride	AsF_5	80.8
Hydrogen fluoride	HF	83.11
Aluminum bromide	Al_2Br_6	87.4
Antimony tribromide	$SbBr_3$	96.8
Sodium	Na	97.8
Iodine	I_2	112.9
Sulfur (monatomic)	S	119
Tin iodide, tetra–	SnI_4	143.4
Indium	In	156.3
Lithium	Li	178.8
Potassium thiocyanate	$KSCN$	179
Argon	Ar	190.2
Aluminum iodide	Al_2I_6	190.9
Aluminum chloride	Al_2Cl_6	192.4
Chromium trioxide	CrO_3	197

Melting Points of Selected Elements and Inorganic Compounds (Continued)
Listed by Melting Point

Compound	Formula	Melting Point °C
Tantalum pentachloride	$TaCl_5$	206.8
Thallium nitrate	$TINO_3$	207
Silver nitrate	$AgNO_3$	209
Selenium	Se	217
Bismuth trichloride	$BiCl_3$	223.8
Tin	Sn	231.7
Tin bromide, di–	$SnBr_2$	231.8
Mercury bromide	$HgBr_2$	241
Tin chloride, di–	$SnCl_2$	247
Lithium nitrate	$LiNO_3$	250
Mercury iodide	HgI_2	250
Sodium chlorate	$NaClO_3$	255
Bismuth	Bi	271
Thallium carbonate	Tl_2CO_3	273
Mercury chloride	$HgCl_2$	276.8
Zincchloride	$ZnCl_2$	283
Rhenium heptoxide	Re_2O_7	296
Thallium	Tl	302.4
Iron (III) chloride	Fe_2Cl_6	303.8
Rubidium nitrate	$RbNO_3$	305
Sodium nitrate	$NaNO_3$	310
Arsenic trioxide	As_4O_6	312.8
Cadmium	Cd	320.8
Sodium hydroxide	NaOH	322
Sodium thiocyanate	NaSCN	323
Tungsten tetrachloride	WCl_4	327
Lead	Pb	327.3
Potassium nitrate	KNO_3	338

Melting Points of Selected Elements and
Inorganic Compounds (Continued)
Listed by Melting Point

Compound	Formula	Melting Point °C
Silver cyanide	AgCN	350
Potassium hydroxide	KOH	360
Cadmium iodide	CdI_2	386.8
Potassium dichromate	$K_2Cr_2O_7$	398
Beryllium chloride	BeCl2	404.8
Cesium nitrate	$CsNO_3$	406.8
Lead iodide	PbI_2	412
Zinc	Zn	419.4
Thallium chloride, mono–	TlCl	427
Copper (I) chloride	CuCl	429
Copper (II) chloride	$CuCl_2$	430
Silver bromide	AgBr	430
Lithium iodide	LiI	440
Thallium iodide, mono–	TlI	440
Boron trioxide	B_2O_3	448.8
Thallium sulfide	Tl_2S	449
Tellurium	Te	453
Silver chloride	AgCl	455
Sodium peroxide	Na_2O_2	460
Thallium bromide, mono–	TlBr	460
Lithium hydroxide	LiOH	462
Copper(l) cyanide	$Cu_2(CN)_2$	473
Beryllium bromide	$BeBr_2$	487.8
Leadbromide	$PbBr_2$	487.8
Potassium peroxide	K_2O_2	490
Antimony trisulfide	Sb_4S_6	546.0
Lithium bromide	LiBr	552
Silver iodide	AgI	557

Melting Points of Selected Elements and Inorganic Compounds (Continued)
Listed by Melting Point

Compound	Formula	Melting Point °C
Calcium nitrate	$Ca(NO_3)_2$	560.8
Sodium cyanide	NaCN	562
Cadmium bromide	$CdBr_2$	567.8
Cadmium chloride	$CdCl_2$	567.8
Phosphorus pentoxide	P_4O_{10}	569.0
Copper (I) iodide	CuI	587
Uranium tetrachloride	UCl_4	590
Barium nitrate	$Ba(NO_3)_2$	594.8
Lithium chloride	LiCl	614
Europium trichloride	$EuCl_3$	622
Potassium cyanide	KCN	623
Antimony	Sb	630
Thallium sulfate	Tl_2SO_4	632
Rubidium iodide	RbI	638
Strontium bromide	$SrBr_2$	643
Magnesium	Mg	650
Manganese dichloride	$MnCl_2$	650
Antimony trioxide	Sb_4O_6	655.0
Silver sulfate	Ag_2SO_4	657
Aluminum	Al	658.5
Sodium iodide	NaI	662
Vanadium pentoxide	V_2O_5	670
Iron (II) chloride	$FeCl_2$	677
Rubidium bromide	RbBr	677
Potassium iodide	KI	682
Sodium molybdate	Na_2MoO_4	687
Sodium tungstate	Na_2WO_4	702
Lithium molybdate	Li_2MoO_4	705

Melting Points of Selected Elements and
Inorganic Compounds (Continued)
Listed by Melting Point

Compound	Formula	Melting Point °C
Barium iodide	BaI_2	710.8
Magnesium bromide	$MgBr_2$	711
Magnesium chloride	$MgCl_2$	712
Rubidium chloride	RbCl	717
Barium	Ba	725
Bismuth trifluoride	BiF_3	726.0
Molybdenum dichloride	$MoCl_2$	726.8
Cobalt (II) chloride	$CoCl_2$	727
Zirconium dichloride	$ZrCl_2$	727
Calcium bromide	$CaBr_2$	729.8
Lithium tungstate	Li_2WO_4	742
Potassium bromide	KBr	742
Sodium bromide	NaBr	747
Strontium	Sr	757
Thorium chloride	$ThCl_4$	765
Potassium chloride	KCl	770
Cerium	Ce	775
Calcium chloride	$CaCl_2$	782
Nickel subsulfide	Ni_3S_2	790
Molybdenum trioxide	MoO_3	795
Sodium chloride	NaCl	800
Chromium (II) chloride	$CrCl_2$	814
Bismuth trioxide	Bi_2O_3	815.8
Arsenic	As	816.8
Lead fluoride	PbF_2	823
Ytterbium	Yb	823
Europium	Eu	826
Rubidium fluoride	RbF	833

Melting Points of Selected Elements and Inorganic Compounds (Continued)
Listed by Melting Point

Compound	Formula	Melting Point °C
Silver sulfide	Ag_2S	841
Barium bromide	$BaBr_2$	846.8
Mercury sulfate	$HgSO_4$	850
Calcium	Ca	851
Sodium carbonate	Na_2CO_3	854
Lithium sulfate	Li_2SO_4	857
Strontium chloride	$SrCl_2$	872
Potassium fluoride	KF	875
Sodium silicate, di–	$Na_2Si_2O_5$	884
Sodium sulfate	Na_2SO_4	884
Lead oxide	PbO	890
Lithium fluoride	LiF	896
Potassium carbonate	K_2CO_3	897
Lanthanum	La	920
Sodium sulfide	Na_2S	920
Praseodymium	Pr	931
Potassium borate, meta–	KBO_2	947
Germanium	Ge	959
Barium chloride	$BaCl_2$	959.8
Silver	Ag	961
Sodium borate, meta–	$NaBO_2$	966
Sodium pyrophosphate	$Na_4P_2O_7$	970
Potassium chromate	K_2CrO_4	984
Sodium phosphate, meta–	$NaPO_3$	988
Titanium oxide	TiO	991
Sodium fluoride	NaF	992
Cadmium sulfate	$CdSO_4$	1000
Neodymium	Nd	1020

Melting Points of Selected Elements and Inorganic Compounds (Continued)
Listed by Melting Point

Compound	Formula	Melting Point °C
Vanadium dichloride	VCl_2	1027
Nickel chloride	$NiCl_2$	1030
Tin oxide	SnO	1042
Actinium227	Ac	1050±50
Gold	Au	1063
Lead molybdate	$PbMoO_4$	1065
Samarium	Sm	1072
Potassium sulfate	K_2SO_4	1074
Copper	Cu	1083
Lead sulfate	$PbSO_4$	1087
Sodium silicate, meta–	Na_2SiO_3	1087
Potassium pyro– phosphate	$K_4P_2O_7$	1092
Sodiumsilicate,aluminum–	$NaAlSi_3O_8$	1107
Cadmium fluoride	CdF_2	1110
Lead sulfide	PbS	1114
Copper (I) sulfide	Cu_2S	1129
Uranium235	U	~1133
Lithium metasilicate	Li_2SiO_3	1177
Iron (II) sulfide	FeS	1195
Manganese	Mn	1220
Magnesium fluoride	MgF_2	1221
Iron carbide	Fe_3C	1226.8
Copper (I) oxide	Cu_2O	1230
Lithium orthosilicate	Li_4SiO_4	1249
Tungsten dioxide	WO_2	1270
Manganese metasilicate	$MnSiO_3$	1274
Beryllium	Be	1278
Calcium carbonate	$CaCO_3$	1282

Melting Points of Selected Elements and Inorganic Compounds (Continued)
Listed by Melting Point

Compound	Formula	Melting Point °C
Barium fluoride	BaF_2	1286.8
Calcium sulfate	$CaSO_4$	1297
Gadolinium	Gd	1312
Magnesium sulfate	$MgSO_4$	1327
Potassium phosphate	K_3PO4	1340
Barium sulfate	$BaSO_4$	1350
Terbium	Tb	1356
Iron (II) oxide	FeO	1380
Calcium fluoride	CaF_2	1382
Strontium fluoride	SrF_2	1400
Dysprosium	Dy	1407
Silicon	Si	1427
Copper (II) oxide	CuO	1446
Nickel	Ni	1452
Holmium	Ho	1461
Tungsten trioxide	WO_3	1470
Cobalt	Co	1490
Erbium	Er	1496
Yttrium	Y	1504
Niobium pentoxide	Nb_2O_5	1511
Calcium metasilicate	$CaSiO_3$	1512
Magnesium silicate	$MgSiO_3$	1524
Iron	Fe	1530.0
Scandium	Sc	1538
Thulium	Tm	1545
Palladium	Pd	1555
Manganese oxide	Mn_3O_4	1590

Melting Points of Selected Elements and
Inorganic Compounds (Continued)
Listed by Melting Point

Compound	Formula	Melting Point °C
Iron oxide	Fe_3O_4	1596
Lutetium	Lu	1651
Barium phosphate	$Ba_3(PO_4)_2$	1727
Zinc sulfide	ZnS	1745
Platinum	Pt	1770
Manganese (II) oxide	MnO	1784
Titanium	Ti	1800
Titanium dioxide	TiO_2	1825
Thorium	Th	1845
Zirconium	Zr	1857
Tantalum pentoxide	Ta_2O_5	1877
Chromium	Cr	1890
Vanadium	V	1917
Barium oxide	BaO	1922.8
Zinc oxide	ZnO	1975
Aluminum oxide	Al_2O_3	2045.0
Vanadium oxide	VO	2077
Hafnium	Hf	2214
Yttrium oxide	Y_2O_3	2227
Chromium (III) sequioxide	Cr_2O_3	2279
Boron	B	2300
Strontium oxide	SrO	2430
Niobium	Nb	2496
Beryllium oxide	BeO	2550.0
Molybdenum	Mo	2622
Magnesium oxide	MgO	2642
Osmium	Os	2700

Melting Points of Selected Elements and Inorganic Compounds (Continued)
Listed by Melting Point

Compound	Formula	Melting Point °C
Calcium oxide	CaO	2707
Zirconium oxide	ZrO_2	2715
Thorium dioxide	ThO_2	2952
Tantalum	Ta	2996 ± 50
Rhenium	Re	3167 ± 60
Tungsten	W	3387

Source: Data from: Weast, R C., Ed., *Handbook of Chemistry and Physics, 55th ed.*, CRC Press, Cleveland, 1974.

Source: Data from: Bolz, R. E. and Tuve, G. L., Eds., *Handbook of Tables for Applied Engineering Science*, 2nd ed., CRC Press, Cleveland, 1973, 479 .

MELTING POINTS OF CERAMICS
Listed by Compound

Compound	(K)	Compound	(K)	Compound	(K)
$AgBr$	703	BaB_4	2543	Bi_2S_3	1020
$AgCl$	728	$BaBr_2$	1123	$CaBr_2$	1003
AgF	708	$BaCl_2$	1235	$CaCl_2$	1055
AgI	831	BaF_2	1627	CaF_2	1675
$AgNO_3$	483	BaI_2	1013	CaI_2	848
Ag_2O	573	$Ba(NO_3)_2$	865	$Ca(NO_3)_2$	623
Ag_2SO_4	933	BaO	2283	Ca_3N_2	1468
Ag_2S	1098	$BaSO_4$	1853	CaO	3183
$AlBr_3$	371	BaS	1473	$CaSO_4$	1723
Al_4C_3	2000	BeB_2	>2243	$CdBr_2$	841
$AlCl_3$	465	$BeBr_2$	793	$CdCl_2$	841
AlF_3	1564	Be_2C	>2375	CdF_2	1373
AlI	464	$BeCl_2$	713	CdI_2	423
AlN	>2475	BeF_2	813	$Cd(NO_3)_2$	834
Al_2O_3	2322	BeI_2	783	CdO	1773
$Al_2(SO_4)_3$	1043	Be_3N_2	2513	$CdSO_4$	1273
Al_2S_3	1373	BeO	2725	CdS	2023
BBr_3	227	$BeSO_4$	848	CeB_6	2463
B_4C	2720	$BiBr_3$	491	$CeCl_3$	1095
BCl_3	166	$BiCl_3$	507	CeF_2	1710
BF_3	146	BiF_3	1000	CeI_3	1025
BN	3000	BiI_3	681	CeO_2	>2873
B_2O_3	723	B_2O_3	1098	CeS	2400
BS_4	663	$Bi(SO_4)_3$	678	$Ce(SO_4)_2$	468

Melting Points of Ceramics (Continued)
Listed by Compound

Compound	(K)	Compound	(K)	Compound	(K)
CrB_2	2123	In_2O_3	2183	Mg_2Si	1375
$Cr3C_2$	2168	In_2S_3	1323	MgS	>2275
CrN	1770	KBr	1008	$MgSO_4$	1397
Cr_2O_3	>2603	KCl	1043	$MnCl_2$	923
$CrSi_2$	1843	KF	1131	MnF_2	1129
$CuBr$	777	KI	958	MnO	1840
$CuCl$	695	KNO_3	610	MoB	2625
CuF_2	1129	K_2O_3	703	Mo_2C	2963
CuI	878	K_2SO_4	1342	MoF_6	290
Cu_3N	573	K_2S	1113	MoI_4	373
Cu_2O	1508	$LiBr$	823	MoO_3	1068
Cu_4Si	1123	$LiCl$	883	$MoSi_2$	2553
Cu_2S	1400	LiF	1119	MoS_2	1458
$FeBr_2$	955	LiI	722	$NaBr$	1023
Fe_3C	2110	$LiNO_3$	527	NaC_2	973
$FeCl_2$	945	Li_3N	1118	$NaCl$	1073
FeF_3	>1275	Li_2O	>1975	NaF	1267
Fe_2O_3	1864	Li_2SO_4	1132	NaI	935
$Fe_2(SO_4)_3$	753	Li_2S	1198	$NaNO_3$	583
FeS	1468	$MgBr_2$	984	Na_2N	573
$InBr_3$	709	$MgCl_2$	987	Na_2SO_4	1157
$InCl$	498	MgF_2	1535	Na_2S	1453
InF_3	1443	MgI_2	<910	NbB	>2270
InI_3	483	MgO	3098	NbC	3770

Melting Points of Ceramics (Continued)
Listed by Compound

Compound	(K)	Compound	(K)	Compound	(K)
NbN	2323	SbF_3	565	SrS	>2275
Nb_2O_5	1764	SbI_3	443	TaB	>2270
$NbSi_2$	2203	Sb_2O_3	928	$TaBr_5$	538
$NiBr_2$	1236	SbS_3	820	TaC	3813
$NiCl_3$	1274	SiC	2970	$TaCl_5$	489
NiF_2	1273	SiF_4	183	TaF_5	370
NiI_2	1070	Si_3N_4	2715	Ta_2N	3360
NiO	2257	SiO_2	1978	Ta_2O_5	2100
$NiSO_4$	1121	$SnBr_2$	488	$TaSi_2$	2670
NiS	1070	$SnCl_2$	581	TaS_4	>1575
$PbBr_2$	643	SnF_4	978	$TeBr_2$	612
$PbCl_2$	771	SnI_2	788	$TeCl_2$	448
PbF_2	1095	SnO	1353	TeO_2	1006
PbI_2	675	$SnSO_4$	>635	ThB_4	>2270
$Pb(NO_3)_2$	743	SnS	1153	$ThBr_4$	883
PbO	1159	SrB_6	2508	ThC	2898
$PbSO_4$	1443	$SrBr_2$	916	$ThCl_4$	1043
PbS	1387	SrC_2	>1970	ThF_4	1375
$PtBr_2$	523	$SrCl_2$	1148	ThN	2903
$PtCl_2$	854	SrF_2	1736	ThO_2	3493
PtI_2	633	SrI_2	593	ThS_2	2198
PtS_2	508	$Sr(NO_3)_2$	643	TiB_2	3253
$SbBr_3$	370	SrO	2933	$TiBr_4$	312
$SbCl_3$	346	$SrSO_4$	1878	TiC	3433

Melting Points of Ceramics (Continued)
Listed by Compound

Compound	(K)	Compound	(K)	Compound	(K)
$TiCl_4$	250	VB_2	2373	$ZnCl_2$	548
TiF_3	1475	VC	3600	ZnF_2	1145
TiI_2	873	VCl_4	245	ZnI_2	719
TiN	3200	VF_3	>1075	ZnO	2248
TiO_2	2113	FI_2	1048	$ZnSO_4$	873
$TiSi_2$	1813	VN	2593	ZrB_2	3313
UB_2	>1770	V_2O_5	947	$ZrBr_2$	>625
UBr_4	789	VSi_2	2023	ZrC	3533
UC	2863	V_2S_3	>875	$ZrCl_2$	623
UCl_4	843	WB	3133	ZrF_4	873
UF_4	1233	WC	2900	ZrI_4	772
UI_4	779	WCl_6	548	ZrN	3250
UN	3123	WO_3	1744	ZrO_2	3123
UO_2	3151	WSi_2	2320	$Zr(SO_4)_2$	683
USi_2	1970	WS_2	1523	ZrS_2	1823
US_2	>1375	$ZnBr_2$	667		

Source: data from: Lynch, Charles T., Ed., *CRC Handbook of Materials Science, Vol. 1*, CRC Press, Boca Raton, 1974, 348.

MELTING POINTS OF CERAMICS
Listed by Melting Point

Compound	(K)	Compound	(K)	Compound	(K)
TaC	3813	CeO_2	>2873	ThB_4	>2270
NbC	3770	UC	2863	TaB	>2270
VC	3600	BeO	2725	NbB	>2270
ZrC	3533	B_4C	2720	NiO	2257
ThO_2	3493	Si_3N_4	2715	ZnO	2248
TiC	3433	$TaSi_2$	2670	BeB_2	>2243
Ta_2N	3360	MoB	2625	$NbSi_2$	2203
ZrB_2	3313	Cr_2O_3	>2603	ThS_2	2198
TiB_2	3253	VN	2593	In_2O_3	2183
ZrN	3250	$MoSi_2$	2553	Cr_3C_2	2168
TiN	3200	BaB_4	2543	CrB_2	2123
CaO	3183	Be_3N_2	2513	TiO_2	2113
UO_2	3151	SrB_6	2508	Fe_3C	2110
WB	3133	AlN	>2475	Ta_2O_5	2100
ZrO_2	3123	CeB_6	2463	VSi_2	2023
UN	3123	CeS	2400	CdS	2023
MgO	3098	Be_2C	>2375	Al_4C_3	2000
BN	3000	VB_2	2373	SiO_2	1978
SiC	2970	NbN	2323	Li_2O	>1975
Mo_2C	2963	Al_2O_3	2322	USi_2	1970
SrO	2933	WSi_2	2320	SrC_2	>1970
ThN	2903	BaO	2283	$SrSO_4$	1878
WC	2900	SrS	>2275	Fe_2O_3	1864
ThC	2898	MgS	>2275	$BaSO_4$	1853

Melting Points of Ceramics (Continued)
Listed by Melting Point

Compound	(K)	Compound	(K)	Compound	(K)
$CrSi_2$	1843	Na_2S	1453	Na_2SO_4	1157
MnO	1840	$PbSO_4$	1443	SnS	1153
ZrS_2	1823	InF_3	1443	$SrCl_2$	1148
$TiSi_2$	1813	Cu_2S	1400	ZnF_2	1145
CdO	1773	$MgSO_4$	1397	Li_2SO_4	1132
UB_2	>1770	PbS	1387	KF	1131
CrN	1770	US_2	>1375	MnF_2	1129
Nb_2O_5	1764	ThF_4	1375	CuF_2	1129
WO_3	1744	Mg_2Si	1375	Cu_4Si	1123
SrF_2	1736	CdF_2	1373	$BaBr_2$	1123
$CaSO_4$	1723	Al_2S_3	1373	$NiSO_4$	1121
CeF_2	1710	SnO	1353	LiF	1119
CaF_2	1675	K_2SO_4	1342	Li_3N	1118
BaF_2	1627	In_2S_3	1323	K_2S	1113
TaS_4	>1575	FeF_3	>1275	B_2O_3	1098
AlF_3	1564	$NiCl_3$	1274	Ag_2S	1098
MgF_2	1535	NiF_2	1273	PbF_2	1095
WS_2	1523	$CdSO_4$	1273	$CeCl_3$	1095
Cu_2O	1508	NaF	1267	VF_3	>1075
TiF_3	1475	$NiBr_2$	1236	NaCl	1073
BaS	1473	$BaCl_2$	1235	NiS	1070
FeS	1468	UF_4	1233	NiI_2	1070
Ca_3N_2	1468	Li_2S	1198	MoO_3	1068
MoS_2	1458	PbO	1159	$CaCl_2$	1055

Melting Points of Ceramics (Continued)
Listed by Melting Point

Compound	(K)	Compound	(K)	Compound	(K)
FI_2	1048	$SrBr_2$	916	BeI_2	783
$ThCl_4$	1043	MgI_2	<910	UI_4	779
KCl	1043	$ThBr_4$	883	CuBr	777
$Al_2(SO_4)_3$	1043	LiCl	883	ZrI_4	772
CeI_3	1025	CuI	878	$PbCl_2$	771
NaBr	1023	V_2S_3	>875	$Fe_2(SO_4)_3$	753
Bi_2S_3	1020	ZrF_4	873	$Pb(NO_3)_2$	743
BaI_2	1013	$ZnSO_4$	873	AgCl	728
KBr	1008	TiI_2	873	B_2O_3	723
TeO_2	1006	$Ba(NO_3)_2$	865	LiI	722
$CaBr_2$	1003	$PtCl_2$	854	ZnI_2	719
BiF_3	1000	CaI_2	848	$BeCl_2$	713
$MgCl_2$	987	$BeSO_4$	848	$InBr_3$	709
$MgBr_2$	984	UCl_4	843	AgF	708
SnF_4	978	$CdCl_2$	841	K_2O_3	703
NaC_2	973	$CdBr_2$	841	AgBr	703
KI	958	$Cd(NO_3)_2$	834	CuCl	695
$FeBr_2$	955	AgI	831	$Zr(SO_4)_2$	683
V_2O_5	947	LiBr	823	BiI_3	681
$FeCl_2$	945	SbS_3	820	$Bi(SO_4)_3$	678
NaI	935	BeF_2	813	PbI_2	675
Ag_2SO_4	933	$BeBr_2$	793	$ZnBr_2$	667
Sb_2O_3	928	UBr_4	789	BS_4	663
$MnCl_2$	923	SnI_2	788	$Sr(NO_3)_2$	643

Melting Points of Ceramics (Continued)
Listed by Melting Point

Compound	(K)	Compound	(K)	Compound	(K)
$PbBr_2$	643	WCl_6	548	SbI_3	443
$SnSO_4$	>635	$TaBr_5$	538	CdI_2	423
PtI_2	633	$LiNO_3$	527	MoI_4	373
$ZrBr_2$	>625	$PtBr_2$	523	$AlBr_3$	371
$ZrCl_2$	623	PtS_2	508	TaF_5	370
$Ca(NO_3)_2$	623	$BiCl_3$	507	$SbBr_3$	370
$TeBr_2$	612	$InCl$	498	$SbCl_3$	346
KNO_3	610	$BiBr_3$	491	$TiBr_4$	312
SrI_2	593	$TaCl_5$	489	MoF_6	290
$NaNO_3$	583	$SnBr_2$	488	$TiCl_4$	250
$SnCl_2$	581	InI_3	483	VCl_4	245
Na_2N	573	$AgNO_3$	483	BBr_3	227
Cu_3N	573	$Ce(SO_4)_2$	468	SiF_4	183
Ag_2O	573	$AlCl_3$	465	BCl_3	166
SbF_3	565	AlI	464	BF_3	146
$ZnCl_2$	548	$TeCl_2$	448		

Source: data from: Lynch, Charles T., Ed., *CRC Handbook of Materials Science, Vol. 1*, CRC Press, Boca Raton, 1974, 348.

HEAT OF FUSION FOR SELECTED ELEMENTS AND INORGANIC COMPOUNDS

Compound	Formula	Melting point °C	Heat of fusion cal/g	Heat of fusion cal/g mole
Actinium[227]	Ac	1050±50	(11.0)	(3400)
Aluminum	Al	658.5	94.5	2550
Aluminum bromide	Al_2Br_6	87.4	10.1	5420
Aluminum chloride	Al_2Cl_6	192.4	63.6	19600
Aluminum iodide	Al_2I_6	190.9	9.8	7960
Aluminum oxide	Al_2O_3	2045.0	(256.0)	(26000)
Antimony	Sb	630	39.1	4770
Antimony pentachloride	$SbCl_5$	4.0	8.0	2400
Antimony tribromide	$SbBr_3$	96.8	9.7	3510
Antimony trichloride	$SbCl_3$	73.3	13.3	3030
Antimony trioxide	Sb_4O_6	655.0	(46.3)	(26990)
Antimony trisulfide	Sb_4S_6	546.0	33.0	11200
Argon	Ar	190.2	7.25	290
Arsenic	As	816.8	(22.0)	(6620)
Arsenic pentafluoride	AsF_5	80.8	16.5	2800
Arsenic tribromide	$AsBr_3$	30.0	8.9	2810
Arsenic trichloride	$AsCl_3$	−16.0	13.3	2420
Arsenic trifluoride	AsF_3	−6.0	18.9	2486
Arsenic trioxide	As_4O_6	312.8	22.2	8000
Barium	Ba	725	13.3	1830
Barium bromide	$BaBr_2$	846.8	21.9	6000
Barium chloride	$BaCl_2$	959.8	25.9	5370
Barium fluoride	BaF_2	1286.8	17.1	3000
Barium iodide	BaI_2	710.8	(17.3)	(6800)
Barium nitrate	$Ba(NO_3)_2$	594.8	(22.6)	(5900)
Barium oxide	BaO	1922.8	93.2	13800
Barium phosphate	$Ba_3(PO_4)_2$	1727	30.9	18600
Barium sulfate	$BaSO_4$	1350	41.6	9700

Heat of Fusion For Selected Elements and Inorganic Compounds (Continued)

Compound	Formula	Melting point °C	Heat of fusion cal/g	cal/g mole
Beryllium	Be	1278	260.0	–
Beryllium bromide	$BeBr_2$	487.8	(26.6)	(4500)
Beryllium chloride	BeCl2	404.8	(30)	(3000)
Beryllium oxide	BeO	2550.0	679.7	17000
Bismuth	Bi	271	12.0	2505
Bismuth trichloride	$BiCl_3$	223.8	8.2	2600
Bismuth trifluoride	BiF_3	726.0	(23.3)	(6200)
Bismuth trioxide	Bi_2O_3	815.8	14.6	6800
Boron	B	2300	(490)	(5300)
Boron tribromide	BBr_3	−48.8	(2.9)	(700)
Boron trichloride	BCl_3	−107.8	(4.3)	(500)
Boron trifluoride	BF_3	−128.0	7.0	480
Boron trioxide	B_2O_3	448.8	78.9	5500
Bromine	Br_2	−7.2	16.1	2580
Bromine pentafluoride	BrF_5	−61.4	7.07	1355
Cadmium	Cd	320.8	12.9	1460
Cadmium bromide	$CdBr_2$	567.8	(18.4)	(5000)
Cadmium chloride	$CdCl_2$	567.8	28.8	5300
Cadmium fluoride	CdF_2	1110	(35.9)	(5400)
Cadmium iodide	CdI_2	386.8	10.0	3660
Cadmium sulfate	$CdSO_4$	1000	22.9	4790
Calcium	Ca	851	55.7	2230
Calcium bromide	$CaBr_2$	729.8	20.9	4180
Calcium carbonate	$CaCO_3$	1282	(126)	(12700)
Calcium chloride	$CaCl_2$	782	55	6100
Calcium fluoride	CaF_2	1382	52.5	4100
Calcium metasilicate	$CaSiO_3$	1512	115.4	13400
Calcium nitrate	$Ca(NO_3)_2$	560.8	31.2	5120

Heat of Fusion For Selected Elements and
Inorganic Compounds (Continued)

Compound	Formula	Melting point °C	Heat of fusion cal/g	cal/g mole
Calcium oxide	CaO	2707	(218.1)	(12240)
Calcium sulfate	$CaSO_4$	1297	49.2	6700
Carbon dioxide	CO_2	−57.6	43.2	1900
Carbon monoxide	CO	−205	7.13	199.7
Cyanogen	C_2N_2	−27.2	39.6	2060
Cyanogen chloride	CNCl	−5.2	36.4	2240
Cerium	Ce	775	27.2	2120
Cesium	Cs	28.3	3.7	500
Cesium chloride	CsCl	38.5	21.4	3600
Cesium nitrate	$CsNO_3$	406.8	16.6	3250
Chlorine	Cl_2	−103+5	22.8	1531
Chromium	Cr	1890	62.1	3660
Chromium (II) chloride	$CrCl_2$	814	65.9	7700
Chromium (III) sequioxide	Cr_2O_3	2279	27.6	4200
Chromium trioxide	CrO_3	197	37.7	3770
Cobalt	Co	1490	62.1	3640
Cobalt (II) chloride	$CoCl_2$	727	56.9	7390
Copper	Cu	1083	49.0	3110
Copper (II) chloride	$CuCl_2$	430	24.7	4890
Copper (I) chloride	CuCl	429	26.4	2620
Copper(l) cyanide	$Cu_2(CN)_2$	473	(30.1)	(5400)
Copper (I) iodide	CuI	587	(13.6)	(2600)
Copper (II) oxide	CuO	1446	35.4	2820
Copper (I) oxide	Cu_2O	1230	(93.6)	(13400)
Copper (I) sulfide	Cu_2S	1129	62.3	5500
Dysprosium	Dy	1407	25.2	4100
Erbium	Er	1496	24.5	4100
Europium	Eu	826	16.4	2500

Heat of Fusion For Selected Elements and
Inorganic Compounds (Continued)

Compound	Formula	Melting point °C	Heat of fusion cal/g	cal/g mole
Europium trichloride	EuCl$_3$	622	(20.9)	(8000)
Fluorine	F$_2$	–219.6	6.4	244.0
Gadolinium	Gd	1312	23.8	3700
Gallium	Ga	29	19.1	1336
Germanium	Ge	959	(114.3)	(8300)
Gold	Au	1063	(15.3)	3030
Hafnium	Hf	2214	(34.1)	(6000)
Holmium	Ho	1461	24.8	4100
Hydrogen	H$_2$	–259.25	13.8	28
Hydrogen bromide	HBr	–86.96	7.1	575.1
Hydrogen chloride	HCl	–114.3	13.0	476.0
Hydrogen fluoride	HF	83.11	54.7	1094
Hydrogen iodide	HI	–50.91	5.4	686.3
Hydrogen nitrate	HNO$_3$	–47.2	9.5	601
Hydrogen oxide (water)	H$_2$O	0	79.72	1436
Deuterium oxide	D$_2$O	3.78	75.8	1516
Hydrogen peroxide	H$_2$O$_2$	–0.7	8.58	2920
Hydrogen selenate	H$_2$SeO$_4$	57.8	23.8	3450
Hydrogen sulfate	H$_2$SO$_4$	10.4	24.0	2360
Hydrogen sulfide	H$_2$S	–85.6	16.8	5683
Hydrogen sulfide, di–	H$_2$S$_2$	–89.7	27.3	1805
Hydrogen telluride	H$_2$Te	–49.0	12.9	1670
Indium	In	156.3	6.8	781
Iodine	I$_2$	112.9	14.3	3650
Iodine chloride (α)	ICl	17.1	16.4	2660
Iodine chloride (β)	ICl	13.8	13.3	2270
Iron	Fe	1530.0	63.7	3560
Iron carbide	Fe$_3$C	1226.8	68.6	12330

Heat of Fusion For Selected Elements and
Inorganic Compounds (Continued)

Compound	Formula	Melting point °C	Heat of fusion cal/g	Heat of fusion cal/g mole
Iron (III) chloride	Fe_2Cl_6	303.8	63.2	20500
Iron (II) chloride	$FeCl_2$	677	61.5	7800
Iron (II) oxide	FeO	1380	(107.2)	(7700)
Iron oxide	Fe_3O_4	1596	142.5	33000
Iron pentacarbonyl	$Fe(CO)_5$	−21.2	16.5	3250
Iron (II) sulfide	FeS	1195	56.9	5000
Lanthanum	La	920	17.4	2400
Lead	Pb	327.3	5.9	1224
Leadbromide	$PbBr_2$	487.8	11 7	4290
Lead chloride	$PbCl_2$	497 8	20.3	5650
Lead fluoride	PbF_2	823	7.6	1860
Lead iodide	PbI_2	412	17.9	5970
Lead molybdate	$PbMoO_4$	1065	70.8	(25800)
Lead oxide	PbO	890	12.6	2820
Lead sulfate	$PbSO_4$	1087	31.6	9600
Lead sulfide	PbS	1114	17.3	4150
Lithium	Li	178.8	158.5	1100
Lithium bromide	$LiBr$	552	33 4	2900
Lithium chloride	$LiCl$	614	75.5	3200
Lithium fluoride	LiF	896	(91.1)	(2360)
Lithium hydroxide	$LiOH$	462	103.3	2480
Lithium iodide	LiI	440	(10.6)	(1420)
Lithium metasilicate	Li_2SiO_3	1177	80.2	7210
Lithium molybdate	Li_2MoO_4	705	24.1	4200
Lithium nitrate	$LiNO_3$	250	87.8	6060
Lithium orthosilicate	Li_4SiO_4	1249	60.5	7430
Lithium sulfate	Li_2SO_4	857	27.6	3040
Lithium tungstate	Li_2WO_4	742	(25.6)	(6700)

Heat of Fusion For Selected Elements and Inorganic Compounds (Continued)

Compound	Formula	Melting point °C	Heat of fusion cal/g	cal/g mole
Lutetium	Lu	1651	26.3	4600
Magnesium	Mg	650	88.9	2160
Magnesium bromide	$MgBr_2$	711	45.0	8300
Magnesium chloride	$MgCl_2$	712	82.9	8100
Magnesium fluoride	MgF_2	1221	94.7	5900
Magnesium oxide	MgO	2642	459.0	18500
Magnesium silicate	$MgSiO_3$	1524	146.4	14700
Magnesium sulfate	$MgSO_4$	1327	28.9	3500
Manganese	Mn	1220	62.7	3450
Manganese dichloride	$MnCl_2$	650	58.4	7340
Manganese metasilicate	$MnSiO_3$	1274	(62.6)	(8200)
Manganese (II) oxide	MnO	1784	183.3	13000
Manganese oxide	Mn_3O_4	1590	(170.4)	(39000)
Mercury	Hg	−39	2.7	557.2
Mercury bromide	$HgBr_2$	241	10.9	3960
Mercury chloride	$HgCl_2$	276.8	15.3	4150
Mercury iodide	HgI_2	250	9.9	4500
Mercury sulfate	$HgSO_4$	850	(4.8)	(1440)
Molybdenum	Mo	2622	(68.4)	(6600)
Molybdenum dichloride	$MoCl_2$	726.8	3.58	6000
Molybdenum hexafluoride	MoF_6	17	11.9	2500
Molybdenum trioxide	MoO_3	795	(17.3)	(2500)
Neodymium	Nd	1020	11.8	1700
Neon	Ne	− 248.6	3.83	77.4
Nickel	Ni	1452	71.5	4200
Nickel chloride	$NiCl_2$	1030	142 5	18470
Nickel subsulfide	Ni_3S_2	790	25.8 1	5800
Niobium	Nb	2496	(68.9)	(6500)

Heat of Fusion For Selected Elements and
Inorganic Compounds (Continued)

Compound	Formula	Melting point °C	Heat of fusion cal/g	Heat of fusion cal/g mole
Niobium pentachloride	$NbCl_5$	21.1	30 8	8400
Niobium pentoxide	Nb_2O_5	1511	91.0	24200
Nitric oxide	NO	−163.7	18.3	549.5
Nitrogen	N_2	−210	6.15	172.3
Nitrogen tetroxide	N_2O_4	−13.2	60.2	5540
Nitrous oxide	N_2O	−90.9	35.5	1563
Osmium	Os	2700	(36.7)	(7000)
Osmium tetroxide (white)	OsO_4	41.8	9.2	2340
Osmium tetroxide (yellow)	OsO_4	55.8	15.5	4060
Oxygen	O_2	−218.8	3.3	106.3
Palladium	Pd	1555	38.6	4120
Phosphoric acid	H_3PO_4	42.3	25.8	2520
Phosphoric acid. hypo–	$H_4P_2O_6$	54.8	51.2	8300
Phosphorus acid, hypo–	H_3PO_2	17.3	35.0	2310
Phosphorus acid, ortho–	H_3PO_3	73.8	37.4	3070
Phosphorus oxychloride	$POCl_3$	1.0	20.3	3110
Phosphorus pentoxide	P_4O_{10}	569.0	60.1	17080
Phosphorus trioxide	P_4O_6	23.7	15.3	3360
Phosphorus, yellow	P_4	44.1	4.8	600
Platinum	Pt	1770	24.1	4700
Potassium	K	63.4	14.6	574
Potassium borate, meta–	KBO_2	947	(69.1)	(5660)
Potassium bromide	KBr	742	42.0	5000
Potassium carbonate	K_2CO_3	897	56.4	7800
Potassium chloride	KCl	770	85.9	6410
Potassium chromate	K_2CrO_4	984	35.6	6920
Potassium cyanide	KCN	623	(53.7)	(3500)
Potassium dichromate	$K_2Cr_2O_7$	398	29.8	8770

Heat of Fusion For Selected Elements and Inorganic Compounds (Continued)

Compound	Formula	Melting point °C	Heat of fusion cal/g	Heat of fusion cal/g mole
Potassium fluoride	KF	875	111.9	6500
Potassium hydroxide	KOH	360	(35.3)	(1980)
Potassium iodide	Kl	682	24.7	4100
Potassium nitrate	KNO_3	338	78.1	2840
Potassium peroxide	K_2O_2	490	55.3	6100
Potassium phosphate	K_3PO4	1340	41.9	8900
Potassium pyro– phosphate	$K_4P_2O_7$	1092	42.4	14000
Potassium sulfate	K_2SO_4	1074	46.4	8100
Potassium thiocyanate	KSCN	179	23.1	2250
Praseodymium	Pr	931	19.0	2700
Rhenium	Re	3167±60	(42.4)	(7900)
Rhenium heptoxide	Re_2O_7	296	30.1	15340
Rhenium hexafluoride	ReF_6	19.0	16.6	5000
Rubidium	Rb	38 .9	6. 1	525
Rubidium bromide	RbBr	677	22.4	3700
Rubidium chloride	RbCl	717	36.4	4400
Rubidium fluoride	RbF	833	39.5	4130
Rubidium iodide	Rbl	638	14.0	2990
Rubidium nitrate	$RbNO_3$	305	9.1	1340
Samarium	Sm	1072	17.3	2600
Scandium	Sc	1538	84.4	3800
Selenium	Se	217	15.4	1220
Seleniumoxychloride	$SeOCl_3$	9.8	6.1	1010
Silane, hexaHuoro–	Si_2F_6	−28.6	22.9	3900
Silicon	Si	1427	337.0	9470
Silicon dioxide (Cristobalite)	SiO_2	1723	35.0	2100
Silicon tetrachloride	$SiCl_4$	−67.7	10.8	1845

Heat of Fusion For Selected Elements and Inorganic Compounds (Continued)

Compound	Formula	Melting point °C	Heat of fusion cal/g	cal/g mole
Silver	Ag	961	25.0	2700
Silver bromide	AgBr	430	11.6	2180
Silver chloride	AgCl	455	22.0	3155
Silver cyanide	AgCN	350	20.5	2750
Silver iodide	AgI	557	9.5	2250
Silver nitrate	$AgNO_3$	209	16.2	2755
Silver sulfate	Ag_2SO_4	657	(13.7)	(4280)
Silver sulfide	Ag_2S	841	13.5	3360
Sodium	Na	97.8	27.4	630
Sodium borate, meta–	$NaBO_2$	966	134.6	8660
Sodium bromide	NaBr	747	59.7	6140
Sodium carbonate	Na_2CO_3	854	66.0	7000
Sodium chlorate	$NaClO_3$	255	49.7	5290
Sodium chloride	NaCl	800	123.5	7220
Sodium cyanide	NaCN	562	(88.9)	(4360)
Sodium fluoride	NaF	992	166.7	7000
Sodium hydroxide	NaOH	322	50.0	2000
Sodium iodide	NaI	662	35.1	5340
Sodium molybdate	Na_2MoO_4	687	17.5	3600
Sodium nitrate	$NaNO_3$	310	44.2	3760
Sodium peroxide	Na_2O_2	460	75.1	5860
Sodium phosphate, meta–	$NaPO_3$	988	(48.6)	(4960)
Sodium pyrophosphate	$Na_4P_2O_7$	970	(51.5)	(13700)
Sodiumsilicate,aluminum–	$NaAlSi_3O_8$	1107	50.1	13150
Sodium silicate, di–	$Na_2Si_2O_5$	884	46.4	8460
Sodium silicate, meta–	Na_2SiO_3	1087	84.4	10300
Sodium sulfate	Na_2SO_4	884	41.0	5830
Sodium sulfide	Na_2S	920	15.4	(1200)

Heat of Fusion For Selected Elements and
Inorganic Compounds (Continued)

Compound	Formula	Melting point °C	Heat of fusion cal/g	Heat of fusion cal/g mole
Sodium thiocyanate	NaSCN	323	54.8	4450
Sodium tungstate	Na_2WO_4	702	19.6	5800
Strontium	Sr	757	25.0	2190
Strontium bromide	$SrBr_2$	643	19.3	4780
Strontium chloride	$SrCl_2$	872	26.5	4100
Strontium fluoride	SrF_2	1400	34.0	4260
Strontium oxide	SrO	2430	161.2	16700
Sulfur (monatomic)	S	119	9.2	295
Sulfur dioxide	SO_2	-73.2	32.2	2060
Sulfur trioxide (α)	SO_3	16.8	25.8	2060
Sulfur trioxide (β)	SO_3	32.3	36.1	2890
Sulfur trioxide (γ)	SO_3	62.1	79.0	6310
Tantalum	Ta	2996 ± 50	34.6 –41.5	(7500)
Tantalum pentachloride	$TaCl_5$	206.8	25.1	9000
Tantalum pentoxide	Ta_2O_5	1877	108.6	48000
Tellurium	Te	453	25.3	3230
Terbium	Tb	1356	24.6	3900
Thallium	Tl	302.4	5.0	1030
Thallium bromide, mono–	TlBr	460	21.0	5990
Thallium carbonate	Tl_2CO_3	273	9.5	4400
Thallium chloride, mono–	TlCl	427	17.7	4260
Thallium iodide, mono–	TlI	440	9.4	3125
Thallium nitrate	$TlNO_3$	207	8.6	2290
Thallium sulfate	Tl_2SO_4	632	10.9	5500
Thallium sulfide	Tl_2S	449	6.8	3000
Thorium	Th	1845	(<19.8)	(<4600)
Thorium chloride	$ThCl_4$	765	61.6	22500
Thorium dioxide	ThO_2	2952	1102.0	291100

Heat of Fusion For Selected Elements and
Inorganic Compounds (Continued)

Compound	Formula	Melting point °C	Heat of fusion cal/g	cal/g mole
Thulium	Tm	1545	26.0	4400
Tin	Sn	231.7	14.4	1720
Tin bromide, di–	$SnBr_2$	231.8	(6.1)	(1720)
Tin bromide, tetra–	$SnBr_4$	29.8	6.8	3000
Tin chloride, di–	$SnCl_2$	247	16.0	3050
Tinchloride,tetra–	$SnCl_4$	–33.3	8.4	2190
Tin iodide, tetra–	SnI_4	143.4	(6.9)	(4330)
Tin oxide	SnO	1042	(46.8)	(6400)
Titanium	Ti	1800	(104.4)	(5000)
Titanium bromide, tetra–	$TiBr_4$	38	(5.6)	(2060)
Titanium chloride, tetra–	$TiCl_4$	–23.2	11.9	2240
Titanium dioxide	TiO_2	1825	(142.7)	(11400)
Titanium oxide	TiO	991	219	14000
Tungsten	W	3387	(45.8)	(8420)
Tungsten dioxide	WO_2	1270	60 1	13940
Tungsten hexafluoride	WF_6	–0.5	6.0	1800
Tungsten tetrachloride	WCl_4	327	18.4	6000
Tungsten trioxide	WO_3	1470	60 1	13940
Uranium[235]	U	~1133	20	3700
Uranium tetrachloride	UCl_4	590	27.1	10300
Vanadium	V	1917	(70)	(4200)
Vanadium dichloride	VCl_2	1027	65.6	8000
Vanadium oxide	VO	2077	224.0	15000
Vanadium pentoxide	V_2O_5	670	85.5	15560
Xenon	Xe	–111.6	5.6	740
Ytterbium	Yb	823	12.7	2200
Yttrium	Y	1504	46.1	4100
Yttrium oxide	Y_2O_3	2227	110.7	25000

Heat of Fusion For Selected Elements and Inorganic Compounds (Continued)

Compound	Formula	Melting point °C	Heat of fusion cal/g	cal/g mole
Zinc	Zn	419.4	24.4	1595
Zinc chloride	$ZnCl_2$	283	(406)	(5540)
Zinc oxide	ZnO	1975	54.9	4470
Zinc sulfide	ZnS	1745	(93.3)	(9100)
Zirconium	Zr	1857	(60)	(5500)
Zirconium dichloride	$ZrCl_2$	727	45.0	7300
Zirconium oxide	ZrO_2	2715	168.8	20800

For heat of fusion in J/kg, multiply values in cal/g by 4184. For heat of fusion in J/mol, multiply values in cal/g-mol (=cal/mol) by 4.184. For melting point in K, add 273.15 to values in °C. Values in parentheses are of uncertain reliability.

Data from: Weast, R C., Ed., *Handbook of Chemistry and Physics, 55th ed.*, CRC Press, Cleveland, (1974); and Bolz, R. E. and Tuve, G. L., Eds., *Handbook of Tables for Applied Engineering Science*, 2nd ed., CRC Press, Cleveland, (1973)

SURFACE TENSION OF LIQUID ELEMENTS

Element	Purity (wt. %)	σ_{mp} (dyn/cm)	Atm.	°C	σ_t (dyn/cm)
Ag	99.99		H_2	1000	916
	–	(785)	vac.		
	99.96		H_2	1000	893
				1150	862
				1250	849
			vac.	1000	908
	99.995		H_2	1000	907
				1100	894
				1200	876
	99.999	(828)	vac.		
	99.99		Ar, H_2	1000	890
	spect. pure	921			
			\multicolumn{3}{l}{$\sigma = 1136 - 0.174\,T$ (valid 1300 to 2200 K)}
		918			
			\multicolumn{3}{l}{$\sigma = 918 - 0.149\,(t - t_{mp})$}
	99.999		Ar	980	905±10
				1108	890±10
	99.99	860±20	Ar		
	99.72		vac.	950	840
	99.7	863±25	Ar		
			\multicolumn{3}{l}{$\sigma = (863+25) - 0.33\,(t - t_{mp})$}
	99.99	865	vac.		
			\multicolumn{3}{l}{$\sigma = 865 - 0.14\,(t - t_{mp})$}
	99.99	(825)	Ar		
			\multicolumn{3}{l}{$\sigma = 825 - 0.05\,(T - 993)$ *}
	99.99	866	He		
			\multicolumn{3}{l}{$\sigma = 866 - 0.15\,(t - t_{mp})$}
	99.999	873	He	1600	725
			\multicolumn{3}{l}{$\sigma = 873 - 0.15\,(t - t_{mp})$}
Au		(754)	vac.		
	99.999	1130	He	1200	1070
				1300	1020

Surface Tension of Liquid Elements

Element	Purity (wt. %)	σ_{mp} (dyn/cm)	Atm.	°C	σ_t (dyn/cm)
	99.999	(731)	vac.		
	99.999		Ar	1108	1130±10
B	99.8	1060±50	vac.		
Ba	–		Ar	720	224
	99.5	276			

$$\sigma = 351 - 0.075\ T$$
(valid 1410 to 1880 K) *

Element	Purity (wt. %)	σ_{mp} (dyn/cm)	Atm.	°C	σ_t (dyn/cm)
Be	99.98		vac.	1500	1100
Bi		376	vac.		
	99.99	376	vac.		
	99.90		H_2	800	343
				1000	328
			vac.	450	(382)
	99.9	380±10	Ar	350	362
			vac.	700	350
	99.98	380±10	Ar		
			–	450	380
			vac.	300	379
		378	vac., Ar, H_2		
	99.99995	375			

$$\sigma = 423 - 0.088\ T$$
(valid 1352 to 1555 K) *

Element	Purity (wt. %)	σ_{mp} (dyn/cm)	Atm.	°C	σ_t (dyn/cm)
	99.999	380±3	Ar		

$$\sigma = 380 - 0.142\ (t - t_{mp})$$
(valid MP to 555°C)

Element	Purity (wt. %)	σ_{mp} (dyn/cm)	Atm.	°C	σ_t (dyn/cm)
Ca 337	–		Ar	850	
	p.a.	360			

$$\sigma = 472 - 0.100\ T$$
(valid 1445 to 1655 K) *

Surface Tension of Liquid Elements

Element	Purity (wt. %)	σ_{mp} (dyn/cm)	Atm.	°C	σ_t (dyn/cm)
Cd			–	450	600
			–	400	600
			–	350	586
	99.9	(550±10)	Ar	390	604
		(525±30)	H_2		
	99.9999	590±5	–		(non linear)
Co			Ar	1550	1836
	99.99		vac., Al_2O_3	1520	1800
	99.99		He,Al_2O_3	1520	(1630)
	99.99		He, BeO	1520	(1640)
	99.99		He, MgO	1520	(1560)
	99.99		H, Al_2O_3	1520	1780
	99.99		He	1520	(1620)
	99.99		H_2	1520	(1590)
			vac.	1500	1870
				1600	(1640)
				1600	(1600)
			vac.	1600	1815
	99.99		vac., Al_2O_3	1600	1812
		(1520)	H_2, He		
	99.99		H_2, He	1550	1845
	99.9983	1880	vac.		
	99.99			1550	1780
Cr	–		vac.	1950	1590±50
	99.9997	1700±50	Ar		
Cs			Ar	62	68.4
			Ar	62	67.5
				146	62.9
	99.95		Ar	39	69.5
				494	42.8
				642	34.6
	99.995	68.6	He		

$$\sigma = 68.6 - 0.047\,(t - t_{mp})$$

Surface Tension of Liquid Elements

Element	Purity (wt. %)	σ_{mp} (dyn/cm)	Atm.	°C	σ_t (dyn/cm)
				(valid 52 to 1100°C)	
Cu			Ar	1120	1269±20
	–	(1150)	vac.		
			Ar	1120	1285±10
	99.99	73.74	Ar		

$$\sigma = 73.74 - 1.791 \cdot 10^{-2}\,(t - t_{mp})$$
$$-9.610 \cdot 10^{-5}\,(t - t_{mp})^2$$
$$+ 6.629 \cdot 10^{-8}\,(t - t_{mp})^3$$
(valid 71 to 1011°C)

Element	Purity (wt. %)	σ_{mp} (dyn/cm)	Atm.	°C	σ_t (dyn/cm)
	99.98		H_2	1100	1301
				1165	1295
				1255	1287
		1270	vac.		
			vac.	1120	1285
				1440	1298
		(1085)	vac.		
	99.99		He	1250	1290
	99.99		H_2	1250	1300
	99.9	(11802±40)	Ar		
	–			1100	1220
				1183	(1130)
	–		vac.	1150	1370
	99.997	1355	He, H_2		
	99.997	1352	vac.		

$$\sigma = 1352 - 0.17\,(t - t_{mp})$$

Element	Purity (wt. %)	σ_{mp} (dyn/cm)	Atm.	°C	σ_t (dyn/cm)
	99.997	1358	Ar		

$$\sigma = 1358 - 0.20\,(t - t_{mp})$$

Element	Purity (wt. %)	σ_{mp} (dyn/cm)	Atm.	°C	σ_t (dyn/cm)
			Ar, He	1120	1285±10
	99.99		H_2, He	1550	1265
	99.99999	1300	vac.		
			vac.	1130	1268±60

Surface Tension of Liquid Elements

Element	Purity (wt. %)	σ_{mp} (dyn/cm)	Atm.	°C	σ_t (dyn/cm)
	99.98		Ar	1600	1230
	99.999		N_2	1100	1341
				1150	1338
				1200	1335
	99.9	(1127)	vac.		
	99.99		Ar, H_2	1100	1320
Fe				1570	(1731)
				1550	1860
		1720	He		
				1580–1760	(880)
				1570	(1632)
	99.99		He	1650	(1610)
	99.99		He	1650	(1430)
	99.99		H_2	1650	(1400)
	–	(1384)	vac.		
		(1700)	vac.		
			vac., He	1550	1865
	99.99		He	1650	(1430)
	99.99		H,	1650	(1400)
				1650	(1640)
		(1650)	He, H_2		
	99.985		Ar	1550	1788
	–	(1560)			
	99.94		vac., Al_2O_3	1560	(1710)
	99.9998	1880	vac.		
	99.93	1860±40	He		
		(1510)	vac.		
	99.97		vac., BeO	1550	1830±6
	Armco		Ar, N_2	1550	1795
			vac.	1550	1754
	99.987		vac.	1550	(1730)
	99.85	(1619)	vac.		
	99.69		He, Al_2O_3	1550	(1727)
	99.69		H_2, Al_2O_3	1550	(1734)

Surface Tension of Liquid Elements

Element	Purity (wt. %)	σ_{mp} (dyn/cm)	Atm.	°C	σ_t (dyn/cm)
	–	1760±20	He, H_2		
			$\sigma = 1760 - 0.35\,(t - t_{mp})$		
	99.9992	1773	He, H_2		
			$\sigma = 773 + 0.65\,t$		
			(valid 1550 to 1780°C)		
Fr	–	–		100	58.4
		725±10	Ar		
			vac.	350	718
	–		He, Al_2O_3	1500	559
	99.9998	718	vac., Al_2O_3		
			$\sigma = 718 - 0.101\,(t - t_{mp})$		
		650	vac.	1200	530
			vac.	1000	650
		632±5	N_2, He		
Hf		(1460)	vac.		
	97.5±2.5	1630	vac.		
	Zr				
Hg			H	20	(542)
			air	16	(410)
			air	20	(435.5)
			vac	20	472
			vac	20	(402)
				25	476
			vac.	20	(436)
			vac	20	(432)
			H_2	25	476
				25	472
				25	(464)
			H_2	19	473
				20	(437)
			vac.	20	480

Surface Tension of Liquid Elements

Element	Purity (wt. %)	σ_{mp} (dyn/cm)	Atm.	°C	σ_t (dyn/cm)
				25	(516)
				25	(435)
			vac.	25	473
				25	488
				25	(498)
			vac.	20	(420)
				25	476
			vac.	20	(410)
			vac.	20	(455)
				25	484±1.5
			vac.	22	(468)
			vac.	20	(465.2)
				103	449.7
				350	387.1
				25	484.9±1.8
			vac.	20	485.5±1.0
			$\sigma = 489.5\text{–}0.20\ t$		
			Ar	20	(454.7)
				21	(350.5)
	99.99	He, H_2	20	475	
	99.9	Ar	20	(500±15)	
		vac.	–1()	487	
		vac.	25	483.5±1.0	
			22	(465)	
			25	485.1	
			16.5	487.3	
			25	485.4±1.2	
			23–25	482.8±9.7	
			$\sigma=468.7\text{–}1.61 \cdot 10^{-1}t\text{–} 1.815 \cdot 10^{-2}\ t^2$		
		vac.	20	484.6±1.3	
		Ar	25	480	
		vac.	20	482.5 ± 3.0	
			$\sigma = 485.5 \text{–} 0.149\ t \text{–} 2.84 \cdot 10^{-4}\ t^2$		
		Ar	21.5	484.9±0.3	

Surface Tension of Liquid Elements

Element	Purity (wt. %)	σ_{mp} (dyn/cm)	Atm.	°C	σ_t (dyn/cm)
In	99.95	559	H_2	600	515
				623	540
	99.995	556.0	Ar, He		
			vac	185	592
			H_2	600	514
				300	541
	99.999		Ar	200	556
				400	535
				550	527.8
	99.9994		vac.	350	539
	99.9999			560±5	

$$\sigma = 568.0 - 0.04\, t - 7.08 \times 10^{-5}\, t^2$$

Element	Purity (wt. %)	σ_{mp} (dyn/cm)	Atm.	°C	σ_t (dyn/cm)
Ir	99.9980	2250	vac.		
K	99.895	101	Ar		
		110.3± 1	–		
		117	vac.	$\sigma = 117 - 0.66\,(t - t_{mp})$	
		–	Ar	87	112
				457	80
				677	64.8
	99.986	116.95	Ar		

$$\sigma = 116.95 - 6.742 \cdot 10^{-2}\,(t - t_{mp})$$
$$-3.836 \cdot 10^{-5}\,(t - t_{mp})^2$$
$$+3.707 \cdot 10^{-8}(t - t_{mp})^3$$
(valid 77 to 983°C)

Element	Purity (wt. %)	σ_{mp} (dyn/cm)	Atm.	°C	σ_t (dyn/cm)
	99.936	(79.2)	He		

$$(\sigma = 76.8 - 70.3 \cdot 10^{-4}\,(t - 400))$$
(valid 600 to 1126°C)

Element	Purity (wt. %)	σ_{mp} (dyn/cm)	Atm.	°C	σ_t (dyn/cm)
		95 ±9.5	–		
	99.97±0.64	111.35	He		

$$\sigma = 115.51 - 0.0653\, t$$

Surface Tension of Liquid Elements

Element	Purity (wt. %)	σ_{mp} (dyn/cm)	Atm.	°C	σ_t (dyn/cm)
				(valid 70 to 713°C)	
Li	99.95		Ar	180	397.5
				300	380
				500	351.5
	99.98		Ar	287	386
				922	275
				1077	253
Mg			Ar	681	563
				789	532
				894	502
	99.8		N_2	670	552
				700	542
				740	528
	99.9		Ar	700	550±15
	99.91	(525±10)	Ar		
	99.5	583			

$$\sigma = 721 - 0.149\ T\ *$$
(valid 1125 to 1326°K)

Element	Purity (wt. %)	σ_{mp} (dyn/cm)	Atm.	°C	σ_t (dyn/cm)
Mn	99.9985	1100 ± 50	Ar		
	99.94		vac.	1550	1030
				1550	1010
Mo	99.7	(1915)	vac.		
		2080	vac.		
	99.9996	2250	vac.		
	99.98	2049	vac.		
	–	2130	vac.		
Na			Ar	110	205.7
				263	198.2
	99.995	191	Ar		
			vac.	123	198
				129	198.5

Surface Tension of Liquid Elements

Element	Purity (wt. %)	σ_{mp} (dyn/cm)	Atm.	°C	σ_t (dyn/cm)
				140	190
		200.2 ±0.6			
		202	vac.		
			$\sigma = 202-0.092(t-t_{mp})$ (valid 100 to 1000°C)		
	p.a.		Ar	617	144
				764	130
				855	120.4
	99.96	210.12	Ar		
			$\sigma = 210.12-8.105 \cdot 10^{-2}(t-t_{mp})$ $-8.064 \cdot 10^{-5}(t-t_{mp})^2$ $+3.380 \cdot 10^{-8}(t-t_{mp})^3$ (valid 141 to 992°C)		
	99.982	187.4	He		
			$\sigma = 144-0.108(t-500)$ (valid 400 to 1125°C)		
Nb, Cb	99.9986	1900	vac.		
	99.99	(1827)	vac.		
	–	2020	vac.		
Nd		688	Ar	1186	674
Ni	99.7		He	1470	(1615)
	99.7				
			H_2		
				1470	(1570)
	99.7		vac.	1470	1735
		1725	vac.		
			vac.	1475	1725
	–		Ar	1550	(1934)
	99.99		vac., Al_2O_3	1520	1740
	99.99		He,Ar,Al_2O_3	1520	1770
	99.99		H_2,Al_2O_3	1520	(1600)
	99.99		He, MgO	1470	(1530)
	99.99		He, BeO	1470	(1500)

Surface Tension of Liquid Elements

Element	Purity (wt. %)	σ_{mp} (dyn/cm)	Atm.	°C	σ_t (dyn/cm)
	99.99		He	1470	(1490)
		1725	vac.		
	99.99			1600	(1600)
			vac.	1500	1720
	99.99		vac., Al_2O_3	1550	1780
				1550	1735
			H_2, He	1470	1700
	99.99		vac.	1640	1705
	–		vac., Al_2O_3	1560	1810
	99.999	1770±13	vac.		
	99.999	1728±10	vac.		
	99.999	1822±8	vac.		
		(1670)	vac.		
		1760	vac.		
		(1687)	vac.		
	–		He	1500	1745
	–	1809±20	H_2, He, Al_2O_3		
			$\sigma = 1770 - 0.39 (t-1550)$		
	99.99975	(1977)	He		
			$\sigma = 1665 + 0.215\,t$ (valid 1475 to 1650°C)		
Os	99.9998	2500	vac.		
P(white)				50	69.7
				68.7	64.95
Pb	99.98		H, N_2	340	448
				390	442
				440	439
			air	360	452
		451	vac.	425	440
		450	He		
				350–450	450
	99.998	480	H_2		
				623	474
			vac.	362	455

Surface Tension of Liquid Elements

Element	Purity (wt. %)	σ_{mp} (dyn/cm)	Atm.	°C	σ_t (dyn/cm)
				700	428
	99.9		H_2	1000	388
	99.9	(410±5)	Ar		
				350	445
	99.98		vac.	340	442
				400	435
	99.9995	470			

$$\sigma = 538\text{--}0.114\ T\ *$$
$$(\text{valid}\ 1440\ \text{to}\ 1970°K)$$

Element	Purity (wt. %)	σ_{mp} (dyn/cm)	Atm.	°C	σ_t (dyn/cm)
	99.9994		vac	450	438
	99.999		He	1600	310
				390	456
	99.999	470	Ar		

$$\sigma = 470\text{--}0.164\ (t\text{--}t_{mp})$$
$$(\text{valid}\ \text{MP to}\ 535°C)$$

Element	Purity (wt. %)	σ_{mp} (dyn/cm)	Atm.	°C	σ_t (dyn/cm)
Pd		1470	vac.		
	99.998	1500	vac.		
	99.998	1460	He		
Pt		1869	CO_2		
	99.84	(1740±20)	vac.		
	99.999		Ar		1800(1699±20)
	99.9980	1865	vac.		
Pu		550±55			
Rb		(77±5)	vac.		
	99.8		Ar	52	84
				477	55
				632	46.8

Surface Tension of Liquid Elements

Element	Purity (wt. %)	σ_{mp} (dyn/cm)	Atm.	°C	σ_t (dyn/cm)
	99.92	91 17	Ar		
			$\sigma = 91.17 - 9.189 \; 10^{-2} \, (t - t_{mp})$ $+7.228 \cdot 10^{-5} \, (t - t_{mp})^2$ $- 3.830 \cdot 10^{-8} \, (t - t_{mp})^3$ (valid 1104 to 1006°C)		
	99.997	85.7	He		
			$\sigma = 85.7 - 0.054 \, (t - t_{mp})$ (valid 53 to 1115°C)		
Re	99.4	2610	vac.		
	99.9999	2700	vac.		
Ru	99.9980	2250	vac.		
Rh		1940	vac.		
	99.9975	2000	vac.		
S	–	60.9	vac.	250	51.1
Sb			H_2	640	349
				700	349
				974	342
			H_2	750	368
				900	361
				1100	348
			vac.	640	367.9
				762	364.9
	99.5	383	H_2, N_2	675	384
				800	380
	99.99	395±20	Ar		
	99.15	395±20	Ar		
	99.999		N_2	800	359
				1000	351

Surface Tension of Liquid Elements

Element	Purity (wt. %)	σ_{mp} (dyn/cm)	Atm.	°C	σ_t (dyn/cm)
				700	347.6
				800	345.0
	99.999		He	1600	320
Se	–		Ar	230–250	88.0±5
Si			He	1450	725
			vac.	1550	720
	99.99		vac.	1550	750
	99.9999		Ar	1500	825
Sn			N_2	275	612
				500	572
				800	520
			air	280	523
				340	520
	99.99	537	vac.	500	524
				600	508
		530	He		
			H_2	489	543
				572	528
				692	503
			–	250	536
			–	450	530
			–	250	545
	99.93		vac.	250	549
				400	539
				600	526
	99.998	566	H,		
			–	623	559
		610	vac.		
			–	800	500
			–	300	538
			–	300	(527)
			–	290	546

Surface Tension of Liquid Elements

Element	Purity (wt. %)	σ_{mp} (dyn/cm)	Atm.	°C	σ_t (dyn/cm)
	99.99		H_2, He	600	530
	99.9	(526±10)	Ar		
			vac.	290	600
	99.965		H_2	740	508
				950	489.5
				1115	479.5
	99.89	543.7			
		562	vac.		
			vac.	300	554
	99.999	590	vac.		
			H_2	290	(520)
			vac.	290	(524)
	99.9999		H_2	246	552.7
	99.9994		vac.	350	537
	99.999	555.8±1.9			
			$\sigma = 566.84 - 4.76 \cdot 10^{-2}\,t$		
	99.96	552	vac.	1000	470
	99.96	552	Ar		
			$\sigma = 552 - 0\,167\,(t-t_{mp}).$ (valid MP to 500°C)		
Sr			Ar	775	288
				830	282
				893	282
	99.5	303			
			$\sigma = 392 - 0.085\,T$ (valid 1152 to 1602 K)		
Ta		2360	vac.		
		2030	vac.		
		1910	vac.		
	99.9983	2150	vac.		
	99.9	(1884)	vac.		
Te	99.4	186±2	Ar		
			vac.	460	178±1.5
	—		vac.	475	(162)

Surface Tension of Liquid Elements

Element	Purity (wt. %)	σ_{mp} (dyn/cm)	Atm.	°C	σ_t (dyn/cm)
		178			
			$\sigma = 178 - 0.024\,(t - t_{mp})$		
Ti	98.7	1510	vac.		
	99.92	1390	Ar		
		1460	vac.		
	99.9991	1650	vac.		
	99.0		vac.	1680	1576
	99.99999		vac.	1680	1588
	99.85	(1880)	vac.		
	99.69	1402	vac.		
Tl		464.5	Ar		
			–	450	452
	–		vac.	450	450
	99.999	467			
			$(\sigma = 536 - 0.119\,T\,)^*$		
			(valid 1270 to 1695°K)		
	99.999		vac.	450	450
U		1500±75			
		1550	Ar		
	99.94	(1294)	vac.		
V	99.9977	1950	vac.		
	–	(1760)	vac.		
W	–	2310	vac.		
	99.9999	2500	vac.		
	99.8	2220	vac.		
	99.9	(2000)	vac.		
Zn	99.9	750 ±20	Ar		
	99.99	757.0±5	vac.		
	99.999	761.0	vac.		

Surface Tension of Liquid Elements

Element	Purity (wt. %)	σ_{mp} (dyn/cm)	Atm.	°C	σ_t (dyn/cm)
	99.9999	767.5	vac.		
Zr		1400	Ar		
	99.5	1411±70	vac.		
	99.9998	1480	vac.		
	99.7	(1533)	vac.		

* T in Kelvin (t in °C).

The data is a compilation of several studies and measurements were obtained from the "sessile drop", "maximum bubble pressure", and the "pendant drop" methods. The accuracy varies with both method and the study.

Values in parentheses are less certain.

Data from: Lang,G.,in *Handbook of Chemistry and Physics, 55th ed.,*Weast, R.C.,Ed., CRC Press, Cleveland, 1974, F-23.

VAPOR PRESSURE OF THE ELEMENTS
(Very Low Pressures)

Element	Melting point (°C)	10^{-5}	10^{-4}	10^{-3}	10^{-2}	10^{-1}	1
Ag	961	767	848	936	1047	1184	1353
Al	660	724	808	889	996	1123	1279
Au	1063	1083	1190	1316	1465	1646	1867
Ba	717	418	476	546	629	730	858
Be	1284	942	1029	1130	1246	1395	1582
Bi	271	474	536	609	698	802	934
C		2129	2288	2471	2681	2926	3214
Cd	321	148	180	220	264	321	
Co	1478	1249	1362	1494	1649	1833	2056
Cr	1900	907	992	1090	1205	1342	1504
CU	1083	946	1035	1141	1273	1432	1628
Fe	1535	1094	1195	1310	1447	1602	1783
Hg	−38.9	−23.9	−5.5	18.0	48.0	82.0	126
In	157	667	746	840	952	1088	1260
Ir	2454	1993	2154	2340	2556	2811	3118
Mg	651	287	331	383	443	515	605
Mn	1244	717	791	878	980	1103	1251
Mo	2622	1923	2095	2295	2533		
Ni	1455	1157	1257	1371	1510	1679	1884
Os	2697	2101	2264	2451	2667	2920	3221
Pb	328	483	548	625	718	832	975
Pd	1555	1156	1271	1405	1566	1759	2000
Pt	1774	1606	1744	1904	2090	2313	2582
Sb	630	466	525	595	678	779	904
Si	1410	1024	1116	1223	1343	1485	1670

Vapor Pressure of the Elements Continued
(Very Low Pressures)

Element	Melting point (°C)	10^{-5}	10^{-4}	10^{-3}	10^{-2}	10^{-1}	1
				Pressure (mm Hg)			
Sn	232	823	922	1042	1189	1373	
	1609						
Ta	2996	2407	2599	2820			
W	3382	2554	2767	3016	3309		
Zn	419	211	248	292	343	405	
Zr	2127	1527	1660	1816	2001	2212	
	2459						

[a]The values given in this table are from a variety of sources that are not always in agreement; for that reason, the table should be used only as a general guide.

To convert pressures to SI units, 1 mm Hg (torr) = 133.3 N/m^2 and 1 atm = 0.1013 MN/m^2.

Source: From Dushman, S., *Scientific Foundations of Vacuum Technique,* John Wiley & Sons, New York, (1949)

VAPOR PRESSURE OF THE ELEMENTS
(Moderate Pressures)

Element	Symbol	mmHg				
		1	10	100	400	760
Aluminum	Al	1540	1780	2080	2320	2467
Antimony	Sb		960	1280	1570	1750
Arsenic	As	380	440	510	580	610
Barium	Ba	860	1050	1300	1520	1640
Beryllium	Be	1520	1860	2300	2770	2970
Bismuth	Bi		1060	1280	1450	1560
Boron	B	2660	3030	3460	3810	4000
Bromine	Br	−60	−30	+9	39	59
Cadmium	Cd	393	486	610	710	765
Calcium	Ca	800	970	1200	1390	1490
Cesium	Cl		373	513	624	690
Chlorine	Cl	−123	−101	−71	−46	−34
Chromium	Cr	1610	1840	2140	2360	2480
Cobalt	Co	1910	2170	2500	2760	2870
Copper	Cu		1870	2190	2440	2600
Fluorine	F			−203	−193	−188
Gallium	Ca	1350	1570	1850	2060	2180
Germanium	Ge		2080	2440	2710	2830
Gold	Au	1880	2160	2520	2800	2940
Indium	In				1960	2080

Vapor Pressure of the Elements Continued
(Moderate Pressures)

Element	Symbol	mmHg				
		1	10	100	400	760
Iodine	I	40	72	115	160	185
Iridium	Ir	2830	3170	3630	3960	4130
Iron	Fe	1780	2040	2370	2620	2750
Lanthanum	La				3230	3420
Lead	Pb	970	1160	1420	1630	1740
Lithium	Li	750	890	1080	1240	1310
Magnesium	Mg	620	740	900	1040	1110
Manganese	Mn		1510	1810	2050	2100
Mercury	Hg			260	330	356.9
Molybdenum	Mo	3300	3770	4200	4580	4830
Neodymium	Nd				2870	3100
Nickel	Ni	1800	2090	2370	2620	2730
Palladium	Pd	1470	2290	2670	2950	3140
Phosphorus	P		127	199	253	283
Platinum	Pt	2600	2940	3360	3650	3830
Polonium	Po	472	587	752	890	960
Potassium	K			590	710	770
Rhodium	Rh	2530	2850	3260	3590	3760
Rubidium	Rb		390	527	640	700
Selenium	Se		429	547	640	685

Vapor Pressure of the Elements Continued
(Moderate Pressures)

Element	Symbol	mmHg				
		1	10	100	400	760
Silver	Ag	1310	1540	1850	2060	2210
Sodium	Na	440	546	700	830	890
Strontium	Sr	740	900	1100	1280	1380
Sulfur	S		246	333	407	445
Tellurium	Te	520	633	792	900	962
Thallium	Tl		1000	1210	1370	1470
Tin	Sn	1610	1890	2270	2580	2750
Titanium	Ti	2180	2480	2860	3100	3260
Tungsten	W	3980	4490	5160	5470	5940
Uranium	U	2450	2800	3270	3620	3800
Vanadium	V	2290	2570	2950	3220	3380
Zinc	Zn		590	730	840	907

This table lists the temperature in degrees Celsius (Centigrade) at which an element has a vapor pressure indicated by the headings of the columns. To convert pressures to SI units, 1 mm Hg (torr) = 133.3 N/m^2 and 1 atm = 0.1013 MN/m^2.

Source: From Loebel, R., in *Handbook of Chemistry and Physics,* 55th ed., Weast, R. C., Ed., CRC Press, Cleveland, (1974)

VAPOR PRESSURE OF THE ELEMENTS
(High Pressures)

Element	Symbol	atm				
		2	5	10	20	40
Aluminum	Al	2610	2850	3050	3270	3530
Antimony	Sb	1960	2490			
Arsenic	As					
Barium	Ba	1790	2030	2230		
Beryllium	Be	3240	3730	4110	4720	5610
Bismuth	Bi	1660	1850	2000	2180	
Boron	B					
Bromine	Br	78	110			
Cadmium	Cd	830	930	1030	1120	1240
Calcium	Ca	1630	1850	2020	2290	
Cesium	Cl					
Chlorine	Cl	−17	+9	30	55	97
Chromium	Cr	2630	2850	3010	3180	
Cobalt	Co	3040	3270			
Copper	Cu	2760	3010	3500	3460	3740
Fluorine	F	−180.7	−169.1	−159.6		
Gallium	Ca	2320	2560	2730		
Germanium	Ge	2970	3200	3430		
Gold	Au	3120	3490	3630	3890	
Indium	In	2230	2440	2600		

Vapor Pressure of the Elements Continued
(High Pressures)

Element	Symbol	atm				
		2	5	10	20	40
Iodine	I	216	265			
Iridium	Ir	4310	4650			
Iron	Fe	2900	3150	3360	3570	
Lanthanum	La	3620	3960	4270		
Lead	Pb	1880	2140	2320	2620	
Lithium	Li	1420	1518			
Magnesium	Mg	1190	1330	1430	1560	
Manganese	Mn	2360	2580	2850		
Mercury	Hg	398	465	517	581	657
Molybdenum	Mo	5050	5340	5680	5980	
Neodymium	Nd	3300	3680	3990		
Nickel	Ni	2880	3120	3300	3310	
Palladium	Pd	3270	3560	3840		
Phosphorus	P	319				
Platinum	Pt	4000	4310	4570	4860	
Polonium	Po	1060	1200	1340		
Potassium	K	850	950	1110	1240	1420
Rhodium	Rh	3930	4230	4440		
Rubidium	Rb					
Selenium	Se	750	850	920	1010	1120

Vapor Pressure of the Elements Continued
(High Pressures)

Element	Symbol	atm 2	5	10	20	40
Silver	Ag	2360	2600	2850	3050	3300
Sodium	Na	980	1120	1230	1370	
Strontium	Sr	1480	1670	1850	2030	
Sulfur	S	493	574	640	720	
Tellurium	Te	1030	1160	1250		
Thallium	Tl	1560	1750	1900	2050	2260
Tin	Sn	2950	3270	3540	3890	
Titanium	Ti	3400	3650	3800		
Tungsten	W	6260	6670	7250	7670	
Uranium	U	4040	4420			
Vanadium	V	3540	3800			
Zinc	Zn	970	1090	1180	1290	

This table lists the temperature in degrees Celsius (Centigrade) at which an element has a vapor pressure indicated by the headings of the columns. To convert pressures to SI units, 1 mm Hg (torr) = 133.3 N/m^2 and 1 atm = 0.1013 MN/m^2.

Source: From Loebel, R., in *Handbook of Chemistry and Physics,* 55th ed., Weast, R. C., Ed., CRC Press, Cleveland, (1974).

SPECIFIC HEAT OF SELECTED ELEMENTS AT 25°C
Listed by Element

Element	C_p (cal $\cdot$ g^{-1} $\cdot$ K^{-1})
Aluminum	0.215
Antimony	0.049
Argon	0.124
Arsenic	0.0785
Barium	0.046
Beryllium	0.436
Bismuth	0.0296
Boron	0.245
Bromine (Br_2)	0.113
Cadmium	0.0555
Calcium	0.156
Carbon, diamond	0.124
Carbon, graphite	0.170
Cerium	0.049
Cesium	0.057
Chlorine (Cl_2)	0.114
Chromium	0.107
Cobalt	0.109
Columbium (see Niobium)	
Copper	0.092

Specific Heat of Selected Elements at 25°C
Listed by Element

Element	C_p (cal $\cdot$ g^{-1} $\cdot$ K^{-1})
Dysprosium	0.0414
Erbium	0.0401
Europium	0.0421
Fluorine (F_2)	0.197
Gadolinium	0.055
Gallium	0.089
Germanium	0.077
Gold	0.0308
Hafnium	0.035
Helium	1.24
Hollnium	0.0393
Hydrogen (H_2)	3.41
Indium	0.056
Iodine (I_2)	0.102
Iridium	0.0317
Iron (α)	0.106
Krypton	0.059
Lanthanum	0.047
Lead	0.038
Lithium	0.85

Specific Heat of Selected Elements at 25°C
Listed by Element

Element	C_p (cal $\cdot$ g^{-1} $\cdot$ K^{-1})
Lutetium	0.037
Magnesium	0.243
Manganese, α	0.114
Manganese, β	1.119
Mercury	0.0331
Molybdenum	0.599
Neodymium	0.049
Neon	0.246
Nickel	0.106
Niobium	0.064
Nitrogen (N$_2$)	0.249
Osmium	0.03127
Oxygen (O$_2$)	0.219
Palladium	0.0584
Phosphorus, white	0.181
Phosphorus, red, triclinic	0.160
Platinum	0.0317
Polonium	0.030
Potassium	0.180
Praseodymium	0.046

Specific Heat of Selected Elements at 25°C
Listed by Element

Element	C_p (cal $\cdot$ g^{-1} $\cdot$ K^{-1})
Promethium	0.0442
Protactinium	0.029
Radium	0.0288
Radon	0.0224
Rhenium	0.0329
Rhodium	0.0583
Rubidium	0.0861
Ruthenium	0.057
Samarium	0.043
Scandium	0.133
Selenium (Se$_2$)	0.0767
Silicon	0.168
Silver	0.0566
Sodium	0.293
Strontium	0.0719
Sulfur, yellow	0.175
Tantalum	0.0334
Technetium	0.058
Tellurium	0.0481
Terbium	0.0437

Specific Heat of Selected Elements at 25°C
Listed by Element

Element	C_p (cal $\cdot$ g^{-1} $\cdot$ K^{-1})
Thallium	0.0307
Thorium	0.0271
Thulium	0.0382
Tin (α)	0.0510
Tin (β)	0.0530
Titanium	0.125
Tungsten	0.0317
Uranium	0.0276
Vanadium	0.116
Xenon	0.0378
Ytterbium	0.0346
Yttrium	0.068
Zinc	0.0928
Zirconium	0.0671

Source: data from Weast, R. C., Ed., *Handbook of Chemistry and Physics, 55th ed.*, CRC Press, Cleveland, 1974, D-144., Kelly, K. K., Bulletin 592, Bureau of Mines, Washington, D. C., 1961.and Hultgren, R., Orr, R L., Anderson, P. D., and Kelly, K. K., Selected Values of *Thermodynamic Properties of Metals and Alloys*, John Wiley & Sons, New York, (1963).

SPECIFIC HEAT OF SELECTED ELEMENTS AT 25°C
Listed by Specific Heat

Element	C_p (cal $\cdot$ g^{-1} $\cdot$ K^{-1})
Radon	0.0224
Thorium	0.0271
Uranium	0.0276
Radium	0.0288
Protactinium	0.029
Bismuth	0.0296
Polonium	0.030
Thallium	0.0307
Gold	0.0308
Osmium	0.03127
Iridium	0.0317
Platinum	0.0317
Tungsten	0.0317
Rhenium	0.0329
Mercury	0.0331
Tantalum	0.0334
Ytterbium	0.0346
Hafnium	0.035
Lutetium	0.037
Xenon	0.0378
Lead	0.038

Specific Heat of Selected Elements at 25°C
Listed by Specific Heat

Element	C_p (cal $\cdot$ g^{-1} $\cdot$ K^{-1})
Thulium	0.0382
Hollnium	0.0393
Erbium	0.0401
Dysprosium	0.0414
Europium	0.0421
Samarium	0.043
Terbium	0.0437
Promethium	0.0442
Barium	0.046
Praseodymium	0.046
Lanthanum	0.047
Tellurium	0.0481
Antimony	0.049
Cerium	0.049
Neodymium	0.049
Tin (α)	0.0510
Tin (β)	0.0530
Gadolinium	0.055
Cadmium	0.0555
Indium	0.056
Silver	0.0566

Specific Heat of Selected Elements at 25°C
Listed by Specific Heat

Element	C_p (cal $\cdot$ g^{-1} $\cdot$ K^{-1})
Cesium	0.057
Ruthenium	0.057
Technetium	0.058
Rhodium	0.0583
Palladium	0.0584
Krypton	0.059
Niobium	0.064
Zirconium	0.0671
Yttrium	0.068
Strontium	0.0719
Selenium (Se$_2$)	0.0767
Germanium	0.077
Arsenic	0.0785
Rubidium	0.0861
Gallium	0.089
Copper	0.092
Zinc	0.0928
Iodine (I$_2$)	0.102
Iron (α)	0.106
Nickel	0.106
Chromium	0.107

Specific Heat of Selected Elements at 25°C
Listed by Specific Heat

Element	C_p (cal $\cdot$ g^{-1} $\cdot$ K^{-1})
Cobalt	0.109
Bromine (Br$_2$)	0.113
Chlorine (Cl$_2$)	0.114
Manganese (α)	0.114
Vanadium	0.116
Argon	0.124
Carbon, diamond	0.124
Titanium	0.125
Scandium	0.133
Calcium	0.156
Phosphorus, red, triclinic	0.160
Silicon	0.168
Carbon, graphite	0.170
Sulfur, yellow	0.175
Potassium	0.180
Phosphorus, white	0.181
Fluorine (F$_2$)	0.197
Aluminum	0.215
Oxygen (O$_2$)	0.219
Magnesium	0.243
Boron	0.245

Specific Heat of Selected Elements at 25°C
Listed by Specific Heat

Element	C_p (cal $\cdot$ g^{-1} $\cdot$ K^{-1})
Neon	0.246
Nitrogen (N$_2$)	0.249
Sodium	0.293
Beryllium	0.436
Molybdenum	0.599
Lithium	0.85
Manganese (β)	1.119
Helium	1.24
Hydrogen (H$_2$)	3.41

Source: data from Weast, R. C., Ed., *Handbook of Chemistry and Physics, 55th ed.*, CRC Press, Cleveland, 1974, D-144., Kelly, K. K., Bulletin 592, Bureau of Mines, Washington, D. C., 1961.and Hultgren, R., Orr, R L., Anderson, P. D., and Kelly, K. K., *Selected Values of Thermodynamic Properties of Metals and Alloys*, John Wiley & Sons, New York, 1963.

HEAT CAPACITY OF SELECTED CERAMICS
Listed by Ceramic

Ceramic	Heat Capacity, C_p (cal/mole/K)

BORIDES

Chromium Diboride (CrB_2)	$9.61 + 10.72 \times 10^{-3}T$ cal/mole at 494-1010K
Hafnium Diboride (HfB_2)	$9.61 + 10.72 \times 10^{-3}T$ cal/mole at 494-1010K
Tantalum Diboride (TaB_2)	0.04 cal/g°C
Titanium Diboride (TiB_2)	$10.93 + 7.08 \times 10^{-3}T$ cal/mole at 420-1180 K
Zirconium Diboride (ZrB_2)	$15.81T + 4.20 \times 10^{-3}T - 3.52 \times 10^{5}T^{-2}$ for 429-1171K

CARBIDES

Hafnium Monocarbide (HfC)	0.05 at room temp. 15 ± 0.15 at 925°C 16 ± 0.16 at 1525°C
Silicon Carbide (SiC)	0.26 at 540°C 0.27 at 700°C 0.30 at 1000°C 0.32 at 1200°C 0.33 at 1350°C 0.35 at 1550°C

Heat Capacity of Selected Ceramics (Continued)
Listed by Ceramic

Ceramic	Heat Capacity, C_p (cal/mole/K)
Titanium Monocarbide (TiC)	0.150-0.170 cal/g at 150°C
	0.170-0.187 cal/g at 300°C
	0.183-0.196 cal/g at 450°C
	0.192-0.201 cal/g at 600°C
	0.20-0.207 cal/g at 750°C
	0.209 cal/g at 900°C
	0.210 cal/g at 1000°C
	0.211 cal/g at 1100°C

NITRIDES

Aluminum Nitride (AlN)	0.1961 cal/g/°C ; 0-100°C
	0.2277 cal/g/°C ; 0-420°C
	0.2399 cal/g/°C ; 0-598°C
Trisilicon tetranitride (Si_3N_4)	0.17 cal/g/°C

OXIDES

Cerium Dioxide (CeO_2)	$14.24T + 5.62 \times 10^{-3}T$ 491-1140K

SILICIDES

Molybdenum Disilicide ($MoSi_2$)	10-14 cal/g/°C; 425-1000°C
Tungsten Disilicide (WSi_2)	8 cal/g/°C; 425-1450°C

Source: Data compiled by J.S. Park from *No. 1 Materials Index*, Peter T.B. Shaffer, Plenum Press, New York, (1964); *Smithells Metals Reference Book*, Eric A. Brandes, ed., in association with Fulmer Research Institute Ltd. 6th ed. London, Butterworths, Boston, (1983); and *Ceramic Abstracts*, American Ceramic Society (1986-1991)

SPECIFIC HEAT OF SELECTED POLYMERS
Listed by Polymer

Polymer	Specific heat (Btu/lb/°F)
ABS Resins; Molded, Extruded	
Medium impact	0.36—0.38
High impact	0.36—0.38
Very high impact	0.36—0.38
Low temperature impact	0.35—0.38
Heat resistant	0.37—0.39
Acrylics; Cast, Molded, Extruded	
Cast Resin Sheets, Rods:	
General purpose, type I	0.35
General purpose, type II	0.35
Moldings:	
Grades 5, 6, 8	0.35
High impact grade	0.34
Thermoset Carbonate	
Allyl diglycol carbonate	0.3
Cellulose Acetate; Molded, Extruded	
ASTM Grade:	
H6—1	0.3—0.42
H4—1	0.3—0.42
H2—1	0.3—0.42
MH—1, MH—2	0.3—0.42
MS—1, MS—2	0.3—0.42
S2—1	0.3—0.42

Specific Heat of Selected Selected Polymers (Continued)
Listed by Polymer

Polymer	Specific heat (Btu/lb/°F)
Cellulose Acetate Butyrate; Molded, Extruded ASTM Grade:	
H4	0.3—0.4
MH	0.3—0.4
S2	0.3—0.4
Cellusose Acetate Propionate; Molded, Extruded ASTM Grade:	
1	0.3—0.4
3	0.3—0.4
6	0.3—0.4
Chlorinated polyvinyl chloride	0.3
Polycarbonate	0.3
Fluorocarbons; Molded,Extruded	
Polytrifluoro chloroethylene (PTFCE)	0.22
Polytetrafluoroethylene (PTFE)	0.25
Fluorinated ethylene propylene(FEP)	0.28
Polyvinylidene— fluoride (PVDF)	0.33
Epoxies; Cast, Molded, Reinforced	
Standard epoxies (diglycidyl ethers of bisphenol A)	
Cast rigid	0.4-0.5
High strength laminate	0.21
Filament wound composite	0.24

Specific Heat of Selected Selected Polymers (Continued)
Listed by Polymer

Polymer	Specific heat (Btu/lb/°F)
Nylons; Molded, Extruded	
Type 6	
General purpose	0.4
Cast	0.4
Type 8	0.4
Type 11	0.58
Type 12	0.28
Nylons; Molded, Extruded	
6/6 Nylon	
General purpose molding	0.3—0.5
General purpose extrusion	0.3—0.5
6/10 Nylon	
General purpose	0.3—0.5
Phenolics; Molded	
Type and filler	
General: woodflour and flock	0.35—0.40
Shock: paper, flock, or pulp	—
High shock: chopped fabric or cord	0.30—0.35
Very high shock: glass fiber	0.28—0.32
Phenolics: Molded	
Arc resistant—mineral	0.27—0.37
Rubber phenolic—woodflour or flock	0.33
PVC–Acrylic Alloy	
PVC–acrylic sheet	0.293

Specific Heat of Selected Selected Polymers (Continued)
Listed by Polymer

Polymer	Specific heat (Btu/lb/°F)
Polymides	
Unreinforced	0.31
Unreinforced 2nd value	0.25—0.35
Glass reinforced	0.15—0.27
Polyacetals	
Standard	0.35
Copolymer:	
Standard	0.35
High flow	0.35
Polyesters: Thermosets	
Cast polyyester	
Rigid	0.30—0.55
Reinforced polyester moldings	
High strength (glass fibers)	0.25—0.35
Sheet molding compounds, general purpose	0.20—0.25
Phenylene oxides (Noryl)	
Standard	0.24
Polypropylene:	
General purpose	0.45
High impact	0.45—0.48
Polyphenylene sulfide:	
Standard	0.26

Specific Heat of Selected Selected Polymers (Continued)
Listed by Polymer

Polymer	Specific heat (Btu/lb/°F)
Polyethylenes; Molded, Extruded	Speciflc heat,
Type I—lower density (0.910—0.925)	
Melt index 0.3—3.6	0.53—0.55
Melt index 6—26	0.53—0.55
Melt index 200	0.53—0.55
Type II—medium density (0.926—0.940)	
Melt index 20	0.53—0.55
Melt index 1.0—1.9	0.53—0.55
Type III—higher density (0.941—0.965)	
Melt index 0.2—0.9	0.46—0.55
Melt Melt index 0.1—12.0	0.46—0.55
Melt index 1.5—15	0.46—0.55
Polystyrenes; Molded	
Polystyrenes	
General purpose	0.30—0.35
Medium impact	0.30—0.35
High impact	0.30—0.35
Glass fiber -30% reinforced	0.256
Styrene acrylonitrile (SAN)	0.33
Polyvinyl Chloride And Copolymers; Molded, Extruded	
Vinylidene chloride	0.32
Silicones; Molded, Laminated	
Woven glass fabric/ silicone laminate	0.246

Source: data compiled by J.S. Park from Charles T. Lynch, *CRC Handbook of Materials Science*, Vol. 3, CRC Press, Boca Raton, Florida and *Engineered Materials Handbook*, Vol.2, Engineering Plastics, ASM International, Metals Park, Ohio, 1988.

PHASE CHANGE THERMODYNAMIC PROPERTIES
FOR SELECTED ELEMENTS

Element	Phase	Transition Temperature (K)	Heat of Transition (kcal • g mole⁻¹)	Entropy of Transition (e.u.)	Entropy at 298K (e.u.)
Ac	solid	(1090)	(2.5)	(2.3)	(13)
	liquid	(2750)	(70)	(25)	–
Ag	solid	1234	2.855	2.313	10.20
	liquid	2485	60.72	24.43	–
Al	solid	931.7	2.57	2.76	6.769
	liquid	2600		26	–
Am	solid	(1200)	(2.4)	(2.0)	(13)
	liquid	2733	51.7	18.9	–
As	solid	883			8.4
Au	solid	1336.16	3.03	2.27	11.32
	liquid	2933	74.21	25.30	–
B	solid	2313	(3.8)	(1.6)	1.42
	liquid	2800	75	27	–
Ba	solid, α	648	0.14	0.22	16
	solid, β	977	1.83	1.87	–
	liquid	1911	35.665	18.63	–
Be	solid	1556	2.919	1.501	2.28
	liquid	–			

Phase Change Thermodynamic Properties for Selected Elements
(Continued)

Element	Phase	Transition Temperature (K)	Heat of Transition (kcal • g mole⁻¹)	Entropy of Transition (e.u.)	Entropy at 298K (e.u.)
Bi	solid	544.2	2.63	4.83	13.6
	liquid	1900	41.1	21.6	–
C	solid	–	–	–	1.3609
Ca	solid, α	723	0.24	0.33	9.95
	solid, β	1123	2.2	1.96	–
	liquid	1755	38.6	22.0	–
Cd	solid			2.46	12.3
	liquid	1040	23.86	22.94	–
Ce	solid	1048	2.1	2.0	13.8
	liquid	2800	73	26	–
Cl_2	gas	–	–	–	53.286
Co	solid, α	723	0.005	0.007	6.8
	solid, β	1398	0.095	0.068	–
	solid, γ	1766	3.7	2.1	–
	liquid	3370	93	28	–
Cr	solid	2173	3.5	1.6	5.68
	liquid	2495	72.97	29.25	–
Cs	solid	301.9	0.50	1.7	19.8
	liquid	963	16.32	17.0	–

Phase Change Thermodynamic Properties for Selected Elements (Continued)

Element	Phase	Transition Temperature (K)	Heat of Transition (kcal • g mole^{-1})	Entropy of Transition (e.u.)	Entropy at 298K (e.u.)
Cu	solid	1356.2	3.11	2.29	7.97
	liquid	2868	72.8	25.4	–
F$_2$	gas	–	–	–	48.58
Fe	solid, α	1033	0.410	0.397	6.491
	solid, β	1180	0.217	0.184	–
	solid, γ	1673	0.15	0.084	–
	solid, δ	1808	3.86	2.14	–
	liquid	3008	84.62	28.1	–
Ga	solid	302.94	1.335	4.407	9.82
	liquid	2700	–	–	–
Ge	solid	1232	8.3	6.7	10.1
	liquid	2980	68	23	–
H$_2$	gas	–	–	–	31.211
Hf	solid	(2600)	(6.0)	(2.3)	13.1
Hg	liquid	629.73	13.985	22.208	18.46
In	solid	430	0.775	1.80	13.88
	liquid	2440	53.8	22.0	–
Ir	solid	2727	6.6	2.4	8.7

Phase Change Thermodynamic Properties for Selected Elements
(Continued)

Element	Phase	Transition Temperature (K)	Heat of Transition (kcal • g mole^{-1})	Entropy of Transition (e.u.)	Entropy at 298K (e.u.)
K	solid	336.4	0.5575	1.657	15.2
	liquid	1052	18.88	17.95	–
La	solid	1153	(2.3)	(2.0)	13.7
	liquid	3000	80	27	–
Li	solid	459	0.69	1.5	6.70
	liquid	1640	32.48	19.81	–
Mg	solid	923	2.2	2.4	7.77
	liquid	1393	31.5	22.6	–
Mn	solid, α	1000	0.535	0.535	7.59
	solid, β	1374	0.545	0.397	–
	solid, γ	1410	0.430	0.305	–
	solid, δ	1517	3.5	2.31	–
	liquid	2368	53.7	22.7	–
Mo	solid	2883	(5.8)	(2.0)	6.83
N$_2$	gas	–	–	–	45.767
Na	solid	371	0.63	1.7	12.31
	liquid	1187		20.1	–
Nb	solid	2760	(5.8)	(2.1)	8.3

Phase Change Thermodynamic Properties for Selected Elements
(Continued)

Element	Phase	Transition Temperature (K)	Heat of Transition (kcal • g mole⁻¹)	Entropy of Transition (e.u.)	Entropy at 298K (e.u.)
Nd	solid	1297	(2.55)	(197)	13.9
	liquid	(2750)	(61)	(22)	–
Ni	solid α	626	0.092	0.15	7.137
	solid β	1728	4.21	2.44	–
	liquid	3110	90.48	29.0	–
Np	solid	913	(2.3)	(2.5)	(14)
	liquid	(2525)	(55)	(22)	–
O_2	gas	–	–	–	49.003
Os	solid	2970	(6.4)	(2.2)	7.8
P_4	solid, white	317.4	0.601	1.89	42.4
	liquid	553	11.9	21.5	–
Pa	solid	(18.25)	(4.0)	(2.2)	(13.5)
	liquid	(4500)	(115)	(26)	–
Pb	solid	600.6	1.141	1.900	15.49
	liquid	2023	42.5	21.0	–
Pd	solid	1828	4.12	2.25	8.9
	liquid	3440	89	26	–
Po	solid	525		(4.6)	13
	liquid	(1235)	(24.6)	(19.9)	–

Phase Change Thermodynamic Properties for Selected Elements (Continued)

Element	Phase	Transition Temperature (K)	Heat of Transition (kcal • g mole^{-1})	Entropy of Transition (e.u.)	Entropy at 298K (e.u.)
Pr	solid	1205	(25)	(2.1)	(13.5)
	liquid	3563	–	–	–
Pt	solid	2042.5	5.2	25	10.0
	liquid	4100	122	29.8	–
Pu	solid	913	(2.26)	(2.48)	(13.0)
	liquid	–			
Ra	solid	1233	(2.3)	(1.9)	(17)
	liquid	(1700)	(35)	(21)	–
Rb	solid	312.0	0.525	1.68	16.6
	liquid	952	18.11	19.0	–
Re	solid	3440	(7.9)	(2.3)	(8.89)
Rh	solid	2240	(5.2)	(2.3)	7.6
	liquid	4150	127	30.7	–
Ru	solid, α	1308	0.034	0.026	6.9
	solid, β	1473	0	–	–
	solid, γ	1773	0.23	0.13	–
	solid, δ	2700	(6.1)	(2.3)	–
S	solid, α	368.6	0.088	0.24	7.62
	solid, β	392	0.293	0.747	–
	liquid	717.76	2.5	3.5	–

Phase Change Thermodynamic Properties for Selected Elements
(Continued)

Element	Phase	Transition Temperature (K)	Heat of Transition (kcal • g mole^{-1})	Entropy of Transition (e.u.)	Entropy at 298K (e.u.)
Sb	solid (α, β, γ)	903.7	4.8	5.3	10.5
	liquid	1713	46.665	27.3	–
Sc	solid	1670	(4.0)	(2.4)	(9.0)
	liquid	3000	80	27	–
Se	solid	490.6	1.25	2.55	10.144
	liquid	1000	14.27	14.27	–
Si	solid	1683	11.1	6.60	4.50
	liquid	2750	71	26	–
Sm	solid	1623	3.7	2.3	(15)
	liquid	(2800)	(70)	(25)	–
Sn	solid, α, β	505.1	1.69	335	12.3
	liquid	2473	(55)	(22)	–
Sr	solid	1043	2.2	2.1	13.0
	liquid	1657	33.61	20.28	–
Ta	solid	3250	7.5	2.3	99
Tc	solid	(2400)	(5.5)	(2.3)	(8.0)
	liquid	(3800)	(120)	(32)	–

Phase Change Thermodynamic Properties for Selected Elements
(Continued)

Element	Phase	Transition Temperature (K)	Heat of Transition (kcal • g mole^{-1})	Entropy of Transition (e.u.)	Entropy at 298K (e.u.)
Te	solid, α	621	0.13	0.21	11.88
	solid, β	723	4.28	5.92	–
	liquid	1360	11.9	8.75	–
Th	solid	2173	(4.6)	(2.1)	12.76
	liquid	4500	(130)	(29)	–
Ti	solid, α	1155	0.950	0.822	7.334
	solid, β	2000	(4.6)	(23)	–
	liquid	3550	(101)	(28)	–
Tl	solid, α	508.3	0.082	0.16	15.4
	solid, β	576.8	1.03	1.79	–
	liquid	1730	38.81	22.4	–
U	solid, α	938	0.665	0.709	12.03
	solid, β	1049	1.165	1.111	–
	solid, γ	1405	(3.0)	(2.1)	–
	liquid	3800	–	–	–
V	solid	2003	(4.0)	(2.0)	7.05
	liquid	3800	–	–	–
W	solid	3650	8.42	2.3	8.0
Y	solid	1750	(4.0)	(2.3)	(11)
	liquid	3500	(90)	(26)	–

Phase Change Thermodynamic Properties for Selected Elements
(Continued)

Element	Phase	Transition Temperature (K)	Heat of Transition (kcal $\cdot$ g mole^{-1})	Entropy of Transition (e.u.)	Entropy at 298K (e.u.)
Zn	solid	692.7	1.595	2.303	9.95
	liquid	1180	27A3	23.24	–
Zr	solid, α	1135	0.920	0.811	9.29
	solid, β	2125	(4.9)	(2.3)	–
	liquid	(3900)	(100)	(26)	–

From Weast, R. C. Ed., *Handbook of Chemistry and Physics, 69th ed.*, CRC Press, Boca Raton, Fla., 1988, D44.

PHASE CHANGE THERMODYNAMIC PROPERTIES
FOR SELECTED OXIDES

Oxide	Phase	Transition Temperature (K)	Heat of Transition (kcal • g mole^{-1})	Entropy of Transition (e.u.)	Entropy at 298K (e.u.)
Ac_2O_3	Solid	(2250)	(20)	(8.9)	(36.5)
	Liquid	–	–	–	–
Ag_2O	Solid	dec. 460	–	–	29.09
Ag_2O_2	Solid	dec.	–	–	(20.4)
Al_2O_3	Solid	2300	26	11	12.186
	Liquid	dec.	–	–	–
Am_2O_3	Solid	(2225)	(17)	(7.6)	(37)
	Liquid	(3400)	(85)	(25)	–
AmO_2	Solid	dec.	–	–	(20)
As_2O_3	Solid, α	503	4.1	8.2	25.6
	Solid, β	586	4.4	7.5	–
	Liquid	730	7.15	9.79	–
AsO_2	Solid	(1200)	(9.0)	(7.5)	(13)
	Liquid	(dec.)	–	–	–
As_2O_5	Solid	dec. >1100	–	–	25.2
Au_2O_3	Solid	dec.	–	–	30
B_2O_3	Solid	723	5.27	7.29	12.91
	Liquid	2520	(55)	(22)	–

Phase Change Thermodynamic Properties for Selected Oxides

Oxide	Phase	Transition Temperature (K)	Heat of Transition (kcal • g mole^{-1})	Entropy of Transition (e.u.)	Entropy at 298K (e.u.)
Ba_2O	Solid	(880)	(5.2)	(5.9)	(23.5)
	Liquid	(1040)	(20)	(19)	–
BaO	Solid	2196	13.8	6.28	16.8
	Liquid	3000	(62)	(21)	–
BaO_2	Solid	723	(5.7)	(7.9)	(18.5)
	Liquid	dec. 1110	–	–	–
BeO	Solid	dec.	–	–	3.37
BiO	Solid	(1175)	(3.7)	(3.1)	(15)
	Liquid	(1920)	(54)	(28)	–
Bi_2O_3	Solid	1090	6.8	6.2	36.2
	Liquid	(dec.)	–	–	–
CO	Gas	–	–	–	47.30
CO_2	Gas	–	–	–	51.06
CaO	Solid	2860	(18)	(6.3)	9.5
CdO	Solid	dec.	–	–	13.1
Ce_2O_3	Solid	1960	(20)	(10)	(33.5)
	Liquid	(3500)	(80)	(23)	–
CeO_2	Solid	3000	(19)	(6.3)	17.7
CoO	Solid	2078	(12)	(5.8)	10.5
	Liquid	(2900)	(61)	(21)	–

Phase Change Thermodynamic Properties for Selected Oxides

Oxide	Phase	Transition Temperature (K)	Heat of Transition (kcal $\cdot$ g mole^{-1})	Entropy of Transition (e.u.)	Entropy at 298K (e.u.)
Co_3O_4	Solid	dec. 1240	–	–	(35.5)
Cr_2O_3	Solid	2538	(25)	(10)	19.4
CrO_2	Solid	dec. 700	–	–	(11.5)
CrO_3	Solid	460	(6.1)	(13)	(17.5)
	Liquid	(1000)	(25)	(25)	–
Cs_2O	Solid	763	(4.58)	(6.0)	(23)
	Liquid	dec.	–	–	–
Cs_2O_2	Solid	867	(5.5)	(6.3)	(40)
	Liquid	dec.	–	–	–
Cs_2O_3	Solid	775	(7.75)	(10)	(47)
	Liquid	dec.	–	–	–
Cu_2O	Solid	1503	13.4	8.92	
	Liquid	dec.	–	–	–
CuO	Solid	1609	(8.9)	(5.5)	10.4
	Liquid	dec.	–	–	–
FeO	Solid	1641	7.5	4.6	12.9
	Liquid	(2700)	(55)	(20)	–
Fe_3O_4	Solid, α	900	(0)	(0)	35.0
	Solid, β	dec.	–	–	–
Fe_2O_3	Solid, α	950	0.16	0.17	21.5
	Solid, β	1050	0	0	–
	Solid, γ	dec.	–	–	–

Phase Change Thermodynamic Properties for Selected Oxides

Oxide	Phase	Transition Temperature (K)	Heat of Transition (kcal • g mole^{-1})	Entropy of Transition (e.u.)	Entropy at 298K (e.u.)
Ga_2O	Solid	(925)	(8.5)	(9.2)	(22.5)
	Liquid	(1000)	(20)	(20)	–
Ga_2O_3	Solid	2013	(22)	(11)	20.23
	Liquid	(2900)	(75)	(26)	–
GeO	Solid	983	(50)	(51)	(12.5)
GeO_2	Solid (α,β)	1389	10.5	7.56	(12.5)
	Liquid	(2625)	(61)	(23)	–
In_2O	Solid	(600)	(4.5)	(7.5)	(28)
	Liquid	(800)	(16)	(20)	–
InO	Solid	(1325)	(4.0)	(3.0)	(14.5)
	Liquid	(2000)	(60)	(30)	–
In_2O_3	Solid	(2000)	(20)	(10)	30.1
	Liquid	(3600)	(85)	(24)	–
Ir_2O_3	Solid	(1450)	(10)	(6.8)	(26.5)
	Liquid	(2250)	(50)	(22)	–
IrO_2	Solid	dec. 1373	–	–	(15.9)
K_2O	Solid	(980)	(6.8)	(6.9)	(23)
	Liquid	dec.	–	–	–
K_2O_2	Solid	763	(7.0)	(9.2)	(27)

Phase Change Thermodynamic Properties for Selected Oxides

Oxide	Phase	Transition Temperature (K)	Heat of Transition (kcal • g mole^{-1})	Entropy of Transition (e.u.)	Entropy at 298K (e.u.)
KO_2	Solid	653	(4.9)	(7.5)	27.9
	Liquid	dec.	–	–	–
La_2O_3	Solid	2590	(18)	(7)	(36.5)
Li_2O	Solid	2000	(14)	(7)	9.06
	Liquid	2600	(56)	(22)	–
Li_2O_2	Solid	dec.470	–	–	(16.5)
MgO	Solid	3075	18.5	5.8	
MgO_2	Solid	dec. 361	–	–	(20.5)
MnO	Solid	2058	13.0	6.32	14.27
	Liquid	dec.	–	–	
Mn_3O_4	Solid, α	1445	4.97	3.44	35.5
	Solid, β	1863	(33)	(18)	–
	Liquid	(2900)	(75)	(26)	–
Mn_2O_3	Solid	dec. 1620	–	–	26A
MnO_2	Solid	dec. 1120	–	–	12.7
MoO_2	Solid	(2200)	(16)	(7.3)	(14.5)
	Liquid	dec. 2250	–	–	–
MoO_3	Solid	1068	12.54	11.74	18.68
	Liquid	1530	33	22	–
N_2O	Gas	–	–	–	52.58

Phase Change Thermodynamic Properties for Selected Oxides

Oxide	Phase	Transition Temperature (K)	Heat of Transition (kcal $\cdot$ g mole^{-1})	Entropy of Transition (e.u.)	Entropy at 298K (e.u.)
Na_2O	Solid	1193	(7.1)	(6.0)	17.4
	Liquid	dec.	–	–	–
Na_2O_2	Solid	dec. 919	–	–	22.6
NaO_2	Solid	(825)	(6.2)	(7.5)	27.7
	Liquid	(1300)	(28)	(22)	–
NbO	Solid	(2650)	(16)	(6.0)	(12)
NbO_2	Solid	(2275)	(16)	(7.0)	(12.7)
	Liquid	(3800)	(85)	(22)	–
Nb_2O_5	Solid	1733	(28)	(16)	32.8
	Liquid	(3200)	(80)	(25)	–
Nd_2O_3	Solid	2545	(22)	(8.8)	(35.3)
NiO	Solid	2230	(12.1)	(5.43)	9.22
	Liquid	dec.	–	–	–
NpO_2	Solid	(2600)	(15)	(5.7)	19.19
Np_2O_5	Solid	dec. 800–900 K	–	–	(43)
OsO_2	Solid	dec. 923	–	–	(14.5)
OsO_4	Solid	313.3	3.41	10.9	34.7
	Liquid	403	9.45	23.4	–

Phase Change Thermodynamic Properties for Selected Oxides

Oxide	Phase	Transition Temperature (K)	Heat of Transition (kcal • g mole⁻¹)	Entropy of Transition (e.u.)	Entropy at 298K (e.u.)
P_2O_3	Liquid	448.5	4.5	10	(34)
PO_2	Solid	(350)	(2.7)	(7.7)	(11.5)
	Liquid	(dec.)	–	–	–
P_2O_5	Solid	631	8.8	13.9	33.5
PaO_2	Solid	(2560)	(20)	(7.8)	(17.8)
Pa_2O_5	Solid	(2050)	(26)	(13)	(37.5)
	Liquid	(3350)	(95)	(28)	–
PbO	Solid, red	762	(0.4)	(0.5)	16.2
	Solid, yellow	1159	2.8	2.4	–
	Liquid	1745	51	29	–
Pb_2O_4	Solid	dec.	–	–	50.5
PbO_2	Solid	dec.	–	–	18.3
PdO	Solid	dec. 1150	–	–	(9.1)
PoO_2	Solid	(825)	(5.5)	(6.7)	(17)
	Liquid	(dec.)	–	–	–
Pr_2O_3	Solid	(2200)	(22)	(10)	(35.5)
	Liquid	(4000	(90)	(23)	–
PrO_2	Solid	dec. 700	–	–	(17)

Phase Change Thermodynamic Properties for Selected Oxides

Oxide	Phase	Transition Temperature (K)	Heat of Transition (kcal $\cdot$ g mole^{-1})	Entropy of Transition (e.u.)	Entropy at 298K (e.u.)
PtO	Solid	dec. 780	–	–	(13.5)
Pt$_3$O$_4$	Solid	(dec.)	–	–	(41)
PtO$_2$	Solid	723	(4.6)	(6.4)	(16.5)
	Liquid	dec. 750	–	–	–
PuO	Solid	(1290)	(7.2)	(5.6)	(20)
	Liquid	(2325)	(47)	(20)	–
Pu$_2$O$_3$	Solid	(1880)	(16)	(8.5)	(38)
	Liquid	(3250)	(75)	(23)	–
PuO$_2$	Solid	(2400)	(15)	(6.2)	(19.7)
	Liquid	(3500)	(90)	(26)	–
RaO	Solid	(>2500)	–	–	(17)
Rb$_2$O	Solid	(910)	(5.7)	(6.3)	(27)
	Liquid	dec.	–	–	–
Rb$_2$O$_2$	Solid	843	(7.3)	(8.7)	(27.5)
	Liquid	(dec.)	–	–	–
Rb$_2$O$_3$	Solid	762	(7.6)	(10)	(32.5)
	Liquid	dec.	–	–	–
RbO$_2$	Solid	685	(4.1)	(6.0)	(21.5)
	Liquid	dec.	–	–	–
ReO$_2$	Solid	(1475)	(12)	(8.1)	(15)
	Liquid	(3250)	(80)	(25)	–
ReO$_3$	Solid	433	5.2	12	19.8
	Liquid	dec.	–	–	–

Phase Change Thermodynamic Properties for Selected Oxides

Oxide	Phase	Transition Temperature (K)	Heat of Transition (kcal • g mole^{-1})	Entropy of Transition (e.u.)	Entropy at 298K (e.u.)
Re_2O_7	Solid	569	15.8	27.8	44
	Liquid	635.5	17.7	27.9	
ReO_4	Solid	420	(4.2)	(10)	(34.5)
	Liquid	(460)	(9.3)	(20)	
Rh_2O	Solid	dec. 1400	–	–	(25.5)
RhO	Solid	dec. 1394	–	–	(12)
Rh_2O_3	Solid	dec. 1388	–	–	(23)
RuO_2	Solid	dec. 1400	–	–	(12.5)
$RuO4$	Solid	300	(3.2)	(11)	(32.5)
	Liquid	dec.	–	–	–
SO_2	Gas	–	–	–	59.40
Sb_2O_3	Solid	928	14.74	15.88	29.4
	Liquid	1698	8.92	5.25	
SbO_2	Solid	dec.	–	–	15.2
Sb_2O_5	Solid	dec.	–	–	29.9
Sc_2O_3	Solid	(2500)	(23)	(9.3)	24.8
SeO	Solid	(1375)	(7.6)	(5.5)	(11)
	Liquid	(2075)	(45)	(22)	
SeO_2	Solid	603	(24.5)	(40.6)	(15)
SiO	Solid	(2550)	(12)	(4.7)	(6.5)

Phase Change Thermodynamic Properties for Selected Oxides

Oxide	Phase	Transition Temperature (K)	Heat of Transition (kcal • g mole^{-1})	Entropy of Transition (e.u.)	Entropy at 298K (e.u.)
SiO_2	Solid,b	856	0.15	0.18	10.06
	Solid, a	1883	2.04	1.08	–
	Liquid	dec. 2250	–	–	–
Sm_2O_3	Solid	(2150)	(20)	(9.3)	(36.5)
	Liquid	(3800)	(80)	(21)	–
SnO	Solid	(1315)	(6.4)	(4.9)	13.5
	Liquid	(1800)	(60)	(33)	–
SnO_2	Solid	1898	(11.39)	(5.95)	12.5
	Liquid	(3200)	(75)	(23)	–
SrO	Solid	2703	16.7	6.2	13.0
SrO_2	Solid	dec.488	–	–	(14.8)
Ta_2O_5	Solid	2150	(16)	(7.4)	34.2
	Liquid	–	–	–	–
TcO_2	Solid	(2400)	(18)	(7.5)	(13.5)
	Liquid	(4000)	(105)	(26)	–
TcO_3	Solid	(dec. <1200)	–	–	(19.5)
Tc_2O_7	Solid	392.7	(11)	(28)	(42.5)
	Liquid	583.8	(14)	(24)	–
TeO	Solid	(1020)	(7.1)	(7.0)	(13)
	Liquid	(1775)	(50)	(28)	–
TeO_3	Solid	1006	3.2	3.2	16.99
	Liquid	dec.	–	–	–

Phase Change Thermodynamic Properties for Selected Oxides

Oxide	Phase	Transition Temperature (K)	Heat of Transition (kcal • g mole^{-1})	Entropy of Transition (e.u.)	Entropy at 298K (e.u.)
TeO$_2$	Solid	(2150)	(13)	(6.0)	(16)
	Liquid	(3250)	(65)	(20)	–
ThO$_2$	Solid	3225	(18)	(5.6)	15.59
TiO	Solid, α	1264	0.82	0.65	8.31
	Solid, β	dec. 2010	–	–	–
Ti$_2$O$_3$	Solid, α	473	0.215	0.455	18.83
	Solid, β	2400	(24)	(10)	
	Liquid	3300			
Ti$_3$O$_5$	Solid, α	450	2.24	4.98	30.92
	Solid, β	(2450)	(50)	(20)	
	Liquid	(3600)	(85)	(24)	
TiO$_2$	Solid	2128	(16)	(7.5)	12.01
	Liquid	dec. 3200			
Ti$_2$O	Solid	573	(5.0)	(8.7)	23.8
	Liquid	773	(17)	(22)	–
Tl$_2$O$_3$	Solid	990	(12.4)	(13)	(33.5)
	Liquid	(dec.)	–	–	–
UO	Solid	(2750)	(14)	(5.1)	(16)
UO$_2$	Solid	3000	–	–	18.63
U$_3$O$_8$	Solid	dec.	–	–	(66)
UO$_3$	Solid	dec. 925	–	–	23.57

Phase Change Thermodynamic Properties for Selected Oxides

Oxide	Phase	Transition Temperature (K)	Heat of Transition (kcal $\cdot$ g mole^{-1})	Entropy of Transition (e.u.)	Entropy at 298K (e.u.)
VO	Solid	(2350)	(15)	(6.4)	9.3
	Liquid	(3400)	(70)	(21)	
V_2O_3	Solid	2240	(24)	(11)	23.58
	Liquid	dec. 3300	–	–	
V_3O_4	Solid	(2100)	(42)	(20)	(32)
	Liquid	(dec.)	–	–	–
VO_2	Solid, α	345	1.02	2.96	12.32
	Solid, β	1818	13.60	7.48	
	Liquid	dec. 3300	–	–	
V_2O_5	Solid	943	15.56	16.50	313
	Liquid	(2325)	(63)	(27)	
WO_2	Solid	(1543)	(11.5)	(7.45)	(15)
	Liquid	dec. 2125	–	–	–
WO_1	Solid	1743	(17)	(9.8)	19.90
	Liquid	(2100)	(43)	(20)	
Y_2O_3	Solid	(2500)	(25)	(10)	(29.5)
ZnO	Solid	dec.	–	–	10.4
ZrO_2	Solid, α	1478	1.420	0.961	12.03
	Solid, β	2950	20.8	7.0	

Source: Data from Weast, R. C., Ed., *Handbook of Chemistry and Physics*, 55th ed., CRC Press, Cleveland, 1974, D-58.

THERMODYNAMIC COEFFICIENTS

Thermodynamic calculations over a wide range of temperatures are generally made with the aid of algebraic equations representing the characteristic properties of the substances being considered. The necessary integrations and differentiations, or other mathematical manipulations, are then most easily effected. The most convenient starting point in making such calculations for a given substance is the heat capacity at constant pressure. From this quantity and a knowledge of the properties of any phase transitions, the other thermodynamic properties may be computed by the well-known equations given in standard texts on thermodynamics. Please note that the units for a, b, c, and d are cal/g mole, whereas those for A are kcal/g mole. The necessary adjustment must be made when the data are substituted into the equations. Empirical heat capacity equations are generated in the form of a power series, with the absolute temperature T as the independent variable:

$$C_p = a' + (b' \times 10^{-3})T + (c' \times 10^{-6})T_2$$

or

$$C_p = a'' + (b'' \times 10^{-3})T + \frac{d \times 10^5}{T^2}.$$

Since both forms are used in the following, let

$$C_p = a + (b \times 10^{-3})T + (c \times 10^{-6})T^2 + \frac{d \times 10^5}{T^2}.$$

The constants a, b, c, and d are to be determined either experimentally or by some theoretical or semi-empirical approach. The heat content, or enthalpy (H), is determined from the heat capacity by a simple integration of the range of temperatures for which the formula for c_p is valid. Thus, if 298K is taken as a reference temperature,

$$H_T - H_{298} = \int_{298}^{T} C_p dT$$

$$= a(T - 298) + \tfrac{1}{2}(b \times 10^{-3})(T_2 - 298^2) + \tfrac{1}{3}(c \times 10^{-6})(T^3 - 298^3) - (d \times 10^5)\left(\frac{1}{T} - \frac{1}{298}\right)$$

$$= aT + \tfrac{1}{2}(b \times 10^{-3})T^2 + \tfrac{1}{3}(c \times 10^{-6})T^3 - \frac{d \times 10^5}{T} - A,$$

where all the constants on the right-hand side of the equation have been incorporated in the term −A.

In general, the enthalpy is given by a sum of terms for each phase of the substance involved in the temperature range considered plus terms that represent the heats of transitions:

$$H_T - H_{298} = \Sigma \int_{T_1}^{T_2} C_p dT + \Sigma \Delta H_{tr.}$$

In a similar manner, the entropy S is obtained by performing the integration

$$S_T - S'_{298} = \int_{298}^{T} (C_p/T)dt$$

$$= a\ln(T/298) + (b \times 10^{-3})(T - 298) + \tfrac{1}{2}(c \times 10^{-6})(T^2 - 298^2) - \tfrac{1}{2}(d \times 10^5)\left(\frac{1}{T^2} - \frac{1}{298^2}\right)$$

$$= a\ln T + (b \times 10^{-3})T + \tfrac{1}{2}(c \times 10^{-6})T^2 - \frac{\tfrac{1}{2}(d \times 10^5)}{T^2} - B'$$

or

$$S_T = 2.303\,a\log T + (b \times 10^{-3})T + \tfrac{1}{2}(c \times 10^{-6})T^2 - \frac{\tfrac{1}{2}(d \times 10^5)}{T^2} - B$$

where

$$B = B' - S_{298}.$$

From the definition of free energy (F):

$$F = H - TS$$

the quantity

$$F_T - H_{298} = (H_T - H_{298}) - TS_T$$

may be written as:

$$F_T - H_{298} = -2.303\,aT\log T - \tfrac{1}{2}(b \times 10^{-3})T^2 - \tfrac{1}{6}(c \times 10^{-6})T^3 - \frac{\tfrac{1}{2}(d \times 10^5)}{T} + (B + a)T - A$$

and also the free energy function

$$\frac{F_T - H_{298}}{T} = -2.303\,a\log T - \tfrac{1}{2}(b \times 10^{-3})T - \tfrac{1}{6}(c \times 10^{-6})T^2 - \frac{\tfrac{1}{2}(d \times 10^5)}{T^2} + (B + a) - \frac{A}{T}$$

Values of these thermodynamic coefficients are given in the following tables. The first column in each table lists the material. The second column gives the phase to which the coefficients are applicable. The remaining columns list the values of the constants a, b, c, d, A, and B required in the thermodynamic equations. All values that represent estimates are enclosed in parentheses. The heat capacities at temperatures beyond the range of experimental determination were estimated by extrapolation. Where no experimental values were found, analogy with compounds of neighboring elements in the periodic table was used.

Source: from Weast, R. C. Ed., *Handbook of Chemistry and Physics, 69th ed.*, CRC Press, Boca Raton, Fla., 1988, D44.

THERMODYNAMIC COEFFICIENTS FOR SELECTED ELEMENTS

Element	Phase	a —	b —	c (cal $\cdot$ g mole^{-1})	d —	A (kcal $\cdot$ g mole^{-1})	B (e.u)
Ac	solid	(5.4)	(3.0)	–	–	(1.743)	(18.7)
	liquid	(8)	–	–	–	(0.295)	(31.3)
Ag	solid	5.09	1.02	–	0.36	1.488	19.21
	liquid	7.30	–	–	–	0.164	30.12
	gas	(4.97)	–			(–66.34)	(–12.52)
Al	solid	4.94	2.96	–	–	1.604	22.26
	liquid	7.0	–	–	–	0.33	30.83
Am	solid	(4.9)	(4.4)	–	–	(1.657)	(16.2)
	liquid	(8.5)	–	–	–	(0.409)	(34.5)
As	solid	5.17	2.34	–	–	1.646	21.8
Au	solid	6.14	–0.175	0.92	–	1.831	23.65
	liquid	7.00	–			0.631	26.99
B	solid	1.54	4.40	–	–	0.655	8.67
	liquid	(6.0)	–			(–4.599)	(31.4)
Ba	solid, α	5.55	4.50	–	–	1.722	16.1
	solid, β	5.55	1.50	–	–	1.582	15.9
	liquid	(7.4)	–	–	–	(0.843)	(25.3)
	gas	(497)	–			(–39.65)	(–11.7)
Be	solid	5.07	1.21	–	–1.15	1.951	27.62
	liquid	5.27	–			–1.611	25.68

Thermodynamic Coefficients for Selected Elements (Continued)

Element	Phase	a –	b –	c (cal • g mole⁻¹)	d –	A (kcal • g mole⁻¹)	B (e.u)
Bi	solid	5.38	2.60	–	–	1.720	17.8
	liquid	7.60	–			–0.087	25.6
	gas	(4.97)	–	–	–	(–46.19)	(–15.9)
C	solid	4.10	1.02	–	–2.10	1.972	23.484
Ca	solid, α	5.24	3.50	–	–	1.718	2095
	solid, β	6.29	1.40	–	–	1.689	26.01
	liquid	7.4	–			–0.147	30.28
	gas	(4.97)	–			(–43.015)	(–9.88)
Cd	solid	5.31	2.94	–	–	1.714	18.8
	liquid	7.10	–	–	–	0.798	26.1
	gas	(4.97)	–			(–25.28)	(–11.7)
Ce	solid	4.40	6.0	–	–	1.579	13.1
	liquid	(7.9)	–			(–0.148)	(29.1)
Cl₂	gas	8.76	0.27	–	–0.65	2.845	–2.929
Co	solid, α	4.72	4.30	–	–	1.598	21.4
	solid, β	3.30	5.86	–	–	0.974	3.1
	solid, γ	9.60	–	–	–	3.961	50.5
	liquid	8.30	–			–2.034	38.7
Cr	solid	5.35	2.36	–	–0.44	1.848	25.75
	liquid	9.40	–	–	–	1.556	50.13
	gas	(4.97)	–			(–82.47)	(–13.8)

Thermodynamic Coefficients for Selected Elements (Continued)

Element	Phase	a –	b –	c (cal • g mole⁻¹)	d –	A (kcal • g mole⁻¹)	B (e.u)
Cs	solid	7.42	–	–	–	2.212	22.5
	liquid	8.00	–	–	–	1.887	24.1
	gas	(4.97)	–			(−17.35)	(−13.6)
Cu	solid	5.41	1.50	–	–	1.680	23.30
	liquid	7.50	–	–	–	0.024	34.05
F₂	gas	8.29	0.44	–	−0.80	2.760	−0.76
Fe	solid, α	3.37	7.10	–	0.43	1.176	14.59
	solid, β	10.40	–	–	–	4.281	55.66
	solid, γ	4.85	3.00	–	–	0.396	19.76
	solid, δ	10.30	–	–	–	4.382	55.11
	liquid	10.00	–			−0.021	50.73
Ga	solid	5.237	3.33	–	–	1.710	21.01
	liquid	(6.645)	–	–	–	(0.648)	(23.64)
Ge	solid	5.90	1.13	–	–	1.764	23.8
	liquid	(7.3)	–			(−5.668)	(25.7)
H₂	gas	6.62	0.81	–	–	2.010	6.75
Hf	solid	(6.00)	(0.52)	–	–	(1.812)	(21.2)
Mg	liquid	–				1.971	19.20
		4.969	–			13.048	−13.54

Thermodynamic Coefficients for Selected Elements (Continued)

Element	Phase	a –	b –	c (cal • g mole⁻¹)	d –	A (kcal • g mole⁻¹)	B (e.u)
In	solid	5.81	2.50	–	–	1.844	19.97
	liquid	7.50	–	–	–	1.564	27.34
	gas	(4.97)	–			(−58.42)	(−14.46)
Ir	solid	5.56	1.42	–	–	1.721	23.4
K	solid	1.3264	19.405	–	–	1.258	−1.86
	liquid	8.8825	4.565	2.9369	–	1.923	32.55
	gas	(4.97)	–			(−19.689)	(−9.46)
La	solid	6.17	1.60	–	–	1.911	21.9
	liquid	(7.3)	–			(−0.15)	(26.0)
Li	solid	3.05	8.60	–	–	1.292	12.92
	liquid	7.0	–	–	–	1.509	32.00
	gas	(4.97)	–			34.30)	(−2.84)
Mg	solid	5.33	2.45	–	−0.103	1.733	23.39
	liquid	(8.0)	–	–	–	0.942	36.967
	gas	(4.97)	–			(−34.78)	(−7.60)
Mn	solid, α	6.70	3.38	–	−0.37	1.974	26.11
	solid, β	8.33	0.66	–	–	2.672	41.02
	solid, γ	10.70	–	–	–	4.760	56.84
	solid, δ	11.30	–	–	–	5.176	60.88
	liquid	11.00	–	–	–	1.221	56.38
	gas	6.26	–			−63.704	−3.13
Mo	solid	5.48	1.30	–	–	1.692	24.78

Thermodynamic Coefficients for Selected Elements (Continued)

Element	Phase	a —	b —	c (cal • g mole^{-1})	d —	A (kcal • g mole^{-1})	B (e.u)
N$_2$	gas	6.76	0.606	0.13	–	2.044	–7.064
Na	solid	5.657	3.252	0.5785	–	1.836	20.92
	liquid	8.954	–4.577	2.540	–	1.924	36.0
		(4.97)	–			(–24.40)	(–8.7)
Nb	solid	5.66	0.96	–	–	1.730	24.24
Nd	solid	5.61	5.34	–	–	1.910	19.7
	liquid	(9.1)	–			(–0.606)	35.8
Ni	solid α	4.06	7.04	–	–	1.523	18.095
	solid β	6.00	1.80	–	–	1.619	27.16
	liquid	9.20	–	–	–	0.251	45.47
Np	solid	(5.3)	(3.4)	–	–	(1.731)	(17.9)
	liquid	(9.0)	–	–	–	(1.392)	(37.5)
O$_2$	gas	8.27	0.258	–	–1.877	3.007	–0.750
Os	solid	5.69	0.88	–	–	1.736	24.9
P$_4$	solid, white	13.62	28.72	–	–	5.338	43.8
	liquid	19.23	0.51	–	–2.98	6.035	66.7
	gas	(19.5)	(–0.4)	(1.3)	–	(–6.32)	(46.1)
Pa	solid	(5.2)	(4.0)	–	–	(1.728)	(17.3)
	liquid	(8.0)	–	–	–	(–3.823)	(28.8)

Thermodynamic Coefficients for Selected Elements (Continued)

Element	Phase	a –	b –	c (cal • g mole^{-1})	d –	A (kcal • g mole^{-1})	B (e.u)
Pb	solid	5.64	2.30	–	–	1.784	17.33
	liquid	7.75	–0.73	–	–	1.362	27.11
	gas	(4.97)	–			(–45.25)	(–13.6)
Pd	solid	5.80	1.38	–	–	1.791	24.6
	liquid	(9.0)	–	–	–	(1.215)	(43.8)
Po	solid	(5.2)	(3.2)	–	–	(1.693)	(17.6)
	liquid	(9.0)	–	–	–	(0.847)	(35.2)
	gas	(4.97)	–			(–28.73)	(–13.5)
Pr	solid	(5.0)	(4.6)	–	–	(1.705)	(16.4)
	liquid	(8.0)	–			(–0.519)	(30.0)
Pt	solid	5.74	1.34	–	0.10	1.737	23.0
	liquid	(9.0)	–	–	–	(0.406)	(42.6)
Pu	solid	(5.2)	(3.6)	–	–	(1.710)	(17.7)
	liquid	(8.0)	–	–	–	(0.506)	(31.0)
Ra	solid	(5.8)	(1.2)	–	–	(1.783)	(16.4)
	liquid	(8.0)	–	–	–	(1.284)	(28.6)
	gas	(4.97)	–			(–38.87)	(–14.5)
Rb	solid	3.27	13.1	–	–	1.557	5.9
	liquid	7.85	–	–	–	1.814	26.5
	gas	(4.97)	–			(–19.04)	(–12.3)
Re	solid	(5.85)	(0.8)	–	–	(1.780)	(24.7)

Thermodynamic Coefficients for Selected Elements (Continued)

Element	Phase	a –	b –	c (cal • g mole^{-1})	d –	A (kcal • g mole^{-1})	B (e.u)
Rh	solid	5.40	2.19	–	–	1.707	23.8
	liquid	(9.0)	–			(–0.923)	(44.4)
Ru	solid, α	5.25	1.50	–	–	1.632	23.5
	solid, β	7.20	–	–	–	2.867	35.5
	solid, γ	7.20	–	–	–	2.867	35.5
	solid, δ	7.50	–	–	–	3.169	37.6
S	solid, α	3.58	6.24	–	–	1.345	14.64
	solid, β	3.56	6.95	–	–	1.298	14.54
	liquid	5.4	5.0	–	–	1.576	24.02
$\frac{1}{2}$ S$_2$	gas	(4.25)	(0.15)	–	(–1.0)	(–2.859)	(9.57)
Sb	solid, α, β, γ	5.51	1.74	–	–	1.720	21.4
	liquid	7.50	–			1.992	28.1
$\frac{1}{2}$ Sb$_2$	gas		4.47	–	–0.11	–53.876	–21.7
Sc	solid	(5.13)	(3.0)	–	–	1.663	21.1
	liquid	(7.50)	–			(–2.563)	31.3
Se	solid	3.30	8.80	–	–	1.375	11.28
	liquid	7.0	–	–	–	0.881	27.34
Si	solid	5.70	1.02	–	–1.06	2.100	28.88
	liquid	7.4	–			7.646	33.17

Thermodynamic Coefficients for Selected Elements (Continued)

Element	Phase	a –	b –	c (cal • g mole⁻¹)	d –	A (kcal • g mole⁻¹)	B (e.u)
Sm	solid	(6.7)	(3.4)	–	–	(2.149)	(24.2)
	liquid	(9.0)	–			(–2.296)	(33.4)
Sn	solid, α, β	4.42	6.30	–	–	1.598	14.8
	liquid	7.30	–	–	–	0.559	26.2
	gas	(4.97)	–			60.21)	(–14.3)
Sr	solid	(5.60)	(1.37)	–	–	(1.731)	(19.3)
	liquid	(7.7)	–	–	–	(0.976)	(30.4)
	gas	(4.97)	–			(37.16)	(–10.2)
Ta	solid	5.82	0.78	–	–	1.770	23.4
Tc	solid	(5.6)	(2.0)	–	–	(1.759)	(24.5)
	liquid					(3.459)	(59.4)
Te	solid, α	4.58	5.25	–	–	1.599	15.78
	solid, β	4.58	5.25	–	–	1.469	15.57
	liquid	9.0	–			–0.988	34.96
$\frac{1}{2}$ Te₂ gas		4.47	–		19.048	–6.47	
Th	solid	8.2	–0.77	2.04	–	2.591	33.64
	liquid	(8.0)	–			(–7.602)	(26.84)
Ti	solid, α	5.25	2.52	–	–	1.677	23.33
	solid, β	7.50	–	–	–	1.645	35.46
	liquid	(7.8)	–			(–2.355)	(35.45)

Thermodynamic Coefficients for Selected Elements (Continued)

Element	Phase	a —	b —	c (cal • g mole^{-1})	d —	A (kcal • g mole^{-1})	B (e.u)
Tl	solid, α	5.26	3.46	–	–	1.722	15.6
	solid, β	7.30	–	–	–	2.230	26.4
	liquid	7.50	–	–	–	1.315	25.9
	gas	(4.97)	–			(–41.88)	(–15.4)
U	solid, α	3.25	8.15	–	0.80	1.063	8.47
	solid, β	10.28	–	–	–	3.493	48.27
	solid, γ	9.12	–	–	–	1.110	39.09
	liquid	(8.99)	–			(–2.073)	36.01
V	solid	5.57	0.97	–	–	1.704	24.97
	liquid	(8.6)	–	–	–	1.827	44.06
W	solid	5.74	0.76	–	–	1.745	24.9
Y	solid	(5.6)	(2.2)	–	–	(1.767)	(21.6)
	liquid	(7.5)	–			2.277)	(29.6)
Zn	solid	5.35	2.40	–	–	1.702	21.25
	liquid	7.50	–	–	–	1.020	31.35
		(4.97)	–			(–29.407)	(–9.81)
Zr	solid, α	6.83	1.12	–	–0.87	2.378	30.45
	solid, β	7.27	–	–	–	1.159	31.43
	liquid	(8.0)	–			(–2.190)	(34.7)

Source: Data from Weast, R. C. Ed., *Handbook of Chemistry and Physics, 69th ed.,* CRC Press, Boca Raton, Fla., 1988, D44.

THERMODYNAMIC COEFFICIENTS FOR SELECTED OXIDES

Oxide	Phase	a —	b (cal • g mole^{-1})	c —	d —	A (kcal • g mole^{-1})	B (e.u)
Ac_2O_3	Solid	(20.0)	(20.4)	–	–	(6.870)	(80.9)
	Liquid	(40)	–	–	–	(–19.767)	(180.5)
Ag_2O	Solid	13.26	7.04	–	–	4.266	48.56
Ag_2O_2	Solid	(16.4)	(12.2)	–	–	(5.432)	(76.7)
Al_2O_3	Solid	26.12	4.388	–	–7.269	10.422	142.03
	Liquid	(33)	–	–	–	(– 11.655)	(174.1)
Am_2O_3	Solid	(20.0)	(15.6)	–	–	(6.657)	(81.6)
	Liquid	(38.5)	–	–	–	(–7.796)	(181.8)
AmO_2	Solid	(14.0)	(6.8)	–	–	(4.477)	(61.8)
As_2O_3	Solid, α	8.37	48.6	–	–	4.656	36.6
	Solid, β	8.37	48.6	–	–	0.556	28.4
	Liquid	(39)	–	–	–	(5.760)	(187.6)
	Gas	(21.5)	–	–	–	(–14.164)	(62.5)
AsO_2	Solid	(8.5)	(9.4)	–	–	(2.952)	(38.2)
	Liquid	(21)	–	–	–	(2.184)	(108.0)
As_2O_5	Solid	(31.1)	(16.4)	–	(–5.4)	(11.813)	(159.9)
Au_2O_3	Solid	(23.5)	(4.8)	–	–	(7.220)	(105.3)
B_2O_3	Solid	8.73	25.40	–	–1.31	4.171	45.04
	Liquid	30.50	–	–	–	7.822	161.59
Ba_2O	Solid	(20.0)	(2.2)	–	–	(6.061)	(91.1)
	Liquid	(22)	–	–	–	(1.769)	(96.8)
	Gas	(15)	–	–	–	(–25.51)	(29.0)

Thermodynamic Coefficients for Selected Oxides (Continued)

Oxide	Phase	a –	b (cal • g mole^{-1})	c –	d –	A (kcal • g mole^{-1})	B (e.u)
BaO	Solid	12.74	1.040	–	–1.984	4.510	57.2
	Liquid	(13.9)	–	–	–	(–9.341)	(57.5)
BaO$_2$	Solid	(13.6)	(2.0)	–	–	(4.144)	(59.6)
	Liquid	(21)	–	–	–	(3.241)	(99.0)
BeO	Solid	8.69	3.65	–	–3.13	3.803	48.99
BiO	Solid	(9.7)	(3.0)	–	–	(3.025)	(41.2)
	Liquid	(14)	–	–	–	(2.306)	(64.9)
	Gas	(8.9)	–	–	–	(–61.49)	(–1.8)
Bi$_2$O$_3$	Solid	23.27	11.05	–	–	7.429	99.7
	Liquid	(35.7)	–	–	–	(7.614)	(168.3)
CO	Gas	6.60	1.2	–	–	2.021	–9.34
CO$_2$	Gas	7.70	5.3	–0.83	–	2.490	–5.64
CaO	Solid	10.00	4.84	–	–1.08	3.559	49.5
CdO	Solid	9.65	2.08	–	–	2.970	42.5
Ce$_2$O$_3$	Solid	(–23.0)	(9.0)	–	–	(7.258)	(100.2)
	Liquid	(37)	–	–	–	(–2.591)	(178.5)
CeO$_2$	Solid	15.0	2.5	–	–	4.579	68.5
CoO	Solid	(9.8)	(2.2)	–	–	(3.020)	(46.0)
	Liquid	(15.5)	–	–	–	(–1.886)	(79.2)
Co$_3$O$_4$	Solid	(29.5)	(17.0)	–	–	(9.551)	(137.6)

Thermodynamic Coefficients for Selected Oxides (Continued)

Oxide	Phase	a —	b (cal • g mole^{-1})	c —	d —	A (kcal • g mole^{-1})	B (e.u)
Cr_2O_3	Solid	28.53	2.20	–	–3.736	9.857	145.9
CrO_2	Solid	(16.1)	(3.0)	–	(–3.0)	(5.946)	(82.8)
CrO_3	Solid	(18.1)	(4.0)	–	(–2.0)	(6.245)	(87.9)
	Liquid	(27)	–	–	–	(3.381)	(127.0)
	Gas	(20)	–	–	–	(–28.62)	(53.6)
Cs_2O	Solid	(16.51)	(5.4)	–	–	(5.160)	(72.6)
	Liquid	(22)	–	–	–	(3.205)	(99.0)
Cs_2O_2	Solid	(21.4)	(11.4)	–	–	(6.887)	(85.3)
	Liquid	(29.5)	–	–	–	(4.125)	(123.8)
Cs_2O_3	Solid	(24.0)	(22.6)	–	–	(8.160)	(96.5)
	Liquid	(35)	–	–	–	(2.148)	(142.2)
Cu_2O	Solid	(13.4)	(8.6)	–	–	(4.378)	(96.0)
	Liquid	(21.5)	–	–	–	(3.721)	(54.9)
CuO	Solid	14.34	6.2	–	–	4.551	61.11
	Liquid	(22)	–	–	–	(–4.339)	(98.91)
FeO	Solid	9.27	4.80	–	–	(2.977)	(43.8)
	Liquid	(14.5)	–	–	–	(–3.721)	(69.2)
Fe_3O_4	Solid, α	12.38	1.62	–	–0.38	3.826	58.3
	Solid, β	(14.5)	–	–	–	(–2.399)	(66.7)
Fe_2O_3	Solid, α						
	Solid, β	21.88	48.20	–	–	8.666	104.0
	Solid, γ	48.00	18.6	–	–	12.652	238.3
Ga_2O	Solid	23.49	18.6	–	–3.55	9.021	119.9
	Liquid	36.00	–	–	–	11.979	187.6

Thermodynamic Coefficients for Selected Oxides (Continued)

Oxide	Phase	a –	b (cal • g mole^{-1})	c –	d –	A (kcal • g mole^{-1})	B (e.u)
	Gas	31.71	1.8	–	–	8.467	159.7
Ga_2O_3	Solid	(13.8)	–	–	–	(4.497)	(58.7)
	Liquid	(21.5)	–	–	–	(–0.559)	(94.1)
GeO	Solid	(14)	–	–	–	(–28.06)	(22.3)
	Gas	11.77	25.2	–	–	(4.630)	(54.35)
GeO_2	Solid (α,β)	(35.5)	–	–	–	(–20.66)	(173.2)
	Liquid	(10.4)	(2.6)	–	(–0.5)	(3.3)	(47.8)
In_2O	Solid	(14.7)	(7.8)	–	–	(4.730)	(58.1)
	Liquid	(22)	–	–	–	(3.206)	(92.6)
	Gas	(15)	–	–	–	(–18.39)	(25.8)
InO	Solid	(10.0)	(3.2)	–	–	(3.124)	(43.4)
	Liquid	(14)	–	–	–	(1.615)	(64.9)
	Gas	(9.0)	–	–	–	(–68.38)	(–3.1)
In_2O_3	Solid	(22.6)	(6.0)	–	–	(7.005)	(100.5)
	Liquid	(35)	–	–	–	(–0.195)	(172.8)
Ir_2O_3	Solid	(21.8)	(14.4)	–	–	(7.140)	(102.0)
	Liquid	(35)	–	–	–	(0.706)	(170.3)
	Gas	(20)	(10)	–	–	(–57.73)	(54.8)
IrO_2	Solid	9.17	15.20	–	–	3.410	40.9
K_2O	Solid	(15.9)	(6.4)	–	–	(5.025)	(69.5)
	Liquid	(22)	–	–	–	(1.130)	(98.3)
K_2O_2	Solid	(20.8)	(5.4)	–	–	(6.442)	(93.1)
	Liquid	(29)	–	–	–	(4.127)	(134.2)
	Gas	(20)	–	–	–	(–57.07)	(41.7)

Thermodynamic Coeffieicnts for Selected Oxides (Continued)

Oxide	Phase	a –	b (cal • g mole⁻¹)	c –	d –	A (kcal • g mole⁻¹)	B (e.u)
K_2O_3	Solid	(19.1)	(23.2)	–	–	(6.750)	(82.2)
	Liquid	(35.5)	–	–	–	(6.447)	(164.7)
	Gas	(20)	(5.0)	–	–	(−31.29)	(37.3)
KO_2	Solid	(15.0)	(12.0)	–	–	(5.006)	(61.1)
	Liquid	(24)	–	–	–	(3.424)	(105.5)
La_2O_3	Solid	28.86	3.076	–	−3.275	9.840	(130.7)
Li_2O	Solid	(11.4)	(5.4)	–	–	(3.639)	(57.5)
	Liquid	(21)	–	–	–	(−1.961)	(112.7)
Li_2O_2	Solid	(17.0)	(5.4)	–	–	(5.309)	(82.0)
MgO	Solid	10.86	1.197	–	−2.087	3.991	57.0
MgO_2	Solid	(12.1)	(2A)	–	–	(3.714)	(49.2)
MnO	Solid	11.11	1.94	–	−0.88	3.689	50.10
	Liquid	(13.5)	–	–	–	(−8.543)	(58.02)
Mn_3O_4	Solid, α	34.64	10.82	–	−2.20	11.312	166.3
	Solid, β	50.20	–	–	–	17.376	260.4
	Liquid	(49)	–	–	–	(−17.86)	(233.4)
Mn_2O_3	Solid	24.73	8.38	–	−3.23	8.829	118.8
MnO_2	Solid	16.60	2.44	–	−3.88	6.359	84.8
MoO_2	Solid	(16.2)	(3.0)	–	(−3.0)	(5.973)	(80.4)
	Liquid	(23)	–	–	–	(−2.463)	(118.4)
MoO_3	Solid	13.6	13.5	–	–	4.655	62.83
	Liquid	(28.4)	–	–	–	(0.222)	(139.88)
	Gas	(18.1)	–	–	–	(−48.54)	(42.8)

Thermodynamic Coeffieicnts for Selected Oxides (Continued)

Oxide	Phase	a —	b (cal • g mole^{-1})	c —	d —	A (kcal • g mole^{-1})	B (e.u)
N_2O	Gas	(10.92)	2.06	–	–2.04	4.032	11.40
Na_2O	Solid	15.70	5.40	–	–	4.921	73.7
	Liquid	(22)	–	–	–	(1.494)	(105.9)
Na_2O_2	Solid	(20.2)	(3.8)	–	–	(6.192)	(93.6)
NaO_2	Solid	(16.2)	(3.6)	–	–	(4.990)	(65.7)
	Liquid	(23)	–	–	–	(3.175)	(100.9)
	Gas	(15)	–	–	–	(–35.22)	(22.0)
NbO	Solid	(9.6)	(4.4)	–	–	(3.058)	(44.0)
NbO_2	Solid	(17.1)	(1.6)	–	(–2.8)	(6.109)	(84.6)
	Liquid	(24)	–	–	–	(1.033)	(127.2)
Nb_2O_5	Solid	21.88	28.2	–	–	7.776	100.3
	Liquid	(44.2)	–	–	–	(–24.09)	(201.6)
Nd_2O_3	Solid	28.99	5.760	–	(–4.159)	10.295	(133.9)
NiO	Solid	13.69	0.83	–	–2.915	5.097	70.67
	Liquid	(14.3)	–	–	–	(–7.861)	(67.91)
NpO_2	Solid	(17.7)	(3.2)	–	(–2.6)	(6.292)	(84.08)
Np_2O_5	Solid	(32.4)	(12.6)	–	–	(10.22)	(145.4)
OsO_2	Solid	(11.5)	(6.0)	–	–	(3.696)	(52.8)
OsO_4	Solid	(16.4)	(23.1)	–	(–2.4)	(6.726)	(67.0)
	Liquid	(33)	–	–	–	(6.612)	(143.0)
	Gas	16.46	8.60	–	–4.6	(–7.644)	(25.3)

Thermodynamic Coeffieicnts for Selected Oxides (Continued)

Oxide	Phase	a –	b (cal • g mole⁻¹)	c –	d –	A (kcal • g mole⁻¹)	B (e.u)
P_2O_3	Liquid	(34.5)	–	–	–	(10.287)	(162.6)
	Gas	(153)	(10)	–	–	(–1.953)	(38.0)
PO_2	Solid	(11.3)	(5.0)	–	–	(3.591)	(54.4)
	Liquid	(20)	–	–	–	(3.640)	(95.9)
P_2O_5	Solid	8.375	5.40	–	–	4.897	30.3
	Gas	36.80	–	–	–	3.284	165.6
PaO_2	Solid	(14.4)	(2.6)	–	–	(4.409)	(65.0)
Pa_2O_5	Solid	(28.4)	(11.4)	–	–	(8.975)	(127.7)
	Liquid	(48)	–	–	–	(–0.800)	(241.1)
PbO	Solid, red	10.60	4.00	–	–	3.338	45.4
	Solid, yellow	9.05	6.40	–	–	2.454	36.4
	Liquid	(14.6)	–	–	–	1.788	65.7
	Gas	(8.1)	(0.4)	–	–	(–59.94)	(–11.0)
Pb_2O_4	Solid	(31.1)	(17.6)	–	–	(10.055)	(132.0)
PbO_2	Solid	12.7	7.80	–	–	4.133	56.4
PdO	Solid	3.30	14.2	–	–	1.615	(13.9)
PoO_2	Solid	(14.3)	(5.6)	–	–	(4.513)	(66.1)
	Liquid	(22)	–	–	–	(3.460)	(106.5)
Pr_2O_3	Solid	(29.0)	(4.0)	–	(–4.0)	(10.166)	(133.2)
	Liquid	(36)	–	–	–	(–6.298)	(168.3)
PrO_2	Solid	(17.6)	(3.4)	–	(–2.8)	(6.338)	(85.9)
PtO	Solid	(9.0)	(6.4)	–	–	(2.968)	(39.7)

Thermodynamic Coeffieicnts for Selected Oxides (Continued)

Oxide	Phase	a –	b (cal • g mole⁻¹)	c –	d –	A (kcal • g mole⁻¹)	B (e.u)
Pt_3O_4	Solid	(30.8)	(17.4)	–	–	(9.957)	(139.7)
PtO_2	Solid	(11.1)	(9.6)	–	–	(3.736)	(49.6)
	Liquid	(21)	–	–	–	(3.785)	(101.5)
PuO	Solid	(12.0)	(2.4)	–	–	(3.685)	(49.1)
	Liquid	(14.5)	–	–	–	(–2.287)	(58.3)
	Gas	(8.9)	–	–	–	(–62.307)	(–5.3)
Pu_2O_3	Solid	(21.2)	(18.2)	–	–	(7.130)	(88.2)
	Liquid	(40)	–	–	–	(–5.691)	(187.2)
PuO_2	Solid	(17.1)	(3.4)	–	(–2.6)	(6.122)	(80.2)
	Liquid	(20.5)	–	–	–	(–10.62)	(92.2)
RaO	Solid	(10.5)	(2.0)	–	–	(3.220)	(43.4)
Rb_2O	Solid	(15.4)	(5.8)	–	–	(4.850)	(62.5)
	Liquid	(22)	–	–	–	(2.754)	(95.9)
Rb_2O_2	Solid	(20.9)	(8.0)	–	–	(6.587)	(94.0)
	Liquid	(29)	–	–	–	(3.273)	(133.2)
Rb_2O_3	Solid	(20.5)	(13.0)	–	–	(6.690)	(88.2)
	Liquid	(34)	–	–	–	(5.603)	(157.8)
RbO_2	Solid	(13.8)	(6.4)	–	–	(4.399)	(59.0)
	Liquid	(21)	–	–	–	(3.720)	(95.7)
ReO_2	Solid	(10.8)	(9.8)	–	–	(3.656)	(49.5)
	Liquid	(24.5)	–	–	–	(1.204)	(127.0)
ReO_3	Solid	(18.0)	(5.8)	–	–	(5.625)	(84.5)
	Liquid	29	–	–	–	(4.644)	(136.8)

Thermodynamic Coeffieicnts for Selected Oxides (Continued)

Oxide	Phase	a	b	c	d	A	B
		–	(cal • g mole^{-1})	–	–	(kcal • g mole^{-1})	(e.u)
Re$_2$O$_7$	Solid	(41.8)	(14.8)	–	(–3.0)	(14.127)	(200.3)
	Liquid	(65.7)	–	–	–	(9.203)	(314.7)
	Gas	(38.2)	–	–	–	(–25.97)	(109.3)
ReO$_4$	Solid	(21.4)	(10.8)	–	(–2.0)	(7.531)	(91.8)
	Liquid	(33)	–	–	–	(6.775)	(146.7)
	Gas	(16.5)	(8.6)	–	(–5.0)	(–8.118)	(30.6)
Rh$_2$O	Solid	15.59	6.47	–	–	4.936	(65.3)
RhO	Solid	(9.84)	(553)	–	–	(3.179)	(45.7)
Rh$_2$O$_3$	Solid	20.73	13.80	–	–	6.794	(99.2)
RuO$_2$	Solid	(11.4)	(6.0)	–	–	3.666	(54.2)
RuO4	Solid	(20)	–	–	–	(5.963)	(81.5)
	Liquid	(33)	–	–	–	(6.663)	(144.9)
SO$_2$	Gas	11.4	1.414	–	–2.045	4.148	7.12
Sb$_2$O$_3$	Solid	19.10	17.1	–	–	6.455	84.5
	Liquid	(36)	–	–	–	(0.035)	(168.2)
	Gas	(20.8)	–	–	–	(–34.70)	(49.9)
SbO$_2$	Solid	11.30	8.1	–	–	3.725	51.6
Sb$_2$O$_5$	Solid	(22.4)	(23.6)	–	–	(7.723)	(104.8)
Sc$_2$O$_3$	Solid	23.17	5.64	–	–	7.159	1089
SeO	Solid	(9.1)	(3.8)	–	–	(2.882)	(42.0)
	Liquid	(15.5)	–	–	–	(0.490)	(77.5)
	Gas	8.20	0.50	–	–0.80	(–58.54)	(0.7)
SeO$_2$	Solid	(12.8)	(6.1)	–	(–0.2)	(4.150)	(59.9)
	Gas	(14.5)	–	–	–	(–20.45)	(26.4)

Thermodynamic Coeffieicnts for Selected Oxides (Continued)

Oxide	Phase	a —	b (cal • g mole^{-1})	c —	d —	A (kcal • g mole^{-1})	B (e.u)
SiO	Solid	(7.3)	(2.4)	–	–	(2.283)	(35.8)
SiO$_2$	Solid, β	11.22	8.20	–	–2.70	4.615	57.83
	Solid, α	14.41	1.94	–	–	4.602	73.67
	Liquid	(20)	–	–	–	(9.649)	(111.08)
Sm$_2$O$_3$	Solid	(25.9)	(7.0)	–	–	(8.033)	(113.2)
	Liquid	(36)	–	–	–	(–6.431)	(166.3)
SnO	Solid	9.40	3.62	–	–	2.964	41.1
	Liquid	(14.5)	–	–	–	(0.141)	(68.1)
	Gas	(9.0)	–	–	–	(–69.76)	(–6.4)
SnO$_2$	Solid	17.66	2.40	–	–5.16	7.103	91.7
	Liquid	(22.5)	–	–	–	(0.304)	(117.7)
SrO	Solid	12.34	1.120	–	–1.806	4.335	58.7
SrO$_2$	Solid	(16.8)	(2.2)	–	(–3.0)	(6.113)	(83.3)
Ta$_2$O$_5$	Solid	29.2	10.0	–	–	9.151	135.2
	Liquid	(46)	–	–	–	(6.158)	(235.1)
TcO$_2$	Solid	(10.4)	(9.2)	–	–	(3.510)	(48.6)
	Liquid	(25)	–	–	–	(–5.946)	(132.7)
TcO$_3$	Solid	(19.4)	(5.2)	–	(–2.0)	(6.686)	(93.7)
Tc$_2$O$_7$	Solid	(39.1)	(18.6)	–	(–2.4)	(13.29)	(187.2)
	Liquid	(64)	–	–	–	(10.02)	(299.8)
	Gas	(25)	(28)	–	–	(–21.98)	(43.8)

Thermodynamic Coeffieicnts for Selected Oxides (Continued)

Oxide	Phase	a —	b (cal • g mole⁻¹)	c —	d —	A (kcal • g mole⁻¹)	B (e.u)
TeO	Solid	(8.6)	(6.2)	–	–	(2.840)	(37.8)
	Liquid	(15.5)	–	–	–	(–0.448)	(72.3)
	Gas	(8.9)	–	–	–	(–62.16)	(–5.2)
TeO$_3$	Solid	13.85	6.87	–	–	4.435	63.97
	Liquid	(20)	–	–	–	(3.940)	(96.4)
TeO$_2$	Solid	(11.0)	(2.4)	–	–	(3.386)	(47.4)
	Liquid	(15)	–	–	–	(–6.561)	(66.9)
ThO$_2$	Solid	16.45	2.346	–	–2.124	5.721	80.03
TiO	Solid, α	10.57	3.60	–	–1.86	3.935	54.03
	Solid, β	11.85	3.00	–	–	4.108	61.71
Ti$_2$O$_3$	Solid, α	7.31	53.52	–	–	4.559	38.78
	Solid, β	34.68	1.30	–	–10.20	13.605	184.48
	Liquid	(37.5)	–	–	–	(–7.796)	(193.2)
Ti$_3$O$_5$	Solid, α	35.47	29.50	–	–	11.887	179.98
	Solid, β	41.60	8.00	–	–	10.230	202.80
	Liquid	(60)	–	–	–	(–18.701)	(306.4)
TiO$_2$	Solid	17.97	0.28	–	–4.35	6.829	92.92
	Liquid	(21.4)	–	–	–	(–2.610)	(111.08
Ti$_2$O	Solid	(15.8)	(6.0)	–	(–0.3)	(5.078)	(68.2)
	Liquid	(22.1)	–	–	–	(2.651)	(96.0)
	Gas	(13.7)	–	–	–	(–20.94)	(18.0)
Tl$_2$O$_3$	Solid	(23.0)	(5.0)	–	–	(7.080)	(99.0)
	Liquid	(35.5)	–	–	–	(4.604)	(167.8)

Thermodynamic Coeffieicnts for Selected Oxides (Continued)

Oxide	Phase	a –	b (cal • g mole^{-1})	c –	d –	A (kcal • g mole^{-1})	B (e.u)
UO	Solid	(10.6)	(2.0)	–	–	(3.249)	(45.0)
UO$_2$	Solid	19.20	1.62	–	–3.957	7.124	93.37
U$_3$O$_8$	Solid	(65)	(7.5)	–	(–10.9)	(23.37)	(312.7)
UO$_3$	Solid	22.09	2.54	–	–2.973	7.969	104.72
VO	Solid	11.32	1.61	–	–1.26	3.869	56.4
	Liquid	(14.5)	–	–	–	(–8.157)	(70.9)
V$_2$O$_3$	Solid	29.35	4.76	–	–5.42	10.780	148.12
	Liquid	(38)	–	–	–	(–6.028)	(193.4)
V$_3$O$_4$	Solid	(36)	(30)	–	–	(12.07)	(182.1)
	Liquid	(55.6)	–	–	–	(–54.72)	(249.1)
VO$_2$	Solid, α	14.96	–	–	–	4.460	72.92
	Solid, β	17.85	1.70	–	–3.94	5.680	89.09
	Liquid	25.50	–	–	–	2.962	135.87
V$_2$O$_5$	Solid	46.54	–390	–	–13.22	18.136	240.2
	Liquid	45.60	–	–	–	2.122	220.1
	Gas	(40)	–	–	–	(–73.90)	(149.6)
WO$_2$	Solid	(17.6)	(4.2)	–	(–4.0)	(6.772)	(88.8)
	Liquid	(24)	–	–	–	(–0.112)	(121.8)
WO$_1$	Solid	17.33	7.74	–	–	5.511	81.15
	Liquid	(30)	–	–	–	(–1.162)	(152.5)
	Gas	(18)	–	–	–	(–69.36)	(40.2)
Y$_2$O$_3$	Solid	(26.0)	(8.2)	–	(–2.2)	(8.846)	(122.3)

Thermodynamic Coefficients for Selected Oxides (Continued)

Oxide	Phase	a —	b (cal • g mole^{-1})	c —	d —	A (kcal • g mole^{-1})	B (e.u)
ZnO	Solid	11.71	1.22	–	–2.18	4.277	57.88
ZrO$_2$	Solid, α	16.64	1.80	–	–3.36	6.168	85.21
	Solid, β	17.80	–	–	–	4.270	89.96

Source: Data from Weast, R. C., Ed., *Handbook of Chemistry and Physics,* 55th ed., CRC Press, Cleveland, 1974, D-58.

THERMAL CONDUCTIVITY OF METALS
AT CRYOGENIC TEMPERATURES

T (K)	Aluminum	Cadmium	Chromium	Copper	Gold
1	7.8	48.7	0.401	28.7	4.4
2	15.5	89.3	0.802	57.3	8.9
3	23.2	104	1.20	85.5	13.1
4	30.8	92.0	1.60	113	17.1
5	38.1	69.0	1.99	138	20.7
6	45.1	44.2	2.38	159	23.7
7	51.5	28.0	2.77	177	26.0
8	57.3	18.0	3.14	189	27.5
9	62.2	12.2	3.50	195	28.2
10	66.1	8.87	3.85	196	28.2
11	69.0	6.91	4.18	193	27.7
12	70.8	5.56	4.49	185	26.7
13	71.5	4.67	4.78	176	25.5
14	71.3	4.01	5.04	166	24.1
15	70.2	3.55	5.27	156	22.6
16	68.4	3.16	5.48	145	20.9
18	63.5	2.62	5.81	124	17.7
20	56.5	2.26	6.01	105	15.0
25	40.0	1.79	6.07	68	10.2
30	28.5	1.56	5.58	43	7.6
35	21.0	1.41	5.03	29	6.1
40	16.0	1.32	4.30	20.5	5.2
45	12.5	1.25	3.67	15.3	4.6
50	10.0	1.20	3.17	12.2	4.2
60	6.7	1.13	2.48	8.5	3.8

T (K)	Aluminum	Cadmium	Chromium	Copper	Gold
70	5.0	1.08	2.08	6.7	3.58
80	4.0	1.06	1.82	5.7	3.52
90	3.4	1.04	1.68	5.14	3.48
100	3.0	1.03	1.58	4.83	3.45

Thermal Conductivity of Aluminum

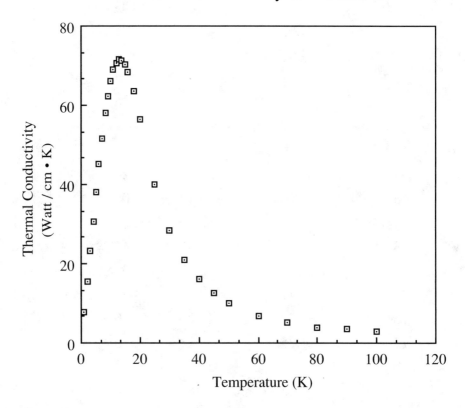

Thermal Conductivity of Metals at Cryogenic Temperatures

Thermal Conductivity of Cadmium

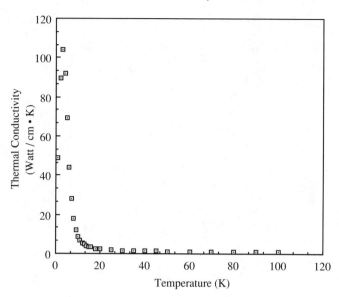

Thermal Conductivity of Chromium

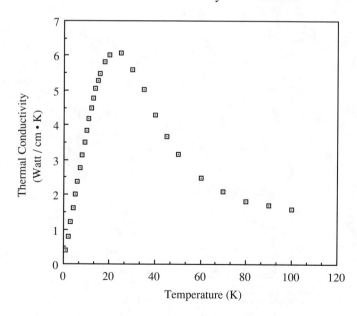

Thermal Conductivity of Metals at Cryogenic Temperatures

Thermal Conductivity of Copper

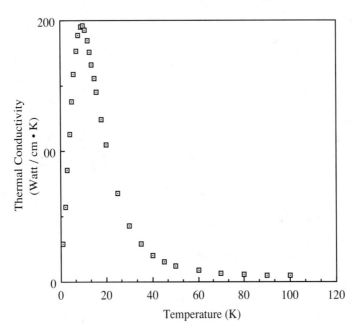

Thermal Conductivity of Gold

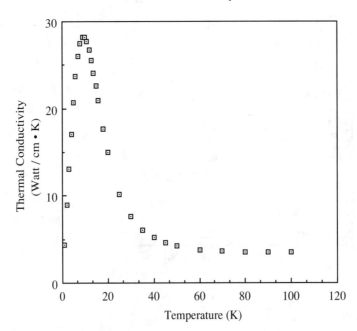

Thermal Conductivity of Metals at Cryogenic Temperatures

T (K)	Iron	Lead	Magnesium	Molybdenum
1	0.75	27.7	1.30	0.146
2	1.49	42.4	2.59	0.292
3	2.24	34.0	3.88	0.438
4	2.97	22.4	5.15	0.584
5	3.71	13.8	6.39	0.730
6	4.42	8.2	7.60	0.876
7	5.13	4.9	8.75	1.02
8	5.80	3.2	9.83	1.17
9	6.45	2.3	10.8	1.31
10	7.05	1.78	11.7	1.45
11	7.62	1.46	12.5	1.60
12	8.13	1.23	13.1	1.74
13	8.58	1.07	13.6	1.88
14	8.97	0.94	14.0	2.01
15	9.30	0.84	14.3	2.15
16	9.56	0.77	14.4	2.28
18	9.88	0.66	14.3	2.53
20	9.97	0.59	13.9	2.77
25	9.36	0.507	12.0	3.25
30	8.14	0.477	9.5	3.55
35	6.81	0.462	7.4	3.62
40	5.55	0.451	5.7	3.51
45	4.50	0.442	4.57	3.26
50	3.72	0.435	3.75	3.00
60	2.65	0.424	2.74	2.60

Thermal Conductivity of Metals at Cryogenic Temperatures

T (K)	Iron	Lead	Magnesium	Molybdenum
70	2.04	0.415	2.23	2.30
80	1.68	0.407	1.95	2.09
90	1.46	0.401	1.78	1.92
100	1.32	0.396	1.69	1.79

Thermal Conductivity of Iron

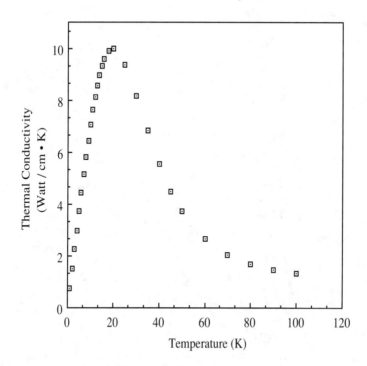

Thermal Conductivity of Metals at Cryogenic Temperatures

Thermal Conductivity of Lead

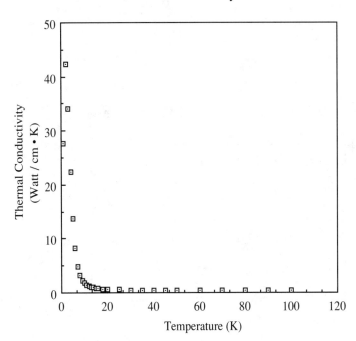

Thermal Conductivity of Magnesium

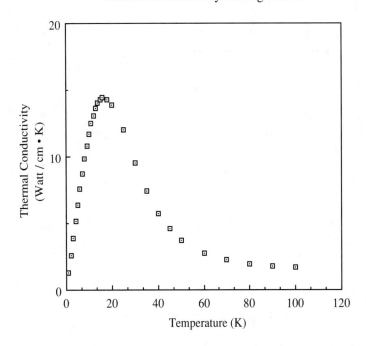

Thermal Conductivity of Metals at Cryogenic Temperatures

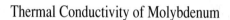

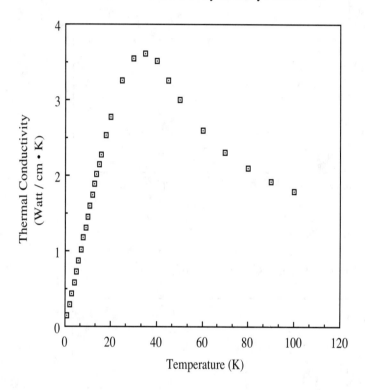

Thermal Conductivity of Metals at Cryogenic Temperatures

T (K)	Nickel	Niobium	Platinum	Silver	Tantalum
1	0.64	0.251	2.31	39.4	0.115
2	1.27	0.501	4.60	78.3	0.230
3	1.91	0.749	6.79	115	0.345
4	2.54	0.993	8.8	147	0.459
5	3.16	1.23	10.5	172	0.571
6	3.77	1.46	11.8	187	0.681
7	4.36	1.67	12.6	193	0.788
8	4.94	1.86	12.9	190	0.891
9	5.49	2.04	12.8	181	0.989
10	6.00	2.18	12.3	168	1.08
11	6.48	2.30	11.7	154	1.16
12	6.91	2.39	10.9	139	1.24
13	7.30	2.46	10.1	124	1.30
14	7.64	2.49	9.3	109	1.36
15	7.92	2.50	8.4	96	1.40
16	8.15	2.49	7.6	85	1.44
18	8.45	2.42	6.1	66	1.47
20	8.56	2.29	4.9	51	1.47
25	8.15	1.87	3.15	29.5	1.36
30	6.95	1.45	2.28	19.3	1.16
35	5.62	1.16	1.80	13.7	0.99
40	4.63	0.97	1.51	10.5	0.87
45	3.91	0.84	1.32	8.4	0.78
50	3.36	0.76	1.18	7.0	0.72
60	2.63	0.66	1.01	5.5	0.651

Thermal Conductivity of Metals at Cryogenic Temperatures

T (K)	Nickel	Niobium	Platinum	Silver	Tantalum
70	2.21	0.61	0.90	4.97	0.616
80	1.93	0.58	0.84	4.71	0.603
90	1.72	0.563	0.81	4.60	0.596
100	1.58	0.552	0.79	4.50	0.592

Thermal Conductivity of Nickel

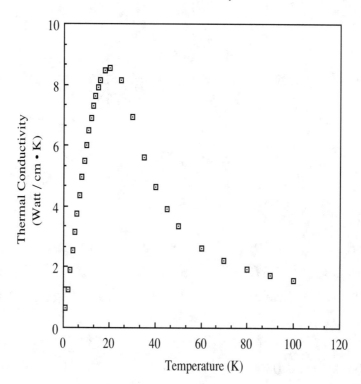

Thermal Conductivity of Metals at Cryogenic Temperatures

Thermal Conductivity of Niobium

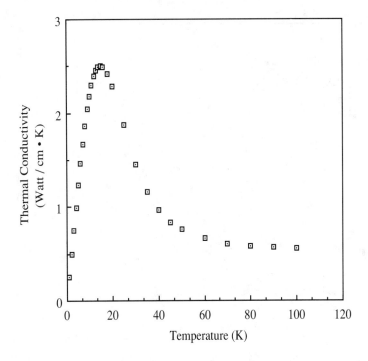

Thermal Conductivity of Platinum

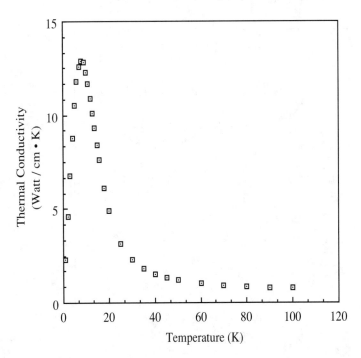

Thermal Conductivity of Metals at Cryogenic Temperatures

Thermal Conductivity of Silver

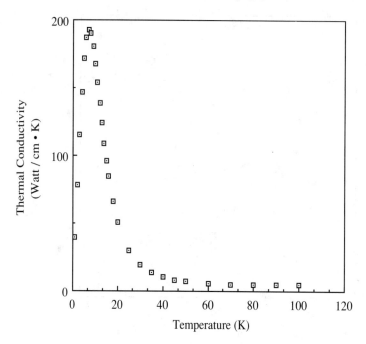

Thermal Conductivity of Tantalum

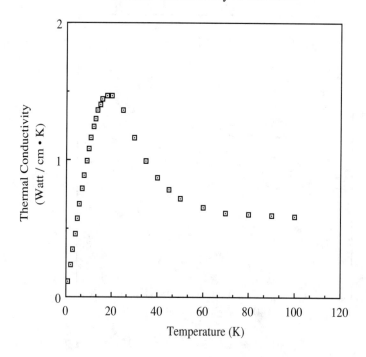

Thermal Conductivity of Metals at Cryogenic Temperatures

T (K)	Tin	Titanium	Tungsten	Zinc	Zirconium
1		0.0144	14.4	19.0	0.111
2		0.0288	28.7	37.9	0.223
3	297	0.0432	42.6	55.5	0.333
4	181	0.0576	55.6	69.7	0.442
5	117	0.0719	67.1	77.8	0.549
6	76	0.0863	76.2	78.0	0.652
7	52	0.101	82.4	71.7	0.748
8	36	0.115	85.3	61.8	0.837
9	26	0.129	85.1	51.9	0.916
10	19.3	0.144	82.4	43.2	0.984
11	14.8	0.158	77.9	36.4	1.04
12	11.6	0.172	72.4	30.8	1.08
13	9.3	0.186	66.4	26.1	1.11
14	7.6	0.200	60.4	22.4	1.13
15	6.3	0.214	54.8	19.4	1.13
16	5.3	0.227	49.3	16.9	1.12
18	4.0	0.254	40.0	13.3	1.08
20	3.2	0.279	32.6	10.7	1.01
25	2.22	0.337	20.4	6.9	0.85
30	1.76	0.382	13.1	4.9	0.74
35	1.50	0.411	8.9	3.72	0.65
40	1.35	0.422	6.5	2.97	0.58
45	1.23	0.416	5.07	2.48	0.535
50	1.15	0.401	4.17	2.13	0.497
60	1.04	0.377	3.18	1.71	0.442

Thermal Conductivity of Metals at Cryogenic Temperatures

T (K)	Tin	Titanium	Tungsten	Zinc	Zirconium
70	0.96	0.356	2.76	1.48	0.403
80	0.91	0.339	2.56	1.38	0.373
90	0.88	0.324	2.44	1.34	0.350
100	0.85	0.312	2.35	1.32	0.332

Values are in watts • cm^{-1} • K^{-1}. These data apply only to metals of purity of at least 99.9%. The third significant figure may not be accurate.

Source: data from Ho, C. Y., Powell, R. W., and Liley, P. *E., Thermal Conductictivity of Selected Materials,* NSRDS–NBS–8 and NSRDS–NBS–1 6, National Standard Reference Data System–National Bureau of Standards, Part 1, 1966; Part 2, 1968.

Thermal Conductivity of Tin

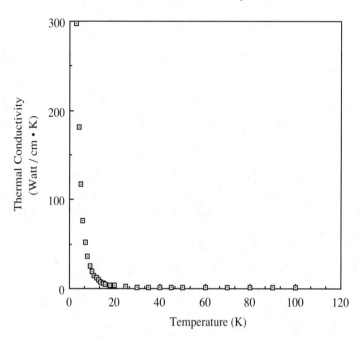

Thermal Conductivity of Metals at Cryogenic Temperatures

Thermal Conductivity of Titanium

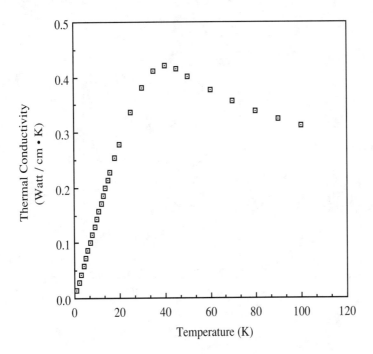

Thermal Conductivity of Tungsten

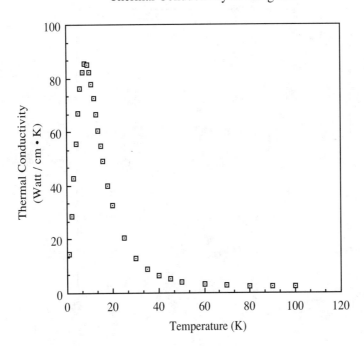

Thermal Conductivity of Metals at Cryogenic Temperatures

Thermal Conductivity of Zinc

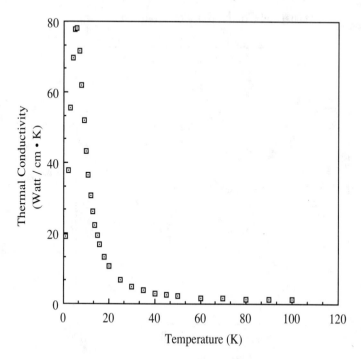

Thermal Conductivity of Zirconium

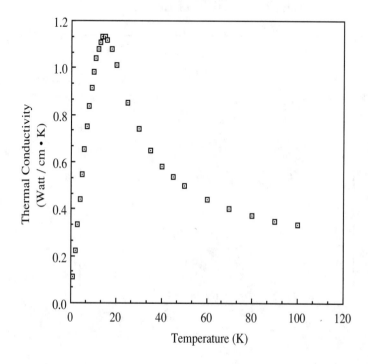

THERMAL CONDUCTIVITY OF METALS
AT 100 TO 3000 K

T (K)	Aluminum	Cadmium	Chromium	Copper	Gold
100	3	1.03	1.58	4.83	3.45
200	2.37	0.993	1.11	4.13	3.27
273	2.36	0.975	0.948	4.01	3.18
300	2.37	0.968	0.903	3.98	3.15
400	2.4	0.947	0.873	3.92	3.12
500	2.37	0.92	0.848	3.88	3.09
600	2.32	(0.42)	0.805	3.83	3.04
700	2.26	(0.49)	0.757	3.77	2.98
800	2.2	(0.559)	0.713	3.71	2.92
900	2.13		0.678	3.64	2.85
1000	(0.93)		0.653	3.57	2.78
1100	(0.96)		0.636	3.5	2.71
1200	(0.99)		0.624	3.42	2.62
1400			0.611		

Thermal Conductivity of Aluminum

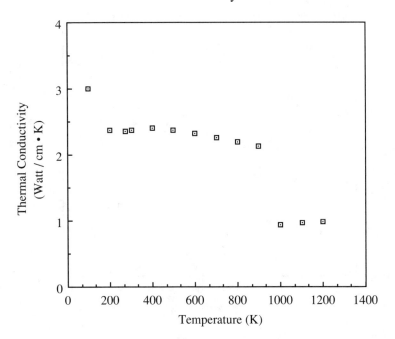

Thermal Conductivity of Metals at 100 to 3000 K

Thermal Conductivity of Cadmium

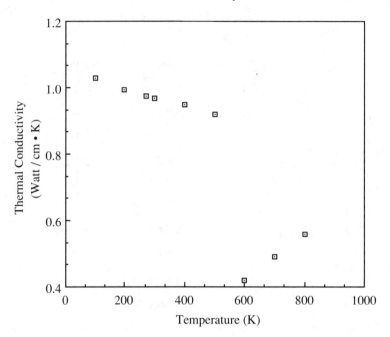

Thermal Conductivity of Chromium

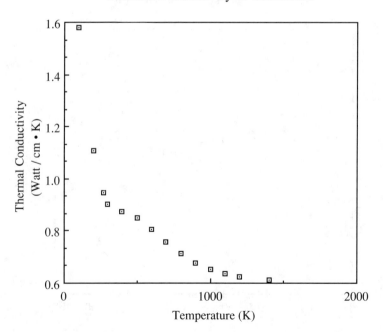

Thermal Conductivity of Metals at 100 to 3000 K

Thermal Conductivity of Copper

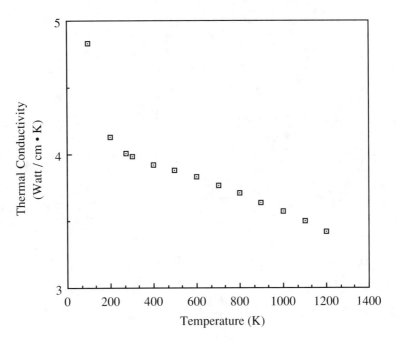

Thermal Conductivity of Gold

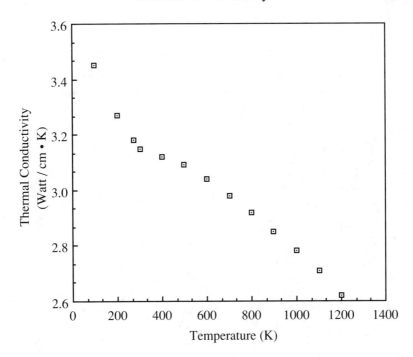

Thermal Conductivity of Metals at 100 to 3000 K

T (K)	Iron	Lead	Magnesium	Mercury	Molybdenum
100	1.32	0.396	1.69		1.79
200	0.94	0.366	1.59		1.43
273	0.835	0.355	1.57	(0.078)	1.39
300	0.803	0.352	1.56	(0.084)	1.38
400	0.694	0.338	1.53	(0.098)	1.34
500	0.613	0.325	1.51	(0.109)	1.3
600	0.547	0.312	1.49	(0.12)	1.26
700	0.487	(0.174)	1.47	(0.127)	1.22
800	0.433	(0.19)	1.46	(0.13)	1.18
900	0.38	(0.203)	1.45		1.15
1000	0.326	(0.215)	(0.84)		1.12
1100	0.297		(0.91)		1.08
1200	0.282		(0.98)		1.05
1400	0.309				0.996
1600	0.327				0.946
1800					0.907
2000					0.88
2200					0.858
2600					0.825

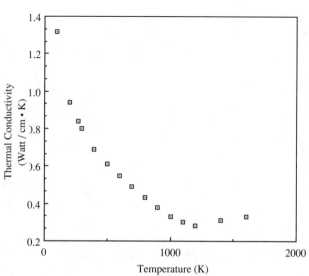

Thermal Conductivity of Iron

Thermal Conductivity of Metals at 100 to 3000 K

Thermal Conductivity of Lead

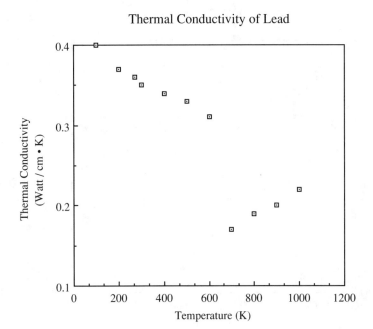

Thermal Conductivity of Magnesium

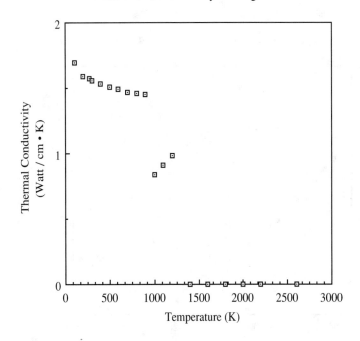

Thermal Conductivity of Metals at 100 to 3000 K

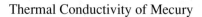

Thermal Conductivity of Mecury

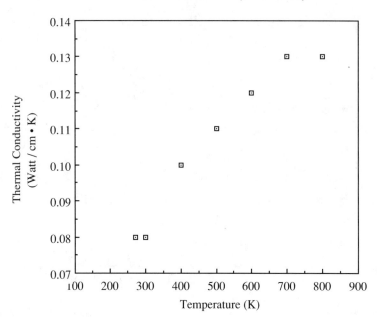

Thermal Conductivity of Molybdenum

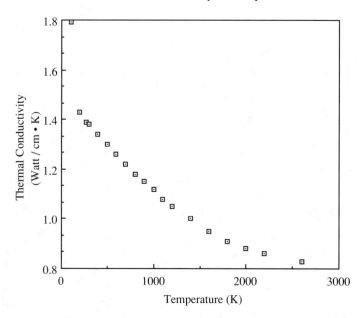

Thermal Conductivity of Metals at 100 to 3000 K

T (K)	Nickel	Niobium	Platinum	Silver	Tantalum
100	1.58	0.552	0.79	4.5	0.592
200	1.06	0.526	0.748	4.3	0.575
273	0.94	0.533	0.734	4.28	0.574
300	0.905	0.537	0.73	4.27	0.575
400	0.801	0.552	0.722	4.2	0.578
500	0.721	0.567	0.719	4.13	0.582
600	0.655	0.582	0.72	4.05	0.586
700	0.653	0.598	0.723	3.97	0.59
800	0.674	0.613	0.729	3.89	0.594
900	0.696	0.629	0.737	3.82	0.598
1000	0.718	0.644	0.748	3.74	0.602
1100	0.739	0.659	0.76	3.66	0.606
1200	0.761	0.675	0.775	3.58	0.610
1400	0.804	0.705	0.807		0.618
1600		0.735	0.842		0.626
1800		0.764	0.877		0.634
2000		0.791	0.913		0.640
2200		0.815			0.647
2600					0.658
3000					0.665

Thermal Conductivity of Nickel

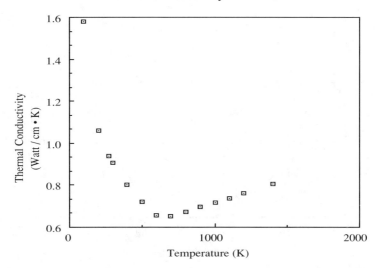

Thermal Conductivity of Metals at 100 to 3000 K

Thermal Conductivity of Niobium

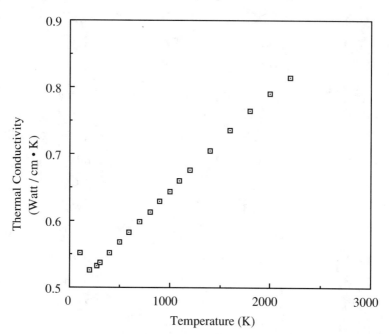

Thermal Conductivity of Platinum

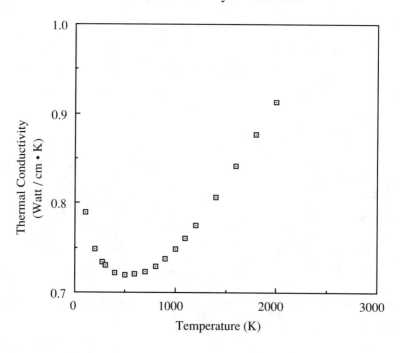

Thermal Conductivity of Metals at 100 to 3000 K

Thermal Conductivity of Silver

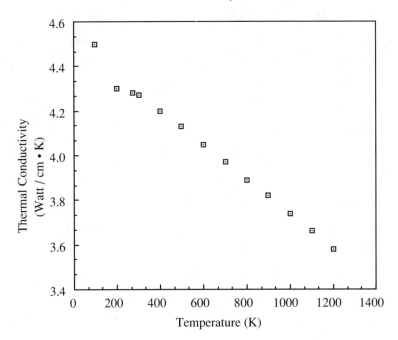

Thermal Conductivity of Tantalum

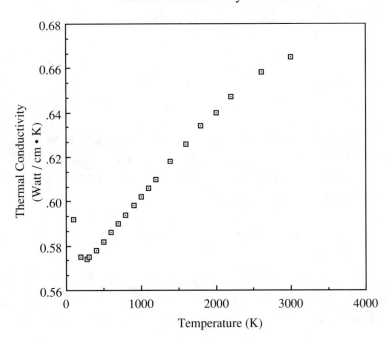

Thermal Conductivity of Metals at 100 to 3000 K

T (K)	Tin	Titanium	Tungsten	Zinc	Zirconium
100	0.85	0.312	2.35	1.32	0.332
200	0.733	0.245	1.97	1.26	0.252
273	0.682	0.224	1.82	1.22	0.232
300	0.666	0.219	1.78	1.21	0.227
400	0.622	0.204	1.62	1.16	0.216
500	0.596	0.197	1.49	1.11	0.210
600	(0.323)	0.194	1.39	1.05	0.207
700	(0.343)	0.194	1.33	(0.499)	0.209
800	(0.364)	0.197	1.28	(0.557)	0.216
900	(0.384)	0.202	1.24	(0.615)	0.226
1000	(0.405)	0.207	1.21	(0.673)	0.237
1100	(0.425)	0.213	1.18	(0.73)	0.248
1200	(0.446)	0.220	1.15		0.257
1400	(0.487)	0.236	1.11		0.275
1600		0.253	1.07		0.290
1800		0.271	1.03		0.302
2000			1.00		0.313
2200			0.98		
2600			0.94		
3000			0.915		

Values in this table are in watts/cm K. To convert to Btu/hr ft °R, multiply the tabular values by 57.818. These data apply only to metals of purity of at least 99.9%. In the table the third significant figure is for smoothness and is not indicative of the degree of accuracy.

Note: Values in parentheses are for liquid state

Source: Data from Ho, C. Y., Powell, R. W., and Liley, P. E., *Thermal Conductivity of Selected Materials*, NSRD-NBS-16, Part 2 National Standard Reference Data System-National Bureau of Standards, February 1968.

Thermal Conductivity of Metals at 100 to 3000 K

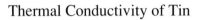

Thermal Conductivity of Tin

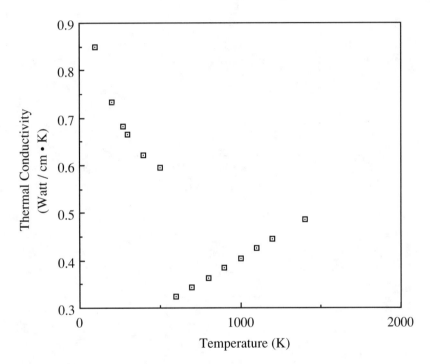

Thermal Conductivity of Metals at 100 to 3000 K

Thermal Conductivity of Titanium

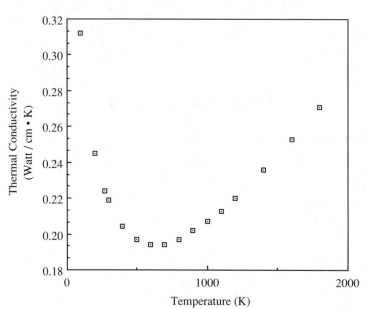

Thermal Conductivity of Tungsten

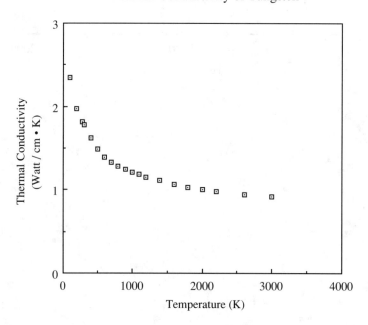

Thermal Conductivity of Metals at 100 to 3000 K

Thermal Conductivity of Zinc

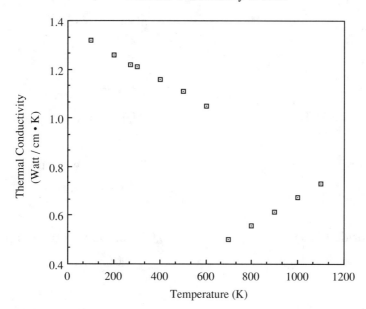

Thermal Conductivity of Zirconium

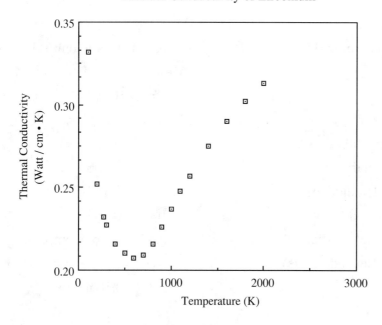

THERMAL CONDUCTIVITY OF SELECTED CERAMICS
Listed by Ceramic

Ceramic	Thermal Conductivity $(cal \bullet cm^{-1} \bullet sec^{-1} \bullet K^{-1})$
BORIDES	
Chromium Diboride (CrB_2)	0.049-0.076 at room temp.
Hafnium Diboride (HfB_2)	0.015 at room temp.
Tantalum Diboride (TaB_2)	0.026 at room temp.
	0.033 at 200 °C.
Titanium Diboride (TiB_2)	0.058-0.062 at room temp.
	0.063 at 200 °C
Zirconium Diboride (ZrB_2)	0.055-0.058 at room temp.
	0.055-0.060 at 200 °C
CARBIDES	
Boron Carbide (B_4C)	0.065-0.069 at room temp.
	0.198 at 425 °C
Hafnium Monocarbide (HfC)	0.053 at room temp.
	$0.15 + 1.20x10$ T watts cm^{-1} K^{-1}
	from 1000-2000K
Silicon Carbide (SiC)	
(with 1 wt% Be addictive)	0.621
(with 1 wt% B addictive)	0.406
(with 1 wt% Al addictive)	0.143
(with 2 wt% BN addictive)	0.263
(with 1.6 wt% BeO addictive)	0.645 at room temp.
(with 3.2 wt% BeO addictive)	0.645 at room temp.
	0.098-0.10 at 20°C

Thermal Conductivity of Selected Ceramics (Continued)
Listed by Ceramic

Ceramic	Thermal Conductivity $(cal \cdot cm^{-1} \cdot sec^{-1} \cdot K^{-1})$
(cubic, CVD)	0.289 at 127°C
	0.049-0.080 at 600°C
	0.061 at 800°C
	0.051 at 1000°C
	0.0059 at 1250°C
	0.0827 at 1327°C
	0.0032 at 1530°C
Tantalum Monocarbide (TaC)	0.053 at room temp.
Titanium Monocarbide (TiC)	0.041-0.074 at room temp.
	0.0135 at 1000 °C
Trichromium Dicarbide (Cr_3C_2)	0.454
Tungsten Monocarbide (WC)	0.201 at 20 °C
(6% Co, 1-3μm grain size)	0.239
(12% Co, 1-3μm grain size)	0.251
(24% Co, 1-3μm grain size)	0.239
(6% Co, 2-4μm grain size)	0.251
(6% Co, 3-6μm grain size)	0.256

Thermal Conductivity of Selected Ceramics (Continued)
Listed by Ceramic

Ceramic	Thermal Conductivity $(cal \cdot cm^{-1} \cdot sec^{-1} \cdot K^{-1})$
Zirconium Monocarbide (ZrC)	0.049 at room temp.
	0.098 at 50°C
	0.069 at 150°C
	0.065 at 188°C
	0.061 at 288°C
	0.080 at 600°C
	0.083 at 800°C
	0.086 at 1000°C
	0.089 at 1200°C
	0.092 at 1400°C
	0.096 at 1600°C
	0.099 at 1800°C
	0.103 at 2000°C
	0.105 at 2200°C
NITRIDES	
Aluminum Nitride (AlN)	0.072 at 25°C
	0.060 at 200°C
	0.053 at 400°C
	0.048 at 600°C
	0.042 at 800°C
Boron Nitride (BN)	
parallel to c axis	0.0687 at 300°C
	0.0646 at 700°C
	0.0637 at 1000°C
parallel to a axis	0.0362 at 300°C
	0.0318 at 700°C
nitride	0.0295 at 1000°C

Thermal Conductivity of Selected Ceramics (Continued)
Listed by Ceramic

Ceramic	Thermal Conductivity $(cal \cdot cm^{-1} \cdot sec^{-1} \cdot K^{-1})$
Titanium Mononitride (TiN)	0.069 at 25 °C
	0.057 at 127 °C
	0.040 at 200 °C
	0.027 at 650 °C
	0.020 at 1000 °C
	0.162 at 1500 °C
	0.136 at 2300 °C
Trisilicon tetranitride (Si_3N_4)	
(pressureless sintered)	0.072 at room temp.
	0.022-0.072 at 127 °C
	0.041 at 200-750 °C
	0.036-0.042 at 500 °C
(pressureless sintered)	0.038 at 1000 °C
	0.033-0.034 at 1200 °C
Zirconium Mononitride (TiN)	0.040 at 200 °C
	0.025 at 425 °C
	0.018 at 650 °C
	0.016 at 875 °C
	0.015 at 1100 °C

Thermal Conductivity of Selected Ceramics (Continued)
Listed by Ceramic

Ceramic	Thermal Conductivity $(\text{cal} \cdot \text{cm}^{-1} \cdot \text{sec}^{-1} \cdot \text{K}^{-1})$
OXIDES	
Aluminum Oxide (Al_2O_3)	0.06 at room temp.
	0.04-0.069 at 100°C
	0.03-0.064 at 200°C
	0.037 at 315°C
	0.02-0.031 at 400°C
	0.035 at 500°C
	0.021-0.022 at 600°C
	0.015-0.017 at 800°C
	0.014-0.016 at 1000°C
	0.013-0.015 at 1200°C
	0.013 at 1400°C
	0.014 at 1600°C
	0.017 at 1800°C
Aluminum Oxide (Al_2O_3) (single crystal)	0.103 at 20°C
	0.047 at 300°C
	0.029 at 800°C

Thermal Conductivity of Selected Ceramics (Continued)
Listed by Ceramic

Ceramic	Thermal Conductivity $(cal \cdot cm^{-1} \cdot sec^{-1} \cdot K^{-1})$
Beryllium Oxide (BeO)	0.038-0.47 at 20°C
	0.032-0.34 at 100°C
	0.14-0.16 at 400°C
	0.089-0.1137 at 600°C
	0.060-0.093 at 800°C
	0.043 at 1100°C
	0.041-0.054 at 1200°C
	0.038 at 1300°C
	0.036 at 1400°C
	0.034 at 1500°C
	0.033-0.039 at 1600°C
	0.033 at 1700°C
	0.036 at 1800°C
	0.036 at 1900°C
	0.036 at 2000°C
Calcium Oxide (CaO)	0.037 at 100°C
	0.027 at 200°C
	0.022 at 400°C
	0.020 at 600°C
	0.019 at 800°C
	0.0186-0.019 at 1000°C
Cerium Dioxide (CeO$_2$)	0.0229 at 400K
	0.00287 at 1400K
Dichromium Trioxide (Cr$_2$O$_3$)	0.0239-0.0788
Hafnium Dioxide (HfO$_2$)	0.0273 at 25-425°C

Thermal Conductivity of Selected Ceramics (Continued)
Listed by Ceramic

Ceramic	Thermal Conductivity $(\text{cal} \cdot \text{cm}^{-1} \cdot \text{sec}^{-1} \cdot \text{K}^{-1})$
Magnesium Oxide (MgO)	0.097 at room temp.
	0.078-0.082 at 100°C
	0.064-0.065 at 200°C
	0.038-0.045 at 400°C
	0.0198-0.026 at 800°C
	0.016-0.020 at 1000°C
	0.0139-0.0148 at 1200°C
	0.012-0.014 at 1400°C
	0.0108-0.016 at 1600°C
	0.0096-0.0191 at 1800°C
Nickel monoxide (NiO)	
(0% porosity)	0.029 at 100°C
(0% porosity)	0.024 at 200°C
(0% porosity)	0.017 at 400°C
(0% porosity)	0.012 at 800°C
(0% porosity)	0.011 at 1000°C
Silicon Dioxide (SiO$_2$)	0.0025 at 200°C
	0.003 at 400°C
	0.004 at 800°C
	0.005 at 1200°C
	0.006 at 1600°C
Thorium Dioxide (ThO$_2$)	
(0% porosity)	0.024 at room temp.
(0% porosity)	0.020 at 100°C
(0% porosity)	0.019 at 200°C
(0% porosity)	0.014 at 400°C
(0% porosity)	0.010 at 600°C

Thermal Conductivity of Selected Ceramics (Continued)
Listed by Ceramic

Ceramic	Thermal Conductivity (cal • cm^{-1} • sec^{-1} • K^{-1})
(0% porosity)	0.008 at 800°C
(0% porosity)	0.007-0.0074 at 1000°C
(0% porosity)	0.006-0.0076 at 1200°C
(0% porosity)	0.006 at 1400°C
Titanium Oxide (TiO$_2$)	
(0% porosity)	0.016 at 100°C
(0% porosity)	0.012 at 200°C
(0% porosity)	0.009 at 400°C
(0% porosity)	0.008 at 600°C
(0% porosity)	0.008 at 800°C
(0% porosity)	0.008 at 1000°C
(0% porosity)	0.008 at 1200°C
Uranium Dioxide (UO$_2$)	
(0% porosity)	0.025 at 100°C
(0% porosity)	0.020 at 200°C
(0% porosity)	0.015 at 400°C
(0% porosity)	0.010 at 600°C
(0% porosity)	0.009 at 800°C
(0% porosity)	0.008 at 1000°C
	0.018 at 100°C
	0.012 at 400°C
	0.008 at 600°C
	0.008 at 700°C
	0.006 at 1000°C
	0.006 at 1200°C

Thermal Conductivity of Selected Ceramics (Continued)
Listed by Ceramic

Ceramic	Thermal Conductivity $(cal \cdot cm^{-1} \cdot sec^{-1} \cdot K^{-1})$
Zirconium Oxide (ZrO_2)	
(stabilized, 0% porosity)	0.005 at 100°C
(stabilized, 0% porosity)	0.005 at 200°C
(stabilized, 0% porosity)	0.005 at 400°C
(stabilized, 0% porosity)	0.0055 at 800°C
(stabilized, 0% porosity)	0.006 at 1200°C
(stabilized, 0% porosity)	0.0065 at 1400°C
(stabilized)	0.004 at 100°C
(stabilized)	0.0044 at 500°C
(stabilized)	0.0048-0.0055 at 1000°C
(stabilized)	0.0049-0.0050 at 1200°C
(MgO stabilized)	0.0076 at room temp.
(Y_2O_3 stabilized)	0.0055 at room temp.
(plasma sprayed)	0.0019-0.0031 at room temp.
(plasma sprayed and coated with Cr_2O_3)	0.0033 at room temp.
(MgO stabilized)	0.0057 at 800°C
(Y_2O_3 stabilized)	0.0053 at 800°C
(plasma sprayed)	0.0019-0.0022 at 800°C
(plasma sprayed and coated with Cr_2O_3)	0.0033 at 800°C
(5-10% CaO stabilized)	0.0045 at 400°C
(5-10% CaO stabilized)	0.0049 at 800°C
(5-10% CaO stabilized)	0.0057 at 1200°C
Cordierite ($2MgO\ 2Al_2O_3\ 5SiO_2$)	
(ρ=2.3g/cm^3)	0.0077 at 20°C
(ρ=2.3g/cm^3)	0.0062 at 300°C
(ρ=2.3g/cm^3)	0.0055 at 500°C
(ρ=2.3g/cm^3)	0.0055 at 800°C

Thermal Conductivity of Selected Ceramics (Continued)
Listed by Ceramic

Ceramic	Thermal Conductivity $(cal \cdot cm^{-1} \cdot sec^{-1} \cdot K^{-1})$
(ρ=2.1g/cm^3)	0.0043 at 20°C
(ρ=2.1g/cm^3)	0.0041 at 300°C
(ρ=2.1g/cm^3)	0.0040 at 500°C
(ρ=2.1g/cm^3)	0.0038 at 800°C
Mullite (3Al$_2$O$_3$ 2SiO$_2$)	
(0% porosity)	0.0145 at 100°C
(0% porosity)	0.013 at 200°C
(0% porosity)	0.011 at 400°C
(0% porosity)	0.010 at 600°C
(0% porosity)	0.0095 at 800°C
(0% porosity)	0.009 at 1000°C
(0% porosity)	0.009 at 1200°C
(0% porosity)	0.009 at 1400°C
Sillimanite (Al$_2$O$_3$ SiO$_2$)	
(0% porosity)	0.0042 at 100°C
(0% porosity)	0.004 at 400°C
(0% porosity)	0.0035 at 800°C
(0% porosity)	0.0035 at 1200°C
(0% porosity)	0.003 at 1500°C
Spinel (Al$_2$O$_3$ MgO)	
(0% porosity)	0.035 at 100°C
(0% porosity)	0.031 at 200°C
(0% porosity)	0.024 at 400°C
(0% porosity)	0.019 at 600°C
(0% porosity)	0.015 at 800°C
(0% porosity)	0.013-0.0138 at 1000°C
(0% porosity)	0.013 at 1200°C

Thermal Conductivity of Selected Ceramics (Continued)
Listed by Ceramic

Ceramic	Thermal Conductivity $(cal \cdot cm^{-1} \cdot sec^{-1} \cdot K^{-1})$
Zircon (SiO_2 ZrO_2)	
(0% porosity)	0.0145 at 100°C
(0% porosity)	0.0135 at 200°C
(0% porosity)	0.012 at 400°C
(0% porosity)	0.010 at 800°C
(0% porosity)	0.0095 at 1200°C
(0% porosity)	0.0095 at 1400°C
SILICIDE	
Molybdenum Disilicide ($MoSi_2$)	0.129 at 150°C
	0.074 at 425°C
	0.053 at 540°C
	0.057 at 650°C
	0.046 at 875°C
	0.041 at 1100°C

Source: Data compiled by J.S. Park from *No. 1 Materials Index*, Peter T.B. Shaffer, Plenum Press, New York, (1964); *Smithells Metals Reference Book*, Eric A. Brandes, ed., in association with Fulmer Research Institute Ltd. 6th ed. London, Butterworths, Boston, (1983); and *Ceramic Abstracts*, American Ceramic Society (1986-1991)

THERMAL CONDUCTIVITY OF SPECIAL CONCRETES

Description; type of aggregate	Thermal Conductivity Btu / (hr • ft • °F)
Frost resisting; 1% $CaCl_2$; normal aggregates	1.0
Frost-resisting porous;6% air entrainment	0.85
Lightweight; with expanded shale or clay	0.25
Lightweight; with foamed slag	0.20
Cinder concrete; fine and coarse	0.25
Pulverized fuel ash	0.25
Lightweight refractory concrete with aluminous cement	0.20
Lightweight; insulating, with perlite	0.15
Lightweight; insulating, with expanded vermiculite	0.10

A great many varieties of aggregates have been used for concrete, dependent largely on the materials available. In general, high density concretes have high strength and high thermal conductivity, although such variables as water/ cement ratio, percentage of fines, and curing conditions may result in wide differences in properties with the same materials.

Source: From Bolz, R. E. and Tuve, C. L., Eds., *Handbook of Tables for Applied Engineering Science, 2nd ed.*, CRC Press, Cleveland, 1973, p.645.

THERMAL CONDUCTIVITY OF CRYOGENIC INSULATION AND SUPPORTS

Class[a]	Material	Thermal Conductivity Range (mW · m^{-1} · K^{-1})	Interspace Pressure (mm Hg)[b]
2	Multilayer	0.04-0.2	10^{-4}
3	Opacified powder	0.26-0.7	10^{-4}
4	Evacuated powder	1.0-2.0	10^{-4}
5	Vacuum flask	5.0	10^{-6}
6	Gas-filled powder	1.7-7.0	760
7	Expanded foam	5.0-35	760
8	Fiber blanket	35-45	760

[a] Classes of cryogenic thermal insulating materials:

1. Liquid and vapor shields - Very low-temperature, valuable, or dangerous liquids such as helium or fluorine are often shielded by an intermediate cryogenic liquid or vapor container that must in turn be insulated by one of the methods described below.
2. Multilayer reflecting shields — Foil or aluminized plastic alternated with paper-thin glass or plastic-fiber sheets; lowest conductivity, low density, and heat storage; good stability; minimum support structure.
3. Opacified evacuated powders - Contain metallic flakes to reduce radiation; conform to irregular shapes.
4. Evacuated dielectric powders - Very fine powders of low-conductivity adsorbent; moderate vacuum requirement; minimum fire hazard in oxygen.
5. Vacuum flasks (Dewar) - Tight shield-space with highly. reflecting walls and high vacuum; minimum heat capacity; rugged; small thickness.
6. Gas-filled powders—Same powders as Class 4 but with air or inert gas; low cost; easy application; no vacuum requirement.
7. Expanded foams - Very light foamed plastic; inexpensive; minimum weight but bulky; self supporting.
8. Porous fiber blankets - Blanket material of fine fibers, usually glass; minimum cost and easy installation but not an adequate insulation for most cryogenic applications.

[b] To convert mm Hg to N · m^{-2} multiply by 133.32.

Thermal Conductivity of Cryogenic Insulation Supports
20–300 K

Material	Mean Thermal Conductivity $(W \cdot m^{-1} \cdot K^{-1})$
Aluminum alloy	86
"K" Monel®	17
Stainbss steel	9.3
Titanhm alloy	6.1
Nylon	0.29
Teflon	0.24

Source: From Boltz, R. E. and Tuve, G. L., Eds., *Handbook of Tables for Applied Engineering Science*, 2nd ed., CRC Press, Cleveland, 1973, 529.

THERMAL CONDUCTIVITY OF
SELECTED POLYMERS
Listed by Polymer

Polymer	Thermal Conductivity (ASTM C177) $(Btu/hr/ft^2/°F/ft)$
ABS Resins; Molded, Extruded	
Medium impact	0.08—0.18
High impact	0.12—0.16
Very high impact	0.01—0.14
Low temperature impact	0.08—0.14
Heat resistant	0.12—0.20
Acrylics; Cast, Molded, Extruded	
Cast Resin Sheets, Rods:	
General purpose, type I	0.12
General purpose, type II	0.12
Moldings:	
Grades 5, 6, 8	0.12
High impact grade	0.12
Thermoset Carbonate	
Allyl diglycol carbonate	1.45
Alkyds; Molded	
Putty (encapsulating)	0.35—0.60
Rope (general purpose)	0.35—0.60
Granular (high speed molding)	0.35—0.60
Glass reinforced (heavy duty parts)	0.20—0.30
Cellulose Acetate; Molded, Extruded	
ASTM Grade:	
H6—1	0.10—0.19

Thermal Conductivity of Selected Polymers (Continued)
Listed by Polymer

Polymer	Thermal Conductivity (ASTM C177) (Btu/hr/ft^2/°F/ft)
H4—1	0.10—0.19
H2—1	0.10—0.19
MH—1, MH—2	0.10—0.19
MS—1, MS—2	0.10—0.19
S2—1	0.10—0.19
Cellulose Acetate Butyrate; Molded, Extruded ASTM Grade:	
H4	0.10—0.19
MH	0.10—0.19
S2	0.10—0.19
Cellulose Acetate Propionate; Molded, Extruded ASTM Grade:	
1	0.10—0.19
3	0.10—0.19
6	0.10—0.19
Chlorinated Polymers	
Chlorinated polyether	0.91
Chlorinated polyvinyl chloride	0.95
Polycarbonates	
Polycarbonate	0.11
Polycarbonate (40% glass fiber reinforced)	0.13

Thermal Conductivity of Selected Polymers (Continued)
Listed by Polymer

Polymer	Thermal Conductivity (ASTM C177) $(Btu/hr/ft^2/°F/ft)$
Fluorocarbons; Molded,Extruded	
Polytrifluoro chloroethylene (PTFCE)	0.145
Polytetrafluoroethylene (PTFE)	0.14
Fluorinated ethylene propylene(FEP)	0.12
Polyvinylidene— fluoride (PVDF)	0.14
Epoxies; Cast, Molded, Reinforced	
Standard epoxies (diglycidyl ethers of bisphenol A)	
Cast rigid	0.1—0.3
Molded	0.1—0.5
High strength laminate	2.35
Melamines; Molded	
Filler & type	
Cellulose electrical	0.17—0.20
Glass fiber	0.28
Nylons; Molded, Extruded	
Type 6	
General purpose	1.2—1.69
Glass fiber (30%) reinforced	1.69—3.27
Cast	1.2—1.7
Type 11	1.5
Type 12	1.7

Thermal Conductivity of Selected Polymers (Continued)
Listed by Polymer

Polymer	Thermal Conductivity (ASTM C177) $(Btu/hr/ft^2/°F/ft)$
6/6 Nylon	
General purpose molding	1.69—1.7
Glass fiber reinforced	1.5— 3.3
General purpose extrusion	1.7
6/10 Nylon	
General purpose	1.5
Glass fiber (30%) reinforced	3.5
Phenolics; Molded	
Type and filler	
General: woodflour and flock	0.097—0.3
Shock: paper, flock, or pulp	0.1—0.16
High shock: chopped fabric or cord	0.097—0.170
Very high shock: glass fiber	0.2
Phenolics: Molded	
Arc resistant—mineral	0.24—0.34
Rubber phenolic—woodflour or flock	0.12
Rubber phenolic—chopped fabric	0.05
Rubber phenolic—asbestos	0.04
ABS–Polycarbonate Alloy	2.46 (per ft)
PVC–Acrylic Alloy	
PVC–acrylic sheet	1.01
PVC–acrylic injection molded	0.98

Thermal Conductivity of Selected Polymers (Continued)
Listed by Polymer

Polymer	Thermal Conductivity (ASTM C177) $(Btu/hr/ft^2/°F/ft)$
Polymides	
Unreinforced	6.78
Unreinforced 2nd value	3.8
Glass reinforced	3.59
Polyacetals	
Homopolymer:	
Standard	0.13
Copolymer:	
Standard	0.16
High flow	1.6
Polyester; Thermoplastic	
Injection Moldings:	
General purpose grade	0.36—0.55
Polyesters: Thermosets	
Cast polyyester	
Rigid	0.10—0.12
Reinforced polyester moldings	
High strength (glass fibers)	1.32—1.68
Phenylene Oxides	
SE—100	1.1
SE—1	1.5
Glass fiber reinforced	1.15,1.1

Thermal Conductivity of Selected Polymers (Continued)
Listed by Polymer

Polymer	Thermal Conductivity (ASTM C177) $(Btu/hr/ft^2/°F/ft)$
Phenylene oxides (Noryl)	
Standard	1.8
Polyarylsulfone	1.1
Polypropylene:	
General purpose	1.21—1.36
High impact	1.72
Polyphenylene sulfide:	
Standard	2
40% glass reinforced	2
Polyethylenes; Molded, Extruded	
Type I—lower density (0.910—0.925)	
Melt index 0.3—3.6	0.19
Melt index 6—26	0.19
Melt index 200	0.19
Type II—medium density (0.926—0.940)	
Melt index 20	0.19
Melt index 1.0—1.9	0.19
Type III—higher density (0.941—0.965)	
Melt index 0.2—0.9	0.19
Melt Melt index 0.1—12.0	0.19
Melt index 1.5—15	0.19
High molecular weight	0.19

Thermal Conductivity of Selected Polymers (Continued)
Listed by Polymer

Polymer	Thermal Conductivity (ASTM C177) $(Btu/hr/ft^2/°F/ft)$
Polystyrenes; Molded	
Polystyrenes	
General purpose	0.058—0.090
Medium impact	0.024—0.090
High impact	0.024—0.090
Glass fiber -30% reinforced	0.117
Polyvinyl Chloride And Copolymers; Molded, Extruded	
Nonrigid—general	0.07—0.10
Nonrigid—electrical	0.07—0.10
Rigid—normal impact	0.07—0.10
Vinylidene chloride	0.053
Silicones; Molded, Laminated	
Fibrous (glass) reinforced silicones	0.18
Granular (silica) reinforced silicones	0.25—0.5
Woven glass fabric/ silicone laminate	0.075—0.125
Ureas; Molded	
Alpha—cellulose filled (ASTM Type l)	0.17—0.244

Source: data compiled by J.S. Park from Charles T. Lynch, *CRC Handbook of Materials Science,* Vol. 3, CRC Press, Boca Raton, Florida, 1975 and *Engineered Materials Handbook*, Vol.2, Engineering Plastics, ASM International, Metals Park, Ohio, 1988.

THERMAL EXPANSION OF SELECTED TOOL STEELS

Type	Thermal Expansion mm/(m•K) from 20 °C to				
	100 °C	200 °C	425°C	540°C	600°C
W1	10.4	11.0	13.1	13.8(a)	14.2(b)
S1	12.4	12.6	13.5	13.9	14.2
S5			12.6	13.3	13.7
S7		12.6	13.3	13.7(a)	13.3
A2	10.7	10.6(c)	12.9	14.0	14.2
H11	11.9	12.4	12.8	12.9	13.3
H13	10.4	11.5	12.2	12.4	13.1
H21	12.4	12.6	12.9	13.5	13.9
H26				12.4	
T1		9.7	11.2	11.7	11.9
T15		9.9	11.0	11.5	
M2	10.1	9.4(c)	11.2	11.9	12.2
L2			14.4	14.6	14.8
L6	11.3	12.6	12.6	13.5	13.7

(a) From 20°C to 500°C
(b) From 20°C to 600°C
(c) From 20°C to 260°C

Source: Data from *ASM Metals Reference Book, Second Edition,* American Society for Metals, Metals Park, Ohio 44073, p242, (1984).

THERMAL EXPANSION AND THERMAL CONDUCTIVITY OF SELECTED ALLOY CAST IRONS

Description	Thermal Expansion Coefficient mm/(m • °C)	Thermal Conductivity W/(m • K)
Abrasion–Resistant White Irons		
Low–C white iron	12[a]	22[b]
Martensitic nickel–chromium iron	8 to 9[a]	30[b]
Corrosion–Resistant Irons		
High– Silicon iron	12.4 to 13.1	
High–chromium iron	9.4 to 9.9	
High–nickel gray iron	8.1 to 19.3	38 to 40
High–nickel ductile iron	12.6 to 18.7	13.4
Heat–Resistant Gray Irons		
Medium–silicon iron	10.8	37
High–chromium iron	9.3 to 9.9	20
High–nickel iron	8.1 to 19.3	37 to 40
Nickel–chromium–silicon iron	12.6 to 16.2	30
High–aluminum iron	15.3	
Heat–Resistant Ductile Irons		
Medium–silicon ductile iron	10.8 to 13.5	
High–nickel ductile (20 Ni)	18.7	13
High–nickel ductile (23 Ni)	18.4	

[a] 10 to 260°C
[b] Estimated.

Source: Data from *ASM Metals Reference Book, Second Edition*, American Society for Metals, Metals Park, Ohio 44073, p172, (1984).

THERMAL EXPANSION OF SELECTED CERAMICS
Listed by Ceramic

Ceramic	Thermal Expansion ($°C^{-1}$)
BORIDES	
Chromium Diboride (CrB_2)	4.6–11.1 x 10^{-6} for 20–1000°C
Hafnium Diboride (HfB_2)	5.5 –5.54 x 10^{-6} for room temp.–1000°C
Tantalum Diboride (TaB_2)	5.1 x 10^{-6} at room temp.
Titanium Diboride (TiB_2)	4.6–8.1 x 10^{-6}
Zirconium Diboride (ZrB_2)	5.69 x 10^{-6} for 25–500°C
	5.5–6.57 x 10^{-6} °C for 25–1000°C
	6.98 x 10^{-6} for 20–1500°C
CARBIDES	
Boron Carbide (B_4C)	4.5 x 10^{-6} for room temp.–800°C
	4.78 x 10^{-6} for 25–500°C
	5.54 x 10^{-6} for 25–1000°C
	6.02 x 10^{-6} for 25–1500°C
	6.53 x 10^{-6} for 25–2000°C
	7.08 x 10^{-6} for 25–2500°C
Hafnium Monocarbide (HfC)	6.27–6.59 x 10^{-6} for 25–650°C
	6.25 x 10^{-6} for 25–1000°C
Silicon Carbide (SiC)	4.63 x 10^{-6} for 25–500°C
	5.12 x 10^{-6} for 25–1000°C
	5.48 x 10^{-6} for 25–1500°C
	5.77 x 10^{-6} for 25–2000°C
	5.94 x 10^{-6} for 25–2500°C
	4.70 x 10^{-6} for 20–1500°C
	4.70 x 10^{-6} for 0–1700°C

Thermal Expansion of Selected Ceramics (Continued)
Listed by Ceramic

Ceramic	Thermal Expansion $(^\circ C^{-1})$
Tantalum Monocarbide (TaC)	6.29–6.32×10^{-6} for 25–500°C
	6.67×10^{-6} for 25–1000°C
	7.12×10^{-6} for 25–1500°C
	7.64×10^{-6} for 25–2000°C
	8.40×10^{-6} for 25–2500°C
	6.50×10^{-6} for 0–1000°C
	6.64×10^{-6} for 0–1200°C
Titanium Monocarbide (TiC)	6.52–7.15×10^{-6} for 25–500°C
	7.18–7.45×10^{-6} for 25–750°C
	7.40–8.82×10^{-6} for 25–1000°C
	9.32×10^{-6} for 25–1250°C
	8.15–9.45×10^{-6} for 25–1500°C
	8.81×10^{-6} for 25–2000°C
	7.90×10^{-6} for 0–2500°C
	7.08×10^{-6} for 0–750°C
	7.85–7.86×10^{-6} for 0–1000°C
	8.02×10^{-6} for 0–1275°C
	8.29×10^{-6} for 0–1400°C
	8.26×10^{-6} for 0–1525°C
	8.40×10^{-6} for 0–1775°C
Trichromium Dicarbide (Cr_3C_2)	8.00×10^{-6} for 25–500°C
	9.95×10^{-6} for 25–500°C
	8.8×10^{-6} for 25–120°C
	10.9×10^{-6} for 150–980°C

Thermal Expansion of Selected Ceramics (Continued)
Listed by Ceramic

Ceramic	Thermal Expansion ($^\circ C^{-1}$)
Tungsten Monocarbide (WC)	4.42 x 10^{-6} for 25–500ºC
	4.84–4.92 x 10^{-6} for 25–1000ºC
	5.35–5.8 x 10^{-6} for 25–1500ºC
	5.82–7.4 x 10^{-6} for 25–2000ºC
Zirconium Monocarbide (ZrC)	6.10x 10^{-6} for 25–500ºC
	6.65x 10^{-6} for 25–800ºC
	6.56x 10^{-6} for 25–1000ºC
	7.06x 10^{-6} for 25–1500ºC
	7.65x 10^{-6} for 25–650ºC
	6.10–6.73 x 10^{-6} for 25–650ºC
	6.32x 10^{-6} for 0–750ºC
	6.46–6.66x 10^{-6} for 0–1000ºC
	6.68x 10^{-6} for 0–1275ºC
	6.83x 10^{-6} for 0–1525ºC
	6.98x 10^{-6} for 0–1775ºC
	9.0x 10^{-6} for 1000–2000ºC
NITRIDES	
Aluminum Nitride (AlN)	4.03 x 10^{-6} for 25 to 200ºC
	4.84 x 10^{-6} for 25 to 500ºC
	4.83 x 10^{-6} for 25 to 600ºC
	5.54–5.64 x 10^{-6} for 25 to 1000ºC
	6.09 x 10^{-6} for 25 to 1350ºC
Boron Nitride (BN)	12.2 x 10^{-6} for 25 to 500ºC
	13.3 x 10^{-6} for 25 to 1000ºC
parallel to c axis	10.15 x 10^{-6} for 25 to 350ºC

Thermal Expansion of Selected Ceramics (Continued)
Listed by Ceramic

Ceramic	Thermal Expansion $(°C^{-1})$
	8.06×10^{-6} for 25 to 700°C
	7.15×10^{-6} for 25 to 1000°C
parallel to a axis	0.59×10^{-6} for 25 to 350°C
	0.89×10^{-6} for 25 to 700°C
	0.77×10^{-6} for 25 to 1000°C
Titanium Mononitride (TiN)	9.35×10^{-6}
Trisilicon tetranitride (Si_3N_4)	2.11×10^{-6} for 25 to 500°C
	2.87×10^{-6} for 25 to 1000°C
	3.66×10^{-6} for 25 to 1500°C
(hot pressed)	$3–3.9 \times 10^{-6}$ for 20 to 1000°C
(sintered)	3.5×10^{-6} for 20 to 1000°C
(reaction sintered)	2.9×10^{-6} for 20 to 1000°C
(pressureless sintered)	3.7×10^{-6} for 40 to 1000°C
Zirconium Mononitride (TiN)	6.13×10^{-6} for 20–450°C
	7.03×10^{-6} for 20–680°C
OXIDES	
Aluminum Oxide (Al_2O_3)	
parallel to c axis	1.95×10^{-6} for 0 to –273°C
	3.01×10^{-6} for 0 to –173°C
	4.39×10^{-6} for 0 to –73°C
	5.31×10^{-6} for 0 to 27°C
	6.26×10^{-6} for 0 to 127°C
	6.86×10^{-6} for 0 to 227°C
	7.31×10^{-6} for 0 to 327°C

Thermal Expansion of Selected Ceramics (Continued)
Listed by Ceramic

Ceramic	Thermal Expansion $(°C^{-1})$
	7.68×10^{-6} for 0 to 427°C
	7.96×10^{-6} for 0 to 527°C
	8.19×10^{-6} for 0 to 627°C
	8.38×10^{-6} for 0 to 727°C
	8.52×10^{-6} for 0 to 827°C
	8.65×10^{-6} for 0 to 927°C
	8.75×10^{-6} for 0 to 1027°C
	8.84×10^{-6} for 0 to 1127°C
	8.92×10^{-6} for 0 to 1227°C
	8.98×10^{-6} for 0 to 1327°C
	9.02×10^{-6} for 0 to 1427°C
	9.08×10^{-6} for 0 to 1527°C
	9.13×10^{-6} for 0 to 1627°C
	9.18×10^{-6} for 0 to 1727°C
Aluminum Oxide (Al_2O_3) (single crystal)	
perpendicular to c axis	1.65×10^{-6} for 0 to –273°C
	2.55×10^{-6} for 0 to –173°C
	3.75×10^{-6} for 0 to –73°C
	4.78×10^{-6} for 0 to 27°C
	5.51×10^{-6} for 0 to 127°C
	6.10×10^{-6} for 0 to 227°C
	6.52×10^{-6} for 0 to 327°C
	6.88×10^{-6} for 0 to 427°C
	7.15×10^{-6} for 0 to 527°C
	7.35×10^{-6} for 0 to 627°C
	7.53×10^{-6} for 0 to 727°C
	7.67×10^{-6} for 0 to 827°C

Thermal Expansion of Selected Ceramics (Continued)
Listed by Ceramic

Ceramic	Thermal Expansion ($°C^{-1}$)
	7.80×10^{-6} for 0 to 927°C
	7.88×10^{-6} for 0 to 1027°C
	7.96×10^{-6} for 0 to 1127°C
	8.05×10^{-6} for 0 to 1227°C
	8.12×10^{-6} for 0 to 1327°C
	8.16×10^{-6} for 0 to 1427°C
	8.20×10^{-6} for 0 to 1527°C
	8.26×10^{-6} for 0 to 1627°C
	8.30×10^{-6} for 0 to 1727°C
Aluminum Oxide (Al_2O_3)	
(polycrystalline)	1.89×10^{-6} for 0 to −273°C
	2.91×10^{-6} for 0 to −173°C
	4.10×10^{-6} for 0 to −73°C
	5.60×10^{-6} for 0 to 27°C
	6.03×10^{-6} for 0 to 127°C
	6.55×10^{-6} for 0 to 227°C
	6.93×10^{-6} for 0 to 327°C
	7.24×10^{-6} for 0 to 427°C
	7.50×10^{-6} for 0 to 527°C
	7.69×10^{-6} for 0 to 627°C
	7.83×10^{-6} for 0 to 727°C
	7.97×10^{-6} for 0 to 827°C
	8.08×10^{-6} for 0 to 927°C
	8.18×10^{-6} for 0 to 1027°C
	8.25×10^{-6} for 0 to 1127°C
	8.32×10^{-6} for 0 to 1227°C
	8.39×10^{-6} for 0 to 1327°C

Thermal Expansion of Selected Ceramics (Continued)
Listed by Ceramic

Ceramic	Thermal Expansion ($°C^{-1}$)
	8.45×10^{-6} for 0 to 1427°C
	8.49×10^{-6} for 0 to 1527°C
	8.53×10^{-6} for 0 to 1627°C
	8.58×10^{-6} for 0 to 1727°C
Beryllium Oxide (BeO) (single crystal) parallel to c axis	6.3×10^{-6} for 28 to 252°C
	6.7×10^{-6} for 28 to 474°C
	7.8×10^{-6} for 28 to 749°C
	8.2×10^{-6} for 28 to 872°C
	8.9×10^{-6} for 28 to 1132°C
Beryllium Oxide (BeO) (single crystal) perpendicular to c axis	7.1×10^{-6} for 28 to 252°C
	7.8×10^{-6} for 28 to 474°C
	8.5×10^{-6} for 28 to 749°C
	9.2×10^{-6} for 28 to 872°C
	9.9×10^{-6} for 28 to 1132°C
Beryllium Oxide (BeO) (single crystal) average for (2a+c)/3	6.83×10^{-6} for 28 to 252°C
	7.43×10^{-6} for 28 to 474°C
	8.27×10^{-6} for 28 to 749°C
	8.87×10^{-6} for 28 to 872°C
	9.57×10^{-6} for 28 to 1132°C

Thermal Expansion of Selected Ceramics (Continued)
Listed by Ceramic

Ceramic	Thermal Expansion ($°C^{-1}$)
Beryllium Oxide (BeO) (polycrystalline)	2.4×10^{-6} for 25–200°C
	$6.3–6.4 \times 10^{-6}$ for 25–300°C
	7.59×10^{-6} for 25–500°C
	$8.4–8.5 \times 10^{-6}$ for 25–800°C
	9.03×10^{-6} for 25–1000°C
	9.18×10^{-6} for 25–1250°C
	10.3×10^{-6} for 25–1500°C
	11.1×10^{-6} for 25–2000°C
	9.40×10^{-6} for 500–1200°C
Cerium Dioxide (CeO_2)	8.22×10^{-6} for 25–500°C
	8.92×10^{-6} for 25–1000°C
	$8.5 + 0.54T$ for 0–1000°C
Dichromium Trioxide (Cr_2O_3)	8.43×10^{-6} for 25–500°C
	8.62×10^{-6} for 25–1000°C
	8.82×10^{-6} for 25–1500°C
	9.55×10^{-6} for 20–1400°C

Hafnium Dioxide (HfO_2) (monoclinic single crystal)

parallel to a axis	
	6.8×10^{-6} for 28–262°C
	6.2×10^{-6} for 28–494°C
	6.7×10^{-6} for 28–697°C
	7.5×10^{-6} for 28–903°C
	7.9×10^{-6} for 28–1098°C

Thermal Expansion of Selected Ceramics (Continued)
Listed by Ceramic

Ceramic	Thermal Expansion ($°C^{-1}$)
parallel to b axis	0 for 28–262°C
	0.9×10^{-6} for 28–494°C
	1.3×10^{-6} for 28–697°C
	1.4×10^{-6} for 28–903°C
	2.1×10^{-6} for 28–1098°C
parallel to c axis	11×10^{-6} for 28–262°C
	11.4×10^{-6} for 28–494°C
	10.8×10^{-6} for 28–697°C
	11.9×10^{-6} for 28–903°C
	12.1×10^{-6} for 28–1098°C
Hafnium Dioxide (HfO_2) (monoclinic polycrystalline)	5.47×10^{-6} for 25–500°C
	5.85×10^{-6} for 25–1000°C
	5.8×10^{-6} for 25–1300°C
	6.30×10^{-6} for 25–1500°C
	6.45×10^{-6} for 20–1700°C
Hafnium Dioxide (HfO_2) (tetragonal polycrystalline)	1.31×10^{-6} for 25–1700°C
	3.03×10^{-6} for 25–2000°C

Thermal Expansion of Selected Ceramics (Continued)
Listed by Ceramic

Ceramic	Thermal Expansion ($°C^{-1}$)
Magnesium Oxide (MgO)	12.83×10^{-6} for 25–500°C
	13.63×10^{-6} for 25–1000°C
	15.11×10^{-6} for 25–1500°C
	15.89×10^{-6} for 25–1800°C
	14.0×10^{-6} for 20–1400°C
	$14.2–14.9 \times 10^{-6}$ for 20–1700°C
	13.3×10^{-6} for 20–1700°C
	13.90×10^{-6} for 0–1000°C
	14.46×10^{-6} for 0–1200°C
	15.06×10^{-6} for 0–1400°C
Silicon Dioxide (SiO_2)	
α quartz	19.35×10^{-6} for 25–500°C
	22.2×10^{-6} for 25–575°C
β quartz	27.8×10^{-6} for 25–575°C
	14.58×10^{-6} for 25–1000°C
α tridymite	18.5×10^{-6} for 25–117°C
$β_1$ tridymite	25.0×10^{-6} for 25–117°C
	27.5×10^{-6} for 25–163°C
$β_2$ tridymite	31.9×10^{-6} for 25–163°C
	19.35×10^{-6} for 25–500°C
	10.45×10^{-6} for 25–1000°C

Thermal Expansion of Selected Ceramics (Continued)
Listed by Ceramic

Ceramic	Thermal Expansion ($°C^{-1}$)
Vitreous	0.527×10^{-6} for 25–500°C
	0.564×10^{-6} for 25–1000°C
	0.5×10^{-6} for 20–1250°C
Thorium Dioxide (ThO_2)	3.67×10^{-6} for 0 to –273°C
	5.32×10^{-6} for 0 to –173°C
	6.47×10^{-6} for 0 to –73°C
	8.10×10^{-6} for 0 to 27°C
	8.06×10^{-6} for 0 to 127°C
	8.31×10^{-6} for 0 to 227°C
	8.53×10^{-6} for 0 to 327°C
	8.71×10^{-6} for 0 to 427°C
	8.87×10^{-6} for 0 to 527°C
	9.00×10^{-6} for 0 to 627°C
	9.14×10^{-6} for 0 to 727°C
	9.24×10^{-6} for 0 to 827°C
	9.34×10^{-6} for 0 to 927°C
	9.42×10^{-6} for 0 to 1027°C
	9.53×10^{-6} for 0 to 1127°C
	9.60×10^{-6} for 0 to 1227°C
	9.68×10^{-6} for 0 to 1327°C
	9.76×10^{-6} for 0 to 1427°C
	9.83×10^{-6} for 0 to 1527°C
	9.91×10^{-6} for 0 to 1627°C
	9.97×10^{-6} for 0 to 1727°C

Thermal Expansion of Selected Ceramics (Continued)
Listed by Ceramic

Ceramic	Thermal Expansion ($^\circ C^{-1}$)
	8.63×10^{-6} for 25 to 500°C
	9.44×10^{-6} for 25 to 1000°C
	10.17×10^{-6} for 25 to 1500°C
	10.43×10^{-6} for 25 to 1700°C
	9.55×10^{-6} for 20 to 800°C
	9.55×10^{-6} for 20 to 1400°C
	7.8×10^{-6} for 27 to 223°C
	8.7×10^{-6} for 27 to 498°C
	8.9×10^{-6} for 27 to 755°C
	9.2×10^{-6} for 27 to 994°C
	9.1×10^{-6} for 27 to 1087°C
	8.96×10^{-6} for 0 to 1000°C
	9.35×10^{-6} for 0 to 1200°C
	9.84×10^{-6} for 0 to 1400°C
α_l (linear expansion coefficient)	$0.6216 \times 10^{-5} + 3.541 \times 10^{-9}T - 0.1124T^{-2}$ from 298–1073K
α_v (volume expansion coefficient)	$1.85 \times 10^{-5} + 10.96 \times 10^{-9}T - 0.3375T^{-2}$ from 298–1073K
Titanium Oxide (TiO$_2$) (polycrystalline)	8.22×10^{-6} for 25–500°C
	8.83×10^{-6} for 25–1000°C
	9.50×10^{-6} for 25–1500°C
	7.8×10^{-6} for 20–600°C
	8.98×10^{-6} for 0–1000°C

Thermal Expansion of Selected Ceramics (Continued)
Listed by Ceramic

Ceramic	Thermal Expansion $(°C^{-1})$
Titanium Oxide (TiO_2) (single crystal) parallel to c axis	9.8×10^{-6} for 26 to 240°C
	10.5×10^{-6} for 26 to 455°C
	10.6×10^{-6} for 26 to 670°C
	10.5×10^{-6} for 26 to 940°C
	10.8×10^{-6} for 26 to 1110°C
Titanium Oxide (TiO_2) (single crystal)) perpendicular to a axis	7.9×10^{-6} for 26 to 240°C
	8.2×10^{-6} for 26 to 455°C
	8.1×10^{-6} for 26 to 670°C
	8.2×10^{-6} for 26 to 940°C
	8.3×10^{-6} for 26 to 1110°C
Titanium Oxide (TiO_2) (single crystal)) average for (2a+c)/3	8.53×10^{-6} for 26 to 240°C
	8.97×10^{-6} for 26 to 455°C
	8.93×10^{-6} for 26 to 670°C
	8.97×10^{-6} for 26 to 940°C
	9.13×10^{-6} for 26 to 1110°C
Uranium Dioxide (UO_2)	9.47×10^{-6} for 25 to 500°C
	11.19×10^{-6} for 25 to 1000°C
	12.19×10^{-6} for 25 to 1200°C
	11.15×10^{-6} for 25 to 1750°C
	9.18×10^{-6} for 27 to 400°C
(heating)	9.07×10^{-6} for 27 to 400°C
(cooling)	9.28×10^{-6} for 27 to 400°C
	10.8×10^{-6} for 400 to 800°C
(heating)	11.1×10^{-6} for 400 to 800°C

Thermal Expansion of Selected Ceramics (Continued)
Listed by Ceramic

Ceramic	Thermal Expansion $(°C^{-1})$
(cooling)	10.8×10^{-6} for 400 to 800°C
	12.6×10^{-6} for 800 to 1250°C
(heating)	13.0×10^{-6} for 800 to 1200°C
(cooling)	12.9×10^{-6} for 800 to 1200°C
Zirconium Oxide (ZrO_2) (monoclinic)	6.53×10^{-6} for 25 to 500°C
	7.59×10^{-6} for 25 to 1000°C
	7.72×10^{-6} for 25 to 1050°C
	8.0×10^{-6} for 25 to 1080°C
Zirconium Oxide (ZrO_2) (tetragonal)	-21.7×10^{-6} for 25 to 1050°C
	-11.11×10^{-6} for 25 to 1500°C
	-9.53×10^{-6} for 25 to 1600°C
	4.0×10^{-6} for 0 to 500°C
	10.5×10^{-6} for 0 to 1000°C
	10.52×10^{-6} for 0 to 1000°C (MgO)
	10.6×10^{-6} for 0 to 1200°C (CaO)
	5.0×10^{-6} for 0 to 1400°C
	11.0×10^{-6} for 0 to 1500°C
	$5.5–5.58 \times 10^{-6}$ for 20 to 1200°C
	7.2×10^{-6} for −10 to 1000°C
	8.64×10^{-6} for −20 to 600°C

Thermal Expansion of Selected Ceramics (Continued)
Listed by Ceramic

Ceramic	Thermal Expansion ($^\circ C^{-1}$)

Zirconium Oxide (ZrO$_2$) (tetragonal, single crystal)

parallel to a axis

8.4×10^{-6} for 27 to 264°C

7.5×10^{-6} for 27 to 504°C

6.8×10^{-6} for 27 to 759°C

7.8×10^{-6} for 27 to 964°C

8.7×10^{-6} for 27 to 1110°C

Zirconium Oxide (ZrO$_2$) (tetragonal, single crystal)

parallel to b axis

3×10^{-6} for 27 to 264°C

2×10^{-6} for 27 to 504°C

1.1×10^{-6} for 27 to 759°C

1.5×10^{-6} for 27 to 964°C

1.9×10^{-6} for 27 to 1110°C

Zirconium Oxide (ZrO$_2$) (tetragonal, single crystal)

parallel to c axis

14×10^{-6} for 27 to 264°C

13×10^{-6} for 27 to 504°C

11.9×10^{-6} for 27 to 759°C

12.8×10^{-6} for 27 to 964°C

13.6×10^{-6} for 27 to 1110°C

Thermal Expansion of Selected Ceramics (Continued)
Listed by Ceramic

Ceramic	Thermal Expansion $(°C^{-1})$
Cordierite $(2MgO\ 2Al_2O_3\ 5SiO_2)$	
$(\rho=2.3g/cm^3)$	2.3×10^{-6} for 25 to 400°C
$(\rho=2.3g/cm^3)$	3.3×10^{-6} for 25 to 700°C
$(\rho=2.3g/cm^3)$	3.7×10^{-6} for 25 to 900°C
$(\rho=2.1g/cm^3)$	2.2×10^{-6} for 25 to 400°C
$(\rho=2.1g/cm^3)$	2.8×10^{-6} for 25 to 700°C
$(\rho=2.1g/cm^3)$	2.8×10^{-6} for 25 to 900°C
$(\rho=1.8g/cm^3)$	0.6×10^{-6} for 25 to 400°C
$(\rho=1.8g/cm^3)$	1.5×10^{-6} for 25 to 700°C
$(\rho=1.8g/cm^3)$	1.7×10^{-6} for 25 to 900°C
$(\rho=2.51g/cm^3)$	2.7×10^{-6} for 25 to 1100°C
(glass)	$3.7–3.8 \times 10^{-6}$ for 25 to 900°C
Mullite $(3Al_2O_3\ 2SiO_2)$	4.5×10^{-6} for 20 to 1325°C
	4.63×10^{-6} for 25 to 500°C
	5.0×10^{-6} for 25 to 800°C
	5.13×10^{-6} for 25 to 1000°C
	5.62×10^{-6} for 20 to 1500°C
Sillimanite $(Al_2O_3\ SiO_2)$	6.58×10^{-6} at 20°C
Spinel $(Al_2O_3\ MgO)$	7.79×10^{-6} for 25 to 500°C
	8.41×10^{-6} for 25 to 1000°C
	9.17×10^{-6} for 25 to 1500°C
	9.0×10^{-6} for 20 to 1250°C

Thermal Expansion of Selected Ceramics (Continued)
Listed by Ceramic

Ceramic	Thermal Expansion ($°C^{-1}$)
Zircon (SiO_2 ZrO_2)	5.5×10^{-6} for 20 to 1200°C
	3.79×10^{-6} for 25 to 500°C
	4.62×10^{-6} for 25 to 1000°C
	5.63×10^{-6} for 20 to 1500°C
SILICIDES	
Molybdenum Disilicide ($MoSi_2$)	7.79×10^{-6} for 25–500°C
	8.51×10^{-6} for 25–1000°C
	9.00–9.18×10^{-6} for 25–1500°C
	8.41×10^{-6} for 0–1000°C
	8.56×10^{-6} for 0–1400°C
Tungsten Disilicide (WSi_2)	7.79×10^{-6} for 25–500°C
	8.31×10^{-6} for 25–1000°C
	8.21×10^{-6} for 0–1000°C
	8.81×10^{-6} for 0–1400°C

Source: Data compiled by J.S. Park from *No. 1 Materials Index*, Peter T.B. Shaffer, Plenum Press, New York, (1964); *Smithells Metals Reference Book*, Eric A. Brandes, ed., in association with Fulmer Research Institute Ltd. 6th ed. London, Butterworths, Boston, (1983); and *Ceramic Abstracts*, American Ceramic Society (1986-1991)

THERMAL EXPANSION COEFFICIENTS FOR MATERIALS USED IN INTEGRATED CIRCUITS

Material	Temperature, K					
	300	400	500	600	700	800
Aluminum	23.2	24.9	26.4	28.3	30.7	33.8
Beryllium oxide	4.7	–	6.0	–	7.0	–
Copper	16.8	17.7	18.3	18.9	19.4	20.0
Germanium	5.7	6.2	6.5	6.7	6.9	7.2
Gold	14.1	14.5	15.	15.4	15.9	16.5
Indium	31.9	38.5	–	–	–	–
Lead	28.9	29.8	32.1	–	–	–
Molybdenum	5.0	5.2	5.3	5.4	5.5	5.7
Nickel	12.7	13.8	15.2	17.2	16.4	16.8
Platinum	8.9	9.2	9.5	9.7	10.0	10.2
Silicon	2.5	3.1	3.5	3.8	4.1	4.3
Silver	19.2	20.0	20.6	21.4	22.3	23.4
Tantalum	6.5	6.6	6.8	6.9	7.0	7.1
Tin	21.2	24.2	27.5	–	–	–
Tungsten	4.5	4.6	4.6	4.7	4.7	4.8
Vitreous silica	.42	.56	.56	.55	.54	.54

Thermal Expansion Coefficients for Materials used in Integrated Circuits (Continued)

Material	Coefficient Range	Temperature Range (°C)
Aluminum oxide ceramic	6.0-7.0	25-300
Brass	17.7-21.2	25-300
Kanthal A	13.9-15.1	20-900
Kovar	5.0	25-300
Pyrex glass	3.2	25-300
Pyroceram (#9608)	420	25-300
Pyroceram cement (Vitreous #45)	4	0-300
Pyroceram cement (Devitrified)	2.4	25-300
Pyroceram cement (#89, #95)	8-10	–
Silicon carbide	4.8	0-1,000
Silicon nitride (a)	2.9	25-1,000
Silicon nitride (β)	2.25	25-1,000
Solder glass (Kimble CV-101)	809	0-300

Coefficient of Linear Thermal Expansion of Selected Materials per K

Note: Multiply all values by 10^{-6}.

Source: From Beadles, R. L., Interconnections and Encapsulation, *Integrated Silicon Device Technology,* Vol. 14, Research Triangle Institute, Research Triangle Park, N. C., 1967. in *CRC Handbook of Materials Science*, Charles T. Lynch, Ed., CRC Press, Cleveland, (1974).

THERMAL EXPANSION OF SELECTED POLYMERS
Listed by Polymer

Polymer	Thermal Expansion Coefficient ASTM D696 ($°F^{-1}$)
ABS Resins; Molded, Extruded	
Medium impact	$3.2—4.8 \times 10^{-6}$
High impact	$5.5—6.0 \times 10^{-6}$
Very high impact	$5.0—6.0 \times 10^{-6}$
Low temperature impact	$5.0—6.0 \times 10^{-6}$
Heat resistant	$3.0—4.0 \times 10^{-6}$
Acrylics; Cast, Molded, Extruded	
Cast Resin Sheets, Rods:	
General purpose, type I	4.5×10^{-6}
General purpose, type II	4.5×10^{-6}
Moldings:	
Grades 5, 6, 8	$3—4 \times 10^{-6}$
High impact grade	$4—6 \times 10^{-6}$
Thermoset Carbonate	
Allyl diglycol carbonate	6×10^{-5}
Alkyds; Molded	
Putty (encapsulating)	1.3×10^{-5}
Rope (general purpose)	1.3×10^{-5}
Granular (high speed molding)	1.3×10^{-5}
Glass reinforced (heavy duty parts)	1.3×10^{-5}
Cellulose Acetate; Molded, Extruded	
ASTM Grade:	
H6—1	$4.4—9.0 \times 10^{-5}$

Thermal Expansion of Selected Selected Polymers (Continued)
Listed by Polymer

Polymer	Thermal Expansion Coefficient ASTM D696 ($°F^{-1}$)
H4—1	4.4—9.0×10^{-5}
H2—1	4.4—9.0×10^{-5}
MH—1, MH—2	4.4—9.0×10^{-5}
MS—1, MS—2	4.4—9.0×10^{-5}
S2—1	4.4—9.0×10^{-5}

Cellulose Acetate Butyrate; Molded, Extruded
ASTM Grade:

H4	6—9×10^{-5}
MH	6—9×10^{-5}
S2	6—9×10^{-5}

Cellusose Acetate Propionate; Molded, Extruded
ASTM Grade:

1	6—9×10^{-5}
3	6—9×10^{-5}
6	6—9×10^{-5}

Chlorinated Polymers

Chlorinated polyether	6.6×10^{-6}
Chlorinated polyvinyl chloride	4.4×10^{-6}

Polycarbonates

Polycarbonate	3.75×10^{-6}
Polycarbonate (40% glass fiber reinforced)	1.0—1.1×10^{-6}

Thermal Expansion of Selected Selected Polymers (Continued)
Listed by Polymer

Polymer	Thermal Expansion Coefficient ASTM D696 ($°F^{-1}$)
Diallyl Phthalates; Molded	
Orlon filled	5.0×10^{-5}
Dacron filled	5.2×10^{-5}
Asbestos filled	4.0×10^{-5}
Glass fiber filled	$2.2.—2.6 \times 10^{-5}$
Fluorocarbons; Molded,Extruded	
Polytrifluoro chloroethylene (PTFCE)	3.88×10^{-5}
Polytetrafluoroethylene (PTFE)	55×10^{-5}
Ceramic reinforced (PTFE)	$1.7—2.0 \times 10^{-5}$
Fluorinated ethylene propylene(FEP)	$8.3—10.5 \times 10^{-5}$
Polyvinylidene— fluoride (PVDF)	8.5×10^{-5}
Epoxies; Cast, Molded, Reinforced	
Standard epoxies (diglycidyl ethers of bisphenol A)	
Cast rigid	3.3×10^{-5}
Cast flexible	$3—5 \times 10^{-5}$
Molded	$1—2 \times 10^{-5}$
General purpose glass cloth laminate	$3.3—4.8 \times 10^{-6}$
High strength laminate	$3.3—4.8 \times 10^{-6}$
Filament wound composite	$2—6 \times 10^{-5}$
Epoxies—Molded, Extruded	
High performance resins (cycloaliphatic diepoxides)	
Molded	$1.7—2.2 \times 10^{-6}$
Epoxy novolacs	
Cast, rigid	$1.6—3.0 \times 10^{-6}$

Thermal Expansion of Selected Selected Polymers (Continued)
Listed by Polymer

Polymer	Thermal Expansion Coefficient ASTM D696 ($°F^{-1}$)
Melamines; Molded	
Filler & type	
Cellulose electrical	$1.11—2.78 \times 10^{-5}$
Glass fiber	0.82×10^{-5}
Nylons; Molded, Extruded	
Type 6	
General purpose	4.8×10^{-5}
Glass fiber (30%) reinforced	1.2×10^{-5}
Cast	4.4×10^{-5}
Type 11	5.5×10^{-5}
Type 12	7.2×10^{-5}
6/6 Nylon	
General purpose molding	$1.69—1.7 \times 10^{-5}$
Glass fiber reinforced	$1.5, 3.3 \times 10^{-5}$
General purpose extrusion	1.7×10^{-5}
6/10 Nylon	
General purpose	1.5×10^{-5}
Glass fiber (30%) reinforced	3.5×10^{-5}
Phenolics; Molded	
Type and filler	
General: woodflour and flock	$1.66—2.50 \times 10^{-5}$
Shock: paper, flock, or pulp	$1.6—2.3 \times 10^{-5}$

Thermal Expansion of Selected Selected Polymers (Continued)
Listed by Polymer

Polymer	Thermal Expansion Coefficient ASTM D696 ($°F^{-1}$)
High shock: chopped fabric or cord	$1.60—2.22 \times 10^{-5}$
Very high shock: glass fiber	0.88×10^{-5}
Phenolics: Molded	
Rubber phenolic—woodflour or flock	$0.83—2.20 \times 10^{-5}$
Rubber phenolic—chopped fabric	1.7×10^{-5}
Rubber phenolic—asbestos	2.2×10^{-5}
ABS–Polycarbonate Alloy	6.12×10^{-5}
PVC–Acrylic Alloy	
PVC–acrylic sheet	3.5×10^{-5}
Polymides	
Unreinforced	2.5×10^{-6}
Unreinforced 2nd value	$3.0—4.5 \times 10^{-6}$
Glass reinforced	0.8×10^{-6}
Homopolymer:	
Standard	4.5×10^{-5}
20% glass reinforced	$2.0—4.5 \times 10^{-5}$
22% TFE reinforced	4.5×10^{-5}
Copolymer:	
Standard	4.7×10^{-5}
25% glass reinforced	$2.2—4.7 \times 10^{-5}$
High flow	4.7×10^{-5}

Thermal Expansion of Selected Selected Polymers (Continued)
Listed by Polymer

Polymer	Thermal Expansion Coefficient ASTM D696 ($°F^{-1}$)
Polyester; Thermoplastic	
Injection Moldings:	
General purpose grade	5.3×10^{-5}
Glass reinforced grades	$2.7—3.3 \times 10^{-5}$
Glass reinforced self extinguishing	3.5×10^{-5}
General purpose grade	$4.9—13.0 \times 10^{-5}$
Polyesters: Thermosets	
Cast polyyester	
Rigid	$3.9—5.6 \times 10^{-5}$
Reinforced polyester moldings	
High strength (glass fibers)	$13—19 \times 10^{-6}$
Phenylene Oxides	
SE—100	3.8×10^{-5}
SE—1	3.3×10^{-5}
Glass fiber reinforced	$2.0,1.4 \times 10^{-5}$
Phenylene oxides (Noryl)	
Standard	3.1×10^{-5}
Glass fiber reinforced	$1.6,1.2 \times 10^{-5}$
Polyarylsulfone	2.6×10^{-5}
Polypropylene:	
General purpose	$3.8—5.8 \times 10^{-5}$
High impact	$4.0—5.9 \times 10^{-5}$
Asbestos filled	$2—3 \times 10^{-5}$

Thermal Expansion of Selected Selected Polymers (Continued)
Listed by Polymer

Polymer	Thermal Expansion Coefficient ASTM D696 ($°F^{-1}$)
Glass reinforced	$1.6—2.4 \times 10^{-5}$
Polyphenylene sulfide:	
Standard	$3.0—4.9 \times 10^{-5}$
40% glass reinforced	4×10^{-5}
Polyethylenes; Molded, Extruded	
Type I—lower density (0.910–0.925)	
Melt index 0.3—3.6	$8.9—11.0 \times 10^{-5}$
Melt index 6—26	$8.9—11.0 \times 10^{-5}$
Melt index 200	11×10^{-5}
Type II—medium density (0.926—0.940)	
Melt index 20	$8.3—16.7 \times 10^{-5}$
Melt index 1.0—1.9	$8.3—16.7 \times 10^{-5}$
Type III—higher density (0.941—0.965)	
Melt index 0.2—0.9	$8.3—16.7 \times 10^{-5}$
Melt Melt index 0.1—12.0	$8.3—16.7 \times 10^{-5}$
Melt index 1.5—15	$8.3—16.7 \times 10^{-5}$
Polystyrenes; Molded	
Polystyrenes	
General purpose	$3.3—4.8 \times 10^{-5}$
Medium impact	$3.3—4.7 \times 10^{-5}$
High impact	$2.2—5.6 \times 10^{-5}$
Glass fiber -30% reinforced	1.8×10^{-5}
Styrene acrylonitrile (SAN)	$3.6—3.7 \times 10^{-5}$
Glass fiber (30%) reinforced SAN	1.6×10^{-5}

Thermal Expansion of Selected Selected Polymers (Continued)
Listed by Polymer

Polymer	Thermal Expansion Coefficient ASTM D696 ($°F^{-1}$)
Polyvinyl Chloride And Copolymers; Molded, Extruded	
Rigid—normal impact	$2.8—3.3 \times 10^{-5}$
Vinylidene chloride	8.78×10^{-5}
Silicones; Molded, Laminated	
Fibrous (glass) reinforced silicones	$3.17—3.23 \times 10^{-5}$
Granular (silica) reinforced silicones	$2.5—5.0 \times 10^{-5}$
Ureas; Molded	
Alpha—cellulose filled (ASTM Type l)	$1.22—1.50 \times 10^{-5}$

Source: data compiled by J.S. Park from Charles T. Lynch, *CRC Handbook of Materials Science*, Vol. 3, CRC Press, Boca Raton, Florida, 1975 and *Engineered Materials Handbook*, Vol.2, Engineering Plastics, ASM International, Metals Park, Ohio, 1988.

VALUES OF THE ERROR FUNCTION

z	erf (z)	z	erf (z)
0.00	0.0000	0.70	0.6778
0.01	0.0113	0.75	0.7112
0.02	0.0226	0.80	0.7421
0.03	0.0338	0.85	0.7707
0.04	0.0451	0.90	0.7969
0.05	0.0564	0.95	0.8209
0.10	0.1125	1.00	0.8427
0.15	0.1680	1.10	0.8802
0.20	0.2227	1.20	0.9103
0.25	0.2763	1.30	0.9340
0.30	0.3286	1.40	0.9523
0.35	0.3794	1.50	0.9661
0.40	0.4284	1.60	0.9763
0.45	0.4755	1.70	0.9838
0.50	0.5205	1.80	0.9891
0.55	0.5633	1.90	0.9928
0.60	0.6039	2.00	0.9953
0.65	0.6420		

Source: From: Handbook of Mathematical Functions, M. Abramowitz and I. A. Stegun, eds., National Bureau of Standards, Applied Mathematics Series 55, Washington, D.C., 1972.

DIFFUSION IN SELECTED METALLIC SYSTEMS

Metal ——— Tracer	Crystalline Form	Purity (%)	Temperature Range (°C)	Activation energy, Q (kcal · mol⁻¹)	Frequency factor, D_0 (cm² · s⁻¹)
Aluminum					
Ag^{110}	S	99.999	371–655	27.83	0.118
Al^{27}	S		450–650	34.0	1.71
Au^{198}	S	99.999	423–609	27.0	0.077
Cd^{115}	S	99.999	441–631	29.7	1.04
Ce^{141}	P	99.995	450–630	26.60	1.9×10^{-6}
Co^{60}	S	99.999	369–655	27.79	0.131
Cr^{51}	S	99.999	422–654	41.74	464
Cu^{64}	S	99.999	433–652	32.27	0.647
Fe^{59}	S	99.99	550–636	46.0	135
Ga^{72}	S	99.999	406–652	29.24	0.49
Ge^{71}	S	99.999	401–653	28.98	0.481
In^{114}	P	99.99	400–600	27.6	0.123
La^{140}	P	99.995	500–630	27.0	1.4×10^{-6}
Mn^{54}	P	99.99	450–650	28.8	0.22
Mo^{99}	P	99.995	400–630	13.1	1.04×10^{-9}
Nb^{95}	P	99.95	350–480	19.65	1.66×10^{-7}
Nd^{147}	P	99.995	450–630	25.0	4.8×10^{-7}
Ni^{63}	P	99.99	360–630	15.7	2.9×10^{-8}
Pd^{103}	P	99.995	400–630	20.2	1.92×10^{-7}
Pr^{142}	P	99.995	520–630	23.87	3.58×10^{-7}
Sb^{124}	P		448–620	29.1	0.09
Sm^{153}	P	99.995	450–630	22.88	3.45×10^{-7}
Sn^{113}	P		400–600	28.5	0.245
V^{48}	P	99.995	400–630	19.6	6.05×10^{-8}
Zn^{65}	S	99.999	357–653	28.86	0.259

Diffusion in Selected Metallic Systems (Continued)

Metal ——— Tracer	Crystalline Form	Purity (%)	Temperature Range (°C)	Activation energy, Q (kcal · mol^{-1})	Frequency factor, D_0 (cm^2 · s^{-1})
Beryllium					
Ag^{110}	S⊥c	99.75	650–900	43.2	1.76
Ag^{110}	S‖c	99.75	650–900	39.3	0.43
Be^{7}	S⊥c	99.75	565–1065	37.6	0.52
Be^{7}	S‖c	99.75	565–1065	39.4	0.62
Fe^{59}	S	99.75	700–1076	51.6	0.67
Ni^{63}	P		800–1250	58.0	0.2
Cadmium					
Ag^{110}	S	99.99	180–300	25.4	2.21
Cd^{115}	S	99.95	110–283	19.3	0.14
Zn^{65}	S	99.99	180–300	19.0	0.0016
Calcium					
C^{14}		99.95	550–800	29.8	3.2 x 10^{-5}
Ca^{45}		99.95	500–800	38.5	8.3
Fe^{59}		99.95	500–800	23.3	2.7 x 10^{-3}
Ni^{63}		99.95	550–800	28.9	1.0 x 10^{-6}
U^{235}		99.95	500–700	34.8	1.1 x 10^{-5}
Carbon					
Ag^{110}	⊥c		750–1050	64.3	9280
C^{14}			2000–2200	163	5

Diffusion in Selected Metallic Systems (Continued)

Metal ——— Tracer	Crystalline Form	Purity (%)	Temperature Range (°C)	Activation energy, Q (kcal $\cdot$ mol^{-1})	Frequency factor, D_0 (cm^2 $\cdot$ s^{-1})
Ni^{63}	$\perp$c		540–920	47.2	102
Ni^{63}	‖c		750–1060	53.3	2.2
Th^{228}	$\perp$c		1400–2200	145.4	1.33 x 10^{-5}
Th^{228}	‖c		1800–2200	114.7	2.48
U^{232}	$\perp$c		140~2200	115.0	6760
U^{232}	‖c		1400 1820	129.5	385

Chromium

C^{14}	P		120~1500	26.5	9.0 x 10^{-3}
Cr^{51}	P	99.98	1030–1545	73.7	0.2
Fe^{59}	P	99.8	980–1420	79.3	0.47
Mo^{99}	P		1100–1420	58.0	2.7 x 10^{-3}

Cobalt

C^{14}	P	99.82	600–1400	34.0	0.21
Co^{60}	P	99.9	1100–1405	67.7	0.83
Fe^{59}	P	99.9	1104–1303	62.7	0.21
Ni^{63}	P		1192–1297	60.2	0.10
S^{35}	P	99.99	1150–1250	5.4	1.3

Copper

Ag^{110}	S, P		580–980	46.5	0.61
As^{76}	P		810–1075	42.13	0.20
Au^{193}	S, P		400–1050	42.6	0.03
Cd^{115}	S	99.98	725–950	45.7	0.935

Diffusion in Selected Metallic Systems (Continued)

Metal ——— Tracer	Crystalline Form	Purity (%)	Temperature Range (°C)	Activation energy, Q (kcal $\cdot$ mol^{-1})	Frequency factor, D_o (cm^2 $\cdot$ s^{-1})
Ce^{141}	P	99.999	766–947	27.6	2.17×10^{-3}
Cr^{51}	S, P		800–1070	53.5	1.02
Co^{60}	S	99.998	701–1077	54.1	1.93
Cu^{67}	S	99.999	698–1061	50.5	0.78
Eu^{152}	P	99.999	750–970	26.85	1.17×10^{-7}
Fe^{59}	S. P		460–1070	52.0	1.36
Ga^{72}			–	45.90	0.55
Ge^{68}	S	99.998	653–1015	44.76	0.397
Hg^{203}	P		–	44.0	0.35
Lu^{177}	P	99.999	857–1010	26.15	4.3×10^{-9}
Mn^{54}	S	99.99	754–950	91.4	10^7
Nb^{95}	P	99.999	807–906	60.06	2.04
Ni^{63}	P		620–1080	53.8	1.1
Pd^{102}	S	99.999	807–1056	54.37	1.71
Pm^{147}	P	99.999	720–955	27.5	3.62×10^{-8}
Pt^{195}	P		843–997	37.5	4.8×10^{-4}
S^{35}	S	99.999	800–1000	49.2	23
Sb^{124}	S	99.999	600–1000	42.0	0.34
Sn^{113}	P		680–910	45.0	0.11
Tb^{160}	P	99.999	770–980	27.45	8.96×10^{-9}
Tl^{204}	S	99.999	785–996	43.3	0.71
Tm^{170}	P	99.999	705–950	24.15	7.28×10^{-9}
Zn^{65}	P	99.999	890–1000	47.50	0.73

Germanium

| Cd^{115} | S | | 750–950 | 102.0 | 1.75×10^9 |
| Fe^{59} | S | | 775–930 | 24.8 | 0.13 |

Diffusion in Selected Metallic Systems (Continued)

Metal ——— Tracer	Crystalline Form	Purity (%)	Temperature Range (°C)	Activation energy, Q (kcal · mol^{-1})	Frequency factor, D_0 (cm^2 · s^{-1})
Ge^{71}	S		766–928	68.5	7.8
In^{114}	S		600–920	39.9	2.9 x 10^{-4}
Sb^{124}	S		720–900	50.2	0.22
Te^{125}	S		770–900	56.0	2.0
Tl^{204}	S		800–930	78.4	1700
Gold					
Ag^{110}	S	99.99	699–1007	40.2	0.072
Au^{198}	S	99.97	850–1050	42.26	0.107
Co^{60}	P	99.93	702–948	41.6	0.068
Fe^{59}	P	99.93	701–948	41.6	0.082
Hg^{203}	S	99.994	600–1027	37.38	0.116
Ni^{63}	P	99.96	880–940	46.0	0.30
Pt^{195}	P, S	99.98	800–1060	60.9	7.6
β–Hafnium					
Hf^{181}	P	97.9	1795–1995	38.7	1.2 x10^{-3}
Indium					
Ag^{110}	S⊥c	99.99	25–140	12.8	0.52
Ag^{110}	S‖c	99.99	25–140	11.5	0.11
Au^{198}	S	99.99	25–140	6.7	9 x 10^{-3}
In^{114}	S⊥c	99.99	44–144	18.7	3.7
In^{114}	S‖c	99.99	44–144	18.7	2.7
Tl^{204}	S	99.99	49–157	15.5	0.049

Diffusion in Selected Metallic Systems (Continued)

Metal ——— Tracer	Crystalline Form	Purity (%)	Temperature Range (°C)	Activation energy, Q (kcal $\cdot$ mol^{-1})	Frequency factor, D$_o$ (cm$^2 \cdot$ s^{-1})
α-Iron					
Ag^{110}	P		748–888	69.0	1950
Au^{198}	P	99.999	800–900	62.4	31
C^{14}	P	99.98	616–844	29.3	2.2
Co^{60}	P	99.995	638–768	62.2	7.19
Cr^{51}	P	99.95	775–875	57.5	2.53
Cu^{64}	P	99.9	800 1050	57.0	0.57
Fe^{55}	P	99.92	809–889	60.3	5.4
K^{42}	P	99.92	500–800	42.3	0.036
Mn^{54}	P	99.97	800–900	52.5	0.35
Mo^{99}	P		750–875	73.0	7800
Ni^{63}	P	99.97	680–800	56.0	1.3
P^{32}	P		860–900	55.0	2.9
Sb^{124}	P		800–900	66.6	1100
V^{48}	P		755–875	55.4	1.43
W^{185}	P		755–875	55.1	0.29
γ-Iron					
Be^{7}	P	99.9	1100–1350	57.6	0.1
C^{14}	P	99.34	800–1400	34.0	0.15
Co^{60}	P	99.98	1138–1340	72.9	1.25
Cr^{51}	P	99.99	950–1400	69.7	10.8
Fe^{59}	P	99.98	1171–1361	67.86	0.49
Hf^{181}	P	99.99	1110–1360	97.3	3600
Mn^{54}	P	99.97	920–1280	62.5	0.16
Ni^{63}	P	99.97	930–2050	67.0	0.77

Diffusion in Selected Metallic Systems (Continued)

Metal ——— Tracer	Crystalline Form	Purity (%)	Temperature Range (°C)	Activation energy, Q (kcal $\cdot$ mol^{-1})	Frequency factor, D$_o$ (cm^2 $\cdot$ s^{-1})
P^{32}	P	99.99	950–1200	43.7	0.01
S^{35}	P		900–1250	53.0	1.7
V^{48}	P	9999	1120–1380	69.3	0.28
W^{185}	P	99.5	1050–1250	90.0	1000
δ-Iron					
Co^{60}	P	99.995	1428–1521	61.4	6.38
Fe^{59}	P	99.95	1428–1492	57.5	2.01
P^{32}	P	99.99	1370–1460	55.0	2.9
Lanthanum					
Au^{198}	P	99.97	600–800	45.1	1.5
La^{140}	P	99.97	690–850	18.1	2.2 x 10^{-2}
Lead					
Ag^{110}	P	99.9	200–310	14.4	0.064
Au^{198}	S	99.999	190–320	10.0	8.7 x 10^{-3}
Cd^{115}	S	99.999	150–320	21.23	0.409
Cu^{64}	S		150–320	14.44	0.046
Pb^{204}	S	99.999	150–320	25.52	0.887
Tl^{205}	P	99.999	207–322	24.33	0.511
Lithium					
Ag^{110}	P	92.5	65–161	12.83	0.37

Diffusion in Selected Metallic Systems (Continued)

Metal ——— Tracer	Crystalline Form	Purity (%)	Temperature Range (°C)	Activation energy, Q (kcal $\cdot$ mol^{-1})	Frequency factor, D$_0$ (cm^2 $\cdot$ s^{-1})
Au^{195}	P	92.5	47–153	10.49	0.21
Bi	P	99.95	141–177	47.3	5.3×10^{13}
Cd^{115}	P	92.5	80–174	16.05	2.35
Cu64	P	99.98	51–120	9.22	0.47
Ga^{72}	P	99.98	58–173	12.9	0.21
Hg^{203}	P	99.98	58–173	14.18	1.04
In^{114}	P	92.5	80–175	15.87	0.39
Li^6	P	99.98	35–178	12.60	0.14
Na^{22}	P	92.5	52–176	12.61	0.41
Pb^{204}	P	99.95	129–169	25.2	160
Sb^{124}	P	99.95	141–176	41.5	1.6×10^{10}
Sn^{113}	P	99.95	108–174	15.0	0.62
Zn^{65}	P	92.5	60–175	12.98	0.57

Magnesium

Tracer	Crystalline Form	Purity (%)	Temperature Range (°C)	Activation energy, Q	Frequency factor, D$_0$
Ag^{110}	P	99.9	476–621	28.50	0.34
Fe^{59}	P	99.95	400–600	21.2	4×10^{-6}
In^{114}	P	99.9	472–610	28.4	5.2×10^{-2}
Mg^{28}	S$\perp$c		467–635	32.5	1.5
Mg^{28}	S||c		467–635	32.2	1.0
Ni^{63}	P	99.95	400 600	22.9	1.2×10^{-5}
U^{235}	P	99.95	500–620	27.4	1.6×10^{-5}
Zn^{65}	P	99.9	467–620	28.6	0.41

Molybdenum

Tracer	Crystalline Form	Purity (%)	Temperature Range (°C)	Activation energy, Q	Frequency factor, D$_0$
C^{14}	P	99.98	1200–1600	41.0	2.04×10^{-2}

Diffusion in Selected Metallic Systems (Continued)

Metal ——— Tracer	Crystalline Form	Purity (%)	Temperature Range (°C)	Activation energy, Q (kcal · mol^{-1})	Frequency factor, D$_o$ (cm^2 · s^{-1})
Co^{60}	P	99.98	1850–2350	106.7	18
Cr^{51}	P		1000–1500	54.0	2.5 x 10^{-4}
Cs^{134}	S	99.99	1000–1470	28.0	8.7 x 10^{-11}
K^{42}	S		800–1100	25.04	5.5 x 10^{-9}
Mo^{99}	P		1850–2350	96.9	0.5
Na^{24}	S		800–1100	21.25	2.95 x 10^{-9}
Nb^{95}	P	99.98	1850–2350	108.1	14
P^{32}	P	99.97	2000–2200	80.5	0.19
Re^{186}	P		1700–2100	94.7	0.097
S^{35}	S	99.97	2220–2470	101.0	320
Ta^{182}	P		1700–2150	83.0	3.5 x 10^{-4}
U^{235}	P	99.98	1500–2000	76.4	7.6 x 10^{-3}
W^{185}	P	99.98	1700–2260	110	1.7

Nickel

Au^{198}	S,P	99.999	700–1075	55.0	0.02
Be^{7}	P	99.9	1020–1400	46.2	0.019
C^{14}	P	99.86	600–1400	34.0	0.012
Co^{60}	P	99.97	1149–1390	65.9	1.39
Cr^{51}	P	99.95	1100–1270	65.1	1.1
Cu^{64}	P	99.95	1050–1360	61.7	0.57
Fe^{59}	P		1020–1263	58.6	0.074
Mo^{99}	P		900–1200	51.0	1.6 x 10^{-3}
Ni^{63}	P	99.95	1042–1404	68.0	1.9
Pu^{238}	P		1025–1125	51.0	0.5
Sb^{124}	P	99.97	1020–1220	27.0	1.8 x 10^{-5}
Sn^{113}	P	99.8	700–1350	58.0	0.83

Diffusion in Selected Metallic Systems (Continued)

Metal ——— Tracer	Crystalline Form	Purity (%)	Temperature Range (°C)	Activation energy, Q (kcal $\cdot$ mol^{-1})	Frequency factor, D_o (cm$^2 \cdot$ s^{-1})
V^{48}	P	99.99	800–1300	66.5	0.87
W^{185}	P	99.95	1100–1300	71.5	2.0
Niobium					
C^{14}	P		800–1250	32.0	1.09×10^{-5}
Co^{60}	P	99.85	1500–2100	70.5	0.74
Cr^{51}	S		943–1435	83.5	0.30
Fe^{51}	P	99.85	1400–2100	77.7	1.5
K^{42}	S		900 1100	22.10	2.38×10^{-7}
Nb^{95}	P, S	99.99	878–2395	96.0	1.1
P^{32}	P	99.0	1300–1800	51.5	5.1×10^{-2}
S^{35}	S	99.9	1100–1500	73.1	2600
Sn^{113}	P	99.85	1850–2400	78.9	0.14
Ta^{182}	P, S	99.997	878–2395	99.3	1.0
Ti^{44}	S		994–1492	86.9	0.099
U^{235}	P	99.55	1500–2000	76.8	8.9×10^{-3}
V^{48}	S	99.99	1000–1400	85.0	2.21
W^{185}	P	99.8	1800–2200	91.7	5×10^{-4}
Palladium					
Pd^{103}	S	99.999	1060–1500	63.6	0.205
Phosphorus					
P^{32}	P		0–44	9.4	1.07×10^{-3}

Diffusion in Selected Metallic Systems (Continued)

Metal ——— Tracer	Crystalline Form	Purity (%)	Temperature Range (°C)	Activation energy, Q (kcal $\cdot$ mol^{-1})	Frequency factor, D$_o$ (cm^2 $\cdot$ s^{-1})
Platinum					
Co^{60}	P	99.99	900–1050	74.2	19.6
Cu^{64}	P		1098–1375	59.5	0.074
Pt^{195}	P	99.99	1325–1600	68.2	0.33
Potassium					
Au^{198}	P	99.95	5.6–52.5	3.23	1.29 x10^{-3}
K^{42}	S	99.7	–52–61	9.36	0.16
Na^{22}	P	99.7	0–62	7.45	0.058
Rb^{86}	P	99.95	0.1–59.9	8.78	0.090
γ–Plutonium					
Pu^{238}	P		190–310	16.7	2.1 x 10^{-5}
δ–Plutonium					
Pu^{238}	P		350–440	23.8	4.5 x 10^{-3}
ε-Plutonium					
Pu^{238}	P		500–612	18.5	2.0 x 10^{-2}
α-Praseodymium					
Ag^{110}	P	99.93	610 730	25.4	0.14
Au^{195}	P	99.93	650–780	19.7	4.3 x 10^{-2}

Diffusion in Selected Metallic Systems (Continued)

Metal ——— Tracer	Crystalline Form	Purity (%)	Temperature Range (°C)	Activation energy, Q (kcal • mol^{-1})	Frequency factor, D_0 (cm^2 • s^{-1})
Co^{60}	P	99.93	660–780	16.4	4.7 x 10^{-2}
Zn^{65}	P	99.96	766–603	24.8	0.18
β-Praseodymium					
Ag^{110}	P	99.93	800–900	21.5	3.2 x 10^{-2}
Au^{195}	P	99.93	800–910	20.1	3.3 x 10^{-2}
Ho^{166}	P	99.96	800–930	26.3	9.5
In^{114}	P	99.96	800–930	28.9	9.6
La^{140}	P	99.96	800–930	25.7	1.8
Pr^{142}	P	99.93	800–900	29.4	8.7
Zn^{65}	P	99.96	822–921	27.0	0.63
Selenium					
Fe^{59}	P		40–100	8.88	—
Hg^{203}	P	99.996	25–100	1.2	—
S^{35}	S⊥c		60–90	29.9	1700
S^{35}	S∥c		60–90	15.6	1100
Se^{75}	P		35–140	11.7	1.4 x 10^{-4}
Silicon					
Au^{198}	S		700–1300	47.0	2.75 x 10^{-3}
C^{14}	P		1070–1400	67.2	0.33
Cu^{64}	P		800–1100	23.0	4 x 10^{-2}
Fe^{59}	S		1000–1200	20.0	6.2 x 10^{-3}
Ni^{63}	P		450–800	97 5	1000

Diffusion in Selected Metallic Systems (Continued)

Metal ——— Tracer	Crystalline Form	Purity (%)	Temperature Range (°C)	Activation energy, Q (kcal $\cdot$ mol^{-1})	Frequency factor, D$_o$ (cm$^2 \cdot$ s^{-1})
P^{32}	S		1100–1250	41.5	–
Sb^{124}	S		1190–1398	91.7	12.9
Si^{31}	S	99.99999	1225–1400	110.0	1800
Silver					
Au^{198}	P	99.99	718–942	48.28	0.85
Ag^{110}	S	99.999	640–955	45.2	0.67
Cd^{115}	S	99.99	592–937	41.69	0.44
Co^{60}	S	99.999	700–940	48.75	1.9
Cu^{64}	P	99.99	717–945	46.1	1.23
Fe^{59}	S	99.99	720–930	49.04	2.42
Ge^{77}	P		640–870	36.5	0.084
Hg^{203}	P	99.99	653–948	38.1	0.079
In^{114}	S	99.99	592–937	40.80	0.41
Ni^{63}	S	99.99	749–950	54.8	21.9
Pb^{210}	P		700–865	38.1	0.22
Pd^{102}	S	99.999	736–939	56.75	9.56
Ru^{103}	S	99.99	793–945	65.8	180
S^{35}	S	99.999	600–900	40.0	1.65
Sb^{124}	P	99.999	780–950	39.07	0.234
Sn^{113}	S	99.99	592–937	39.30	0.255
Te^{125}	P		770–940	38.90	0.47
Tl^{204}	P		640–870	37.9	0.15
Zn^{65}	S	99.99	640–925	41.7	0.54

Diffusion in Selected Metallic Systems (Continued)

Metal ——— Tracer	Crystalline Form	Purity (%)	Temperature Range (°C)	Activation energy, Q (kcal · mol^{-1})	Frequency factor, D_o (cm^2 · s^{-1})
Sodium					
Au^{198}	P	99.99	1.0–77	2.21	3.34 x 10^{-4}
K^{42}	P	99.99	0–91	8.43	0.08
Na^{22}	P	99.99	0–98	10.09	0.145
Rb^{86}	P	99.99	0–85	8.49	0.15
Tantalum					
C^{14}	P		1450–2200	40.3	1.2 x 10^{-2}
Fe^{59}	P		930–1240	71.4	0.505
Mo^{99}	P		1750–2220	81.0	1.8 x 10^{-3}
Nb^{95}	P, S	99.996	921–2484	98.7	0.23
S^{35}	P	99.0	1970–2110	70.0	100
Ta^{182}	P, S	99.996	1250–2200	98.7	1.24
Tellurium					
Hg^{203}	P		270–440	18.7	3.14 x 10^{-5}
Se^{75}	P		320–440	28.6	2.6 x 10^{-2}
Tl^{204}	P		360–430	41.0	320
Te^{127}	S⊥c	99.9999	300–400	46.7	3.91 x 10^{4}
Te^{127}	S∥c	99.9999	300–400	35.5	130
α-Thallium					
Ag^{110}	P⊥c	99.999	80–250	11.8	3.8 x 10^{-2}
Ag^{110}	P∥c	99.999	80–250	11.2	2.7 x 10^{-2}

Diffusion in Selected Metallic Systems (Continued)

Metal ——— Tracer	Crystalline Form	Purity (%)	Temperature Range (°C)	Activation energy, Q (kcal • mol^{-1})	Frequency factor, D_o (cm^2 • s^{-1})
Au^{198}	P⊥c	99.999	110–260	2.8	2.0 x 10^{-5}
Au^{198}	P‖c	99.999	110–260	5.2	5.3 x 10^{-4}
Tl^{204}	S⊥c	99.9	135–230	22.6	0.4
Tl^{204}	S‖c	99.9	135–230	22.9	0.4

β-Thallium

Ag^{110}	P	99.999	230–310	11.9	4.2 x 10^{-2}
Au^{198}	P	99.999	230–310	6.0	5.2 x 10^{-4}
Tl^{204}	S	99.9	230–280	20.7	0.7

α-Thorium

Pa^{231}	P	99.85	770–910	74.7	126
Th^{228}	P	99.85	720–880	716	395
U^{233}	P	99.85	700–880	79.3	2210

Tin

Ag^{110}	S⊥c		135–225	18.4	0.18
Ag^{110}	S‖c		135–225	12.3	7.1 x 10^{-3}
Au^{198}	S⊥c		135–225	17.7	0.16
Au^{198}	S‖c		135–225	11.0	5.8 x 10^{-3}
Co^{60}	S,P		140–217	22.0	5.5
In^{114}	S⊥c	99.998	181–221	25.8	34.1
In^{114}	S‖c	99.998	181–221	25.6	12.2

Diffusion in Selected Metallic Systems (Continued)

Metal ——— Tracer	Crystalline Form	Purity (%)	Temperature Range (°C)	Activation energy, Q (kcal • mol^{-1})	Frequency factor, D_o (cm^2 • s^{-1})
Sn^{113}	S⊥c	99.999	160–226	25.1	10.7
Sn^{113}	S‖c	99.999	160–226	25.6	7.7
Tl^{204}	P	99.999	137–216	14.7	1.2 x 10^{-3}
α-Titanium					
Ti^{44}	P	99.99	700–850	35.9	8.6 x 10^{-6}
β-Titanium					
Ag^{110}	P	99.95	940 1570	43.2	3 x 10^{-3}
Be^{7}	P	99.96	915–1300	40.2	0.8
C^{14}	P	99.62	1100–1600	20.0	3.02 x 10^{-3}
Cr^{51}	P	99.7	950–1600	35.1	5 x 10^{-3}
Co^{60}	P	99.7	900–1600	30.6	1.2 x 10^{-2}
Fe^{59}	P	99.7	900–1600	31.6	7.8 x 10^{-3}
Mo^{99}	P	99.7	900–1600	43.0	8.0 x 10^{-3}
Mn^{54}	P	99.7	900–1600	33.7	6.1 x 10^{-3}
Nb^{95}	P	99.7	1000–1600	39.3	5.0 x 10^{-3}
Ni^{63}	P	99.7	925–1600	29.6	9.2 x 10^{-3}

Diffusion in Selected Metallic Systems (Continued)

Metal ——— Tracer	Crystalline Form	Purity (%)	Temperature Range (°C)	Activation energy, Q (kcal $\cdot$ mol^{-1})	Frequency factor, D$_0$ (cm^2 $\cdot$ s^{-1})
P^{32}	P	99.7	950–1600	24.1	3.62×10^{-3}
Sc^{46}	P	99.95	940–1590	32.4	4.0×10^{-3}
Sn^{113}	P	99.7	950–1600	31.6	3.8×10^{-4}
Ti^{44}	P	99.95	900–1540	31.2	3.58×10^{-4}
U^{235}	P	99.9	900–400	29.3	5.1×10^{-4}
V^{48}	P	99.95	900–1545	32.2	3.1×10^{-4}
W^{185}	P	99.94	900–1250	43.9	3.6×10^{-3}
Zr^{95}	P	98.94	920–1500	35.4	4.7×10^{-3}

Tungsten

C^{14}	P	99.51	1200–1600	53.5	8.91×10^{-3}
Fe^{59}	P		940–1240	66.0	1.4×10^{-2}
Mo^{99}	P		1700–2100	101.0	0.3
Nb^{95}	P	99.99	1305–2367	137.6	3.01
Re^{186}	S		2100–2400	141.0	19.5
Ta^{182}	P	99.99	1305–2375	139.9	3.05
W^{185}	P	99.99	1800–2403	140.3	1.88

α–Uranium

U^{234}	P		580–650	40.0	2×10^{-3}

Diffusion in Selected Metallic Systems (Continued)

Metal —— Tracer	Crystalline Form	Purity (%)	Temperature Range (°C)	Activation energy, Q (kcal · mol^{-1})	Frequency factor, D_o (cm^2 · s^{-1})
β–Uranium					
Co^{60}	P	99.999	692–763	27.45	1.5 x 10^{-2}
U^{235}	P		690–750	44.2	2.8 x10^{-3}
γ-Uranium					
Au^{195}	P	99.99	785–1007	30.4	4.86 x 10^{-3}
Co^{60}	P	99.99	783–989	12.57	3.51 x 10^{-4}
Cr^{51}	P	99.99	797–1037	24.46	5.37 X 10^{-3}
Cu^{64}	P	99.99	787–1039	24.06	1.96 x 10^{-3}
Fe^{55}	P	99.99	787–990	12.0	2.69 x 10^{-4}
Mn^{54}	P	99.99	787–939	13.88	1.81 x 10^{-4}
Nb^{95}	P	99.99	791–1102	39.65	4.87 x 10^{-2}
Ni^{63}	P	99.99	787–1039	15.66	5.36 x10^{-4}
U^{233}	P	99.99	800–1070	28.5	2.33 x 10^{-3}
Zr^{95}	P		800–1000	16.5	3.9 x 10^{-4}
Vanadium					
C^{14}	P	99.7	845–1130	27.3	4.9 x 10^{-3}
Cr^{51}	P	99.8	960–1200	64.6	9.54 x10^{-3}
Fe^{59}	P		960–1350	71.0	0.373
P^{32}	P	99.8	1200–1450	49.8	2.45 x 10^{-2}
S^{35}	P	99.8	1320 1520	34.0	3.1 x 10^{-2}
V^{48}	S,P	99.99	880–1360	73.65	0.36
V^{48}	S,P	99.99	1360–1830	94.14	214.0

Diffusion in Selected Metallic Systems (Continued)

Metal —— Tracer	Crystalline Form	Purity (%)	Temperature Range (°C)	Activation energy, Q (kcal $\cdot$ mol^{-1})	Frequency factor, D_0 (cm^2 $\cdot$ s^{-1})
Yttrium					
Y^{90}	S⊥c		900–1300	67.1	5.2
Y^{90}	S‖c		900–1300	60.3	0.82
Zinc					
Ag^{110}	S⊥c	99.999	271–413	27.6	0.45
Ag^{110}	S‖c	99.999	271–413	26.0	0.32
Au^{198}	S⊥c	99.999	315–415	29.72	0.29
Au^{198}	S‖c	99.999	315–415	29.73	0.97
Cd^{115}	S⊥c	99.999	225–416	20.12	0.117
Cd^{115}	S‖c	99.999	225–416	20.54	0.114
Cu64	S⊥c	99.999	338–415	2.0	2.0
Cu64	S‖c	99.999	338–415	29.53	2.22
Ga^{72}	S⊥c		240–403	18.15	0.018
Ga^{72}	S‖c		240 403	18.4	0.016
Hg^{203}	S⊥c		260–413	20.18	0.073
Hg^{203}	S‖c		260–413	19.70	0.056
In^{114}	S⊥c		271–413	19.60	0.14
In^{114}	S‖c		271–413	19.10	0.062
Sn^{113}	S⊥c		298–400	18.4	0.13
Sn^{113}	S‖c		298–400	19.4	0.15
Zn^{65}	S⊥c	99.999	240–418	23.0	0.18
Zn^{65}	S‖c	99.999	240–418	21.9	0.13

Diffusion in Selected Metallic Systems (Continued)

Metal ——— Tracer	Crystalline Form	Purity (%)	Temperature Range (°C)	Activation energy, Q (kcal $\cdot$ mol^{-1})	Frequency factor, D_o (cm$^2 \cdot$ s^{-1})
α-Zirconium					
Cr^{51}	P	99.9	700–850	18.0	1.19×10^{-8}
Fe^{55}	P		750–840	48.0	2.5×10^{-2}
Mo^{99}	P		600–850	24.76	6.22×10^{-8}
Nb^{95}	P	99.99	740–857	31.5	6.6×10^{-6}
Sn^{113}	P		300–700	22.0	1.0×10^{-8}
Ta^{182}	P	99.6	700–800	70.0	100
V^{48}	P	99.99	600–850	22.9	1.12×10^{-8}
Zr^{95}	P	99.95	750–850	45.5	5.6×10^{-4}
β–Zirconium					
Be^{7}	P	99.7	915–1300	31.1	8.33×10^{-2}
C^{14}	P	96.6	1100–1600	34.2	3.57×10^{-2}
Ce^{141}	P		880–1600	41.4	3.16
Co^{60}	P	99.99	920–1600	21.82	3.26×10^{-3}
Cr^{51}	P	99.9	700–850	18.0	1.19×10^{-8}
Fe^{55}	P		750–840	48.0	2.5×10^{-2}
Mo^{99}	P		900–1635	35.2	1.99×10^{-6}
Nb^{95}	P		1230–1635	36.6	7.8×10^{-4}
P^{32}	P	99.94	950–1200	33.3	0.33
Sn^{113}	P		300–700	22.0	1×10^{-8}
Ta^{182}	P	99.6	900–1200	27.0	5.5×10^{-5}

Diffusion in Selected Metallic Systems (Continued)

Metal ——— Tracer	Crystalline Form	Purity (%)	Temperature Range (°C)	Activation energy, Q (kcal $\cdot$ mol^{-1})	Frequency factor, D_o (cm^2 $\cdot$ s^{-1})
U^{235}	P		900–1065	30.5	5.7 x 10^{-4}
V^{48}	P	99.99	870–1200	45.8	7.59 x 10^{-3}
V^{48}	P	99.99	1200–1400	57.7	0.32
W^{185}	P	99.7	900–1250	55.8	0.41
Zr^{95}	P		1100–1500	30.1	2.4 x 10^{-4}

The diffusion coefficient D_T at a temperature T(K) is given by the following:

$$D_T = D_o\, e^{-Q/RT}$$

Abbreviations:

P= polycrystalline
S = single crystal
$\perp$ c = perpendicular to c direction
$\parallel$ c = parallel to c direction

Source: Data from Askill, J.,in *Handbook of Chemistry and Physics*, 55th ed.,Weast, R.C., Ed., CRC Press, Cleveland,1974, F61.

DIFFUSIVITY VALUES OF METALS
INTO METALS

Diffusing Metal	Matrix Metal	Temperature (°C)	Diffusion Coefficient (cm$^2 \cdot$ hr^{-1})
Ag	Al	466	6.84–8.1 x10^{-7}
		500	7.2–3.96 x 10^{-8}
		573	1.26 x 10^{-5}
	Pb	220	5.40 x 10^{-5}
		250	1.08 x 10^{-4}
		285	3.29 x 10^{-4}
	Sn	500	1.73 x 10^{-1}
Al	Cu	500	6.12 x 10^{-9}
		850	7.92 x 10^{-6}
As	Si		0.32 e$^{-82,000/RT}$
Au	Ag	456	1.76 x 10^{-9}
		491	0.92–2.38 x 10^{-13}
		585	3.6 x 10^{-8}
		601	3.96 x 10^{-8}
		624	2.5–5 x 10^{-11}
		717	1.04–2.25 x 10^{-9}
		729	1.76 x 10^{-9}
		767	1.15 x 10^{-6}
		847	2.30 x 10^{-6}
		858	3.63 x 10^{-8}
		861	3.92 x 10^{-8}
		874	3.92 x 10^{-8}
		916	5.40 x 10^{-6}
		1040	1.17 x 10^{-6}

Diffusivity Values of Metals into Metals (Continued)

Diffusing Metal	Matrix Metal	Temperature (°C)	Diffusion Coefficient (cm$^2 \cdot$ hr^{-1})
		1120	2.29 x 10^{-5}
		1189	5.42 x 10^{-6}
	Au	800	1.17 x 10^{-8}
		900	9 x 10^{-8}
		1020	5.4 x 10^{-7}
	Bi	500	1.88 x 10^{-1}
	Cu	970	5.04 x 10^{-6}
	Hg	11	3 x10^{-2}
	Pb	100	8.28 x 10^{-8}
		150	1.80 x 10^{-4}
		200	3.10 x 10^{-4}
		240	1.58 x 10^{-3}
		300	5.40 x 10^{-3}
		500	1.33 x 10^{-1}
			0.001e$^{-25,000/RT}$
	Sn	500	1.94 x 10^{-1}
B	Si		10.5 e$^{-85,000/RT}$
Ba	Hg	7.8	2.17 x 10^{-2}
Bi	Si		1030e$^{-107,000/RT}$
	Pb	220	1.73 x 10^{-7}
		250	1.33 x 10^{-6}
		285	1.58 x 10^{-6}
C	W	1700	1.87 x 10^{-3}
	Fe	930	7.51–9.18 x 10^{-9}

Diffusivity Values of Metals into Metals (Continued)

Diffusing Metal	Matrix Metal	Temperature (°C)	Diffusion Coefficient ($cm^2 \cdot hr^{-1}$)
Ca	Hg	10.2	2.25×10^{-2}
Cd	Ag	650	9.36×10^{-7}
		800	4.68×10^{-6}
		900	2.23×10^{-5}
	Hg	8.7	6.05×10^{-2}
		15	6.51×10^{-2}
		20	5.47×10^{-2}
		99.1	1.23×10^{-1}
	Pb	200	4.59×10^{-7}
		252	3.10×10^{-6}
Cd, 1 atom%	Pb	167	1.66×10^{-7}
Ce	W	1727	3.42×10^{-6}
Cs	Hg	7.3	1.88×10^{-2}
	W	27	4.32×10^{-3}
		227	5.40×10^{-4}
		427	2.88×10^{-2}
		540	1.44×10^{-1}
Cu	Al	440	1.8×10^{-7}
		457	2.88×10^{-7}
		540	5.04×10^{-6}
		565	$4.68–5.00 \times 10^{-4}$

Diffusivity Values of Metals into Metals (Continued)

Diffusing Metal	Matrix Metal	Temperature (°C)	Diffusion Coefficient ($cm^2 \cdot hr^{-1}$)
	Ag	650	1.04×10^{-6}
		760	1.30×10^{-6}
		895	3.38×10^{-6}
	Au	301	5.40×10^{-10}
		443	8.64×10^{-9}
		560	3.38×10^{-7}
		604	5.10×10^{-7}
		616	7.92×10^{-7}
		740	3.35×10^{-6}
	Cu	650	1.15×10^{-5}
		750	2.34×10^{-8}
		830	$1 .44 \times 10^{-7}$
		850	9.36×10^{-7}
		950	2.30×10^{-6}
		1030	1.01×10^{-5}
	Ge	700–900	$1.01 \pm 0.1 \times 10^{-1}$
	Pt	1041	7.83–9×10^{-8}
		1213	5.04×10^{-7}
		1401	6.12×10^{-6}
Fe	Au	753	1.94×10^{-6}
		1003	2.70×10^{-5}
			$0.0062 \, e^{-20,000/RT}$
Ga	Si		$3.6 \, e^{-81,000/RT}$
Ge	Al	630	3.31×10^{-1}

Diffusivity Values of Metals into Metals (Continued)

Diffusing Metal	Matrix Metal	Temperature (°C)	Diffusion Coefficient (cm$^2 \cdot$ hr^{-1})
	Au	529	1.84×10^{-1}
		563	2.80×10^{-1}
	Ge	766–928	$7.8\ e^{-68,509/RT}$
		1060–1200°K	$87\ e^{-73,000/RT}$
Hg	Cd	156	9.36×10^{-7}
		176	2.55×10^{-6}
		202	9×10^{-6}
	Pb	177	8.34×10^{-8}
		197	2.09×10^{-5}
In	Ag	650	1.04×10^{-6}
		800	6.84×10^{-6}
		895	4.68×10^{-5}
			$16.5\ e^{-90,000/RT}$
K	Hg	10.5	2.21×10^{-2}
	W	207	2.05×10^{-2}
		317	3.6×10^{-1}
		507	$1.1 \times 10^{+1}$
Li	Hg	8.2	2.75×10^{-2}
Mg	Al	365	3.96×10^{-8}
		395	$1.98–2.41 \times 10^{-7}$
		420	$2.38–2.74 \times 10^{-7}$
		440	1.19×10^{-7}

Diffusivity Values of Metals into Metals (Continued)

Diffusing Metal	Matrix Metal	Temperature (°C)	Diffusion Coefficient (cm$^2 \cdot$ hr^{-1})
		447	9.36×10^{-7}
		450	6.84×10^{-6}
		500	$3.96–7.56 \times 10^{-6}$
		577	1.58×10^{-5}
	Pb	220	4.32×10^{-7}
Mn	Cu	400	7.2×10^{-10}
		850	4.68×10^{-7}
Mo	W	1533	9.36×10^{-10}
		1770	4.32×10^{-9}
		2010	7.92×10^{-8}
		2260	2.81×10^{-7}
Na	W	20	2.88×10^{-2}
		227	1.80
		417	9.72
		527	1.19×10^{-1}
Ni	Au	800	2.77×10^{-6}
		1003	2.48×10^{-5}
	Cu	550	2.56×10^{-9}
		950	7.56×10^{-7}
		320	1.26×10^{-6}
	Pt	1043	1.81×10^{-8}
		1241	1.73×10^{-6}

Diffusivity Values of Metals into Metals (Continued)

Diffusing Metal	Matrix Metal	Temperature (°C)	Diffusion Coefficient ($cm^2 \cdot hr^{-1}$)
		1401	5.40×10^{-6}
Ni, 1 atom %	Pb	285	8.34×10^{-7}
Ni, 3 atom%	Pb	252	1.25×10^{-7}
Pb	Cd	252	2.88×10^{-8}
	Pb	250	5.42×10^{-8}
		285	2.92×10^{-7}
	Sn	500	1.33×10^{-1}
Pb, 2 atom %	Hg	9.4	6.46×10^{-9}
		15.6	5.71×10^{-2}
		99.2	8×10^{-2}
Pd	Ag	444	4.68×10^{-9}
		571	1.33×10^{-7}
		642	4.32×10^{-7}
		917	4.32×10^{-6}
	Au	727	2.09×10^{-8}
		970	1.15×10^{-6}
	Cu	490	3.24×10^{-9}
		950	$9.0–10.44 \times 10^{-7}$
Po	Au	470	4.59×10^{-11}
	Al	20	1.08×10^{-9}
		500	1.80×10^{-7}
	Bi	150	1.80×10^{-7}

Diffusivity Values of Metals into Metals (Continued)

Diffusing Metal	Matrix Metal	Temperature (°C)	Diffusion Coefficient ($cm^2 \cdot hr^{-1}$)
		200	1.80×10^{-6}
	Pb	150	4.59×10^{-11}
		200	4.59×10^{-9}
		310	5.41×10^{-7}
Pt	Au	740	1.69×10^{-8}
		986	$6.12–10.08 \times 10^{-7}$
	Cu	490	2.01×10^{-9}
		960	$3.96–8.28 \times 10^{-7}$
	Pb	490	7.04×10^{-2}
Ra	Au	470	1.42×10^{-8}
	Pt	470	3.42×10^{-8}
Ra($\beta+\gamma$)	Ag	470	1.57×10^{-8}
Rb	Hg	7.3	1.92×10^{-9}
Rh	Pb	500	1.27×10^{-1}
Sb	Ag	650	1.37×10^{-6}
		760	5.40×10^{-6}
		895	1.55×10^{-5}
			$5.6\, e^{-91,000/RT}$
Si	Al	465	1.22×10^{-6}
		510	7.2×10^{-6}

Diffusivity Values of Metals into Metals (Continued)

Diftusing Metal	Matrix Metal	Temperature (°C)	Diffusion Coefficient ($cm^2 \cdot hr^{-1}$)
		600	3.35×10^{-5}
		667	1.44×10^{-1}
		697	3.13×10^{-1}
	Fe+C	1400–1600	3.24–5.4×10^{-2}
Sn	Ag	650	2.23×10^{-6}
		895	2.63×10^{-6}
	Cu	400	1.69×10^{-9}
		650	2.48×10^{-7}
		850	1.40×10^{-5}
	Hg	10.7	6.38×10^{-2}
	Pb	245	1.12×10^{-7}
		250	1.83×10^{-7}
		285	5.76×10^{-7}
Sr	Hg	9.4	1.96×10^{-2}
Th	Mo	1615	1.30×10^{-6}
		2000	3.60×10^{-3}
	Tl	285	8.76×10^{-7}
	W	1782	3.96×10^{-7}
		2027	4.03×10^{-6}
		2127	1.29×10^{-5}
		2227	2.45×10^{-5}
Th (β)	Pb	165	2.54×10^{-12}
		260	2.54×10^{-8}

Diffusivity Values of Metals into Metals (Continued)

Diftusing Metal	Matrix Metal	Temperature (°C)	Diffusion Coefficient (cm^2 • hr^{-1})
		324	5.84×10^{-6}
Tl	Hg	11.5	3.63×10^{-2}
	Pb	220	1.01×10^{-7}
		250	7.92×10^{-7}
		270	3.96×10^{-7}
		285	1.12×10^{-6}
		315	2.09×10^{-6}
			$16.5\, e^{-85,000/RT}$
U	W	1727	4.68×10^{-8}
Y	W	1727	6.55×10^{-5}
Zn	Ag	750	1.66×10^{-5}
		850	4.37×10^{-5}
	Al	415	9×10^{-7}
		473	1.91×10^{-6}
		500	$7.2–13.68 \times 10^{-6}$
		555	1.8×10^{-5}
	Hg	11.5	9.09×10^{-2}
		15	8.72×10^{-2}
		99.2	1.20×10^{-1}
	Pb	285	5.84
Zr	W	1727	1.17×10^{-5}

Source: Data from Loebel, R., in *Handbook of Chemistry and Physics*, 51st ed., Weast, R. C., Ed., Chemical Rubber, Cleveland, 1970, F-55.

DIFFUSION IN SOME NON-METALLIC SYSTEMS

Solvent	Solute	D_0 (m$^2 \cdot$ s^{-1})		Q (kJ/mol)	Q (kcal/mol)
Al$_2$O$_3$	Al	2.8	x 10^{-3}	477	114
Al$_2$O$_3$	O	0.19		636	152
MgO	Mg	24.9	x 10^{-6}	330	79
MgO	O	4.3	x 10^{-9}	344	82.1
MgO	Ni	1.8	x 10^{-9}	202	48.3
Si	Si	0.18		460	110
Ge	Ge	1.08	x 10^{-3}	291	69.6
Ge	B	1.1	x 10^3	439	105

Source: Data from P. Kofstad, *Nonstoichiometry, Diffusion, and Electrical Conductivity in Binary Metal Oxides*, John Wiley & Sons, Inc., New York, 1972; and S. M. Hu, in *Atomic Diffusion in Semiconoductors*, D. Shaw, ed., Plenum Press, New York, 1973.

DIFFUSION IN SEMICONDUCTORS

Semiconductor and diffusing element	D_0 (cm$^2 \cdot$ s^{-1})	ΔE (eV)	Temperature Range of Validity (°C)
Aluminum antimonide (AlSb)			
Al		~1.8	
Cu	3.5×10^{-3}	0.36	150–500
Sb		~1.5	
Zn	0.33±.15	1.93±0.04	660–860
Cadmium selenide (CdSe)			
Se	2.6×10^{-3}	1.55	700–1800
Cadmium sulfide (CdS)			
Ag	$2.5 \times 10^{+1}$	1.2	250–500
Cd	3.4	2.0	750–1000
Cu	1.5×10^{-3}	0.76	450–750
Cadmium telluride (CdTe)			
Au	$6.7 \times 10^{+1}$	2.0	600–1000
In	4.1×10^{-1}	1.6	450–1000
Calcium ferrate (III) (CaFe$_2$O$_4$)			
Ca	30	3.7	
Fe	0.4	3.1	
α-Calcium metasilicate (CaSiO$_3$)			
Ca	$7.4 \times 10^{+4}$	4.8	
Gallium antimonide (GaSb)			
Ga	$3.2 \times 10^{+3}$	3.15	650–700
In	1.2×10^{-7}	0.53	400–650
Sb	$3.4 \times 10^{+4}$	3.44	650–700
	$8.7 \times 10^{+2}$	1.13	470–570
Sn	2.4×10^{-5}	0.80	320–570
Te	3.8×10^{-4}	1.2	400–650

Diffusion in Semiconductors (Continued)

Semiconductor and diffusing element	D_0 $(cm^2 \cdot s^{-1})$	ΔE (eV)	Temperature Range of Validity (°C)
Gallium arsenide (GaAs)			
Ag	2.5×10^{-3}	1.5	
	4×10^{-4}	0.8 ± 0.05	500–1160
As	4×10^{21}	10.2 ± 1.2	1200–1250
Au	10^{-3}	1.0 ± 0.2	740–1024
Cd	0.05 ± 0.04	2.43 ± 0.06	868–1149
	$^a 50 \times 10^{-2}$	2.8^a	
Cu	0.03	0.52	100–600
Ga	$1 \times 10^{+7}$	5.60 ± 0.32	1125–1250
Li	0.53	1.0	250–400
Mg	1.4×10^{-4}	1.89	
	2.3×10^{-2}	2.6	740–1024
	$^a 2.6 \times 10^{-2}$	2.7^a	
	$^a 6.5 \times 10^{-1}$	2.49^a	
	8.5×10^{-3}	1.7	740–1024
S	1.2×10^{-4}	1.8	
	$^a 1.6 \times 10^{-5}$	1.63^a	
	2.6×10^{-5}	1.86	
	4×10^3	4.04 ± 0.15	1000–1200
Se	3×10^3	4.16 ± 0.16	1000–1200
Sn	$^a 3.8 \times 10^{-2}$	2.7	
	6×10^{-4}	2.5	1069–1215
Zn	$^a 2.5 \times 10^{-1}$	3.0^a	
	3.0×10^{-7}	1.0	
	6.0×10^{-7}	0.6	
	15 ± 7	2.49 ± 0.05	800

Diffusion in Semiconductors (Continued)

Semiconductor and diffusing element	D_0 $(cm^2 \cdot s^{-1})$	ΔE (eV)	Temperature Range of Validity (˚C)
Gallium phosphide (GaP)			
Zn	1.0	2.1	700–1300
Germanium (Ge)			
Ag	4.4×10^{-2}	1.0	700–900
As	6.3	2.4	600–850
Au	2.2×10^{-2}	2.5	
B	1.6×10^{-9}	4.6	600–850
Cu	1.9×10^{-4}	0.18	600–850
Fe	1.3×10^{-1}	1.1	750–850
Ga	$4.0 \times 10^{+1}$	3.1	600–850
Ge	$8.7 \times 10^{+1}$	3.2	750–920
He	6.1×10^{-3}	0.69	750–850
In	3×10^{-2}	2.4	600–850
Li	1.3×10^{-4}	0.47	200–600
Ni	8×10^{-1}	0.9	700–875
P	2.5	2.5	600–850
Pb	–	3.6	600–850
Sb	4.0	2.4	600–850
Sn	1.7×10^{-2}	1.9	600–850
Zn	$1.0 \times 10^{+1}$	2.5	600–850

Diffusion in Semiconductors (Continued)

Semiconductor and diffusing element	D_0 (cm$^2 \cdot$ s^{-1})	ΔE (eV)	Temperature Range of Validity (°C)
Indium antimonide (InSb)			
Ag	1.0×10^{-7}	0.25	
Au	$^a7 \times 10^{-4}$	0.32^a	140–510
Cd	$^a1.0 \times 10^{-5}$	1.1^a	250–500
	1.23×10^{-9}	0.52	442–519
	1.26	1.75	
	1.3×10^{-4}	1.2	360–500
Co	2.7×10^{-11}	0.39	
	10^{-7}	0.25	440–510
Cu	3.0×10^{-5}	0.37	
	$^a9.0 \times 10^{-4}$	1.08^a	
Fe	10^{-7}	0.25	440–510
HB	$^a4.0 \times 10^{-6}$	1.17^a	
In	0.05	1.81	450–500
	1.8×10^{-9}	0.28	
Ni	10^{-7}	0.25	440–510
Sb	0.05	1.94	450–500
	1.4×10^{-6}	0.75	
Sn	5.5×10^{-8}	0.75	390–512
Te	1.7×10^{-7}	0.57	300–500
Zn	0.5	1.35	360–500
	1.6×10^{-6}	2.3±0.3	360–500
	5.5	1.6	360–500
(Polycrystal)	1.7×10^{-7}	0.85	390–512
	$^a5.3 \times 10^{+7}$	2.61	
(High concentration)	$6.3 \times 10^{+8}$	2.61	
	$^a3.7 \times 10^{-10}$	0.7^a	
(Conc. = 2.2 x 10^{20} cm^{-3})	9.0×10^{-10}	0	
(Single crystal)	1.4×10^{-7}	0.86	390–512

Diffusion in Semiconductors (Continued)

Semiconductor and diffusing element	D_0 $(cm^2 \cdot s^{-1})$	ΔE (eV)	Temperature Range of Validity (°C)
Indium arsenide (InAs)			
Cd	4.35×10^{-4}	1.17	600–900
Cu		0.52a	
Ge	3.74×10^{-6}	1.17	600–900
Mg	1.98×10^{-6}	1.17	600–900
S	6.78	2.20	600–900
Se	12.55	2.20	600–900
Sn	1.49×10^{-6}	1.17	600–900
Te	3.43×10^{-5}	1.28	600–900
Zn	3.11×10^{-3}	1.17	600–900
Indium phosphide (InP)			
In	$1 \times 10^{+5}$	3.85	850–1000
P	$7 \times 10^{+10}$	5.65	850–1000
Iron oxide (Fe_3O_4)			
Fe	5.2	2.4	
Lead metasilicate ($PbSiO_3$)			
Pb	85	2.6	
Lead orthosilicate ($PbSiO_4$)			
Pb	8.2	2.0	
Mercury selenide (HgSe)			
Sb	6.3×10^{-5}	0.85	540–630
Nickel aluminate ($NiAl_2O_4$)			
Cr	1.17×10^{-3}	2.2	
Fe	1.33	3.5	

Diffusion in Semiconductors (Continued)

Semiconductor and diffusing element	D_0 ($cm^2 \cdot s^{-1}$)	ΔE (eV)	Temperature Range of Validity (°C)
Nickel chromate (III) ($NiCr_2O_4$)			
Cr	0.74	3.1	
Cr	2.03×10^{-5}	1.9	
Fe	1.35×10^{-3}	2.6	
Ni	0.85	3.2	
Selenium (Se) (amorphous)			
Fe	1.1×10^{-5}	0.38	300–400
Ge	9.4×10^{-6}	0.39	300–400
In	5.2×10^{-6}	0.32	300–400
Sb	2.8×10^{-8}	0.29	300–400
Se	7.6×10^{-10}	0.14	300–400
Sn	4.8×10^{-8}	0.39	300–400
Te	5.4×10^{-6}	0.53	300–400
Tl	1.4×10^{-6}	0.35	300–400
Zn	3.8×10^{-7}	0.29	300–400
Silicon (Si)			
Al	8.0	3.5	1100–1400
Ag	2×10^{-3}	1.6	1100–1350
As	3.2×10^{-1}	3.5	1100–1350
Au	1.1×10^{-3}	1.1	800–1200
B	$1.0 \times 10^{+1}$	3.7	950–1200
Bi	$1.04 \times 10^{+3}$	4.6	1100–1350
Cu	4×10^{-1}	1.0	800–1100
Fe	6.2×10^{-3}	0.86	1000–1200
Ga	3.6	3.5	1150–1350
Hl	9.4×10^{-3}	0.47	1000–1200
He	1.1×10^{-1}	0.86	1000–1200
In	$1.65 \times 10^{+1}$	3.9	1100–1350
Li	9.4×10^{-3}	0.78	100–800
P	$1.0 \times 10^{+1}$	3.7	1100–1350
Sb	5.6	3.9	1100–1350
Tl	$1.65 \times 10^{+1}$	3.9	1100–1350

Diffusion in Semiconductors (Continued)

Semiconductor and diffusing element	D_0 $(cm^2 \cdot s^{-1})$	ΔE (eV)	Temperature Range of Validity (°C)
Silicon carbide (SiC)			
Al	2.0×10^{-1}	4.9	1800–2250
B	$1.6 \times 10^{+1}$	5.6	1850–2250
Cr	2.3×10^{-1}	4.8	1700–1900
Sulfur (S)			
S	$2.8 \times 10^{+13}$	2.0	>100
Tin zinc oxide (SnZn$_2$O$_4$)			
Sn	$2 \times 10^{+5}$	4.7	
Zn	37	3.3	
Zinc aluminate (ZnAl$_2$O$_4$)			
Zn	$2.5 \times 10^{+2}$	3.4	
Zinc chromate			
Cr	8.5	3.5	
Zn	60	3.7	
Zinc ferrate (III) (ZnFe$_2$O$_4$)			
Fe	$8.5 \times 10^{+2}$	3.5	
Zn	$8.8 \times 10^{+2}$	3.7	
Zinc selenide (ZnSe)			
Cu	1.7×10^{-5}	0.56	200–570
Zinc sulfide (ZnS)			
Zn	$1.0 \times 10^{+16}$	6.50	>1030
	$1.5 \times 10^{+4}$	3.25	940–1030
	3.0×10^{-4}	1.52	<940

$D = D_0 e^{-\Delta E/kT}$

[a] Values obtained at the low concentration limit.

Source: From Bolz, R. E. and Tuve, G. L., Eds., *Handbook of Tables for Applied Engineering Science*, 2nd ed., CRC Press, Cleveland, 1973, 251.

TEMPER DESIGNATION SYSTEM FOR ALUMINUM ALLOYS

Temper	Definition
F	As fabricated
O	Annealed
H 1	Strain-hardened only
H2	Strain-hardened and partially annealed
H3	Strain-hardened and stabilized (mechanical properties stabilized by low- temperature thermal treatment)
T1	Cooled from an elevated-temperature shaping process and naturally aged to a substantially stable condition
T2	Cooled from an elevated temperature shaping process, cold-worked, and naturally aged to a substantially stable condition
T3	Solution heat-treated, cold-worked, and naturally aged to a substantially stable condition
T4	Solution heat-treated and naturally aged to a substantially stable condition
T5	Cooled from an elevated-temperature shaping process and artificially aged
T6	Solution heat-treated and artificially aged
T7	Solution heat-treated and stabilized
T8	Solution heat-treated, cold-worked, and artificially aged
T9	Solution heat-treated, artificially aged, and cold-worked
T10	Cooled from an elevated temperature shaping process, cold-worked, and artificially aged

Source: Data from *Metals Handbook*, 9th ed., Vol. 2, American Society for Metals, Metals Park, Ohio, 1979, 24-27.

Structure, Compositions
and
Phase Diagram Sources

THE SEVEN CRYSTAL SYSTEMS

System	Axial Lengths and Angles	Unit Cell Geometry
Cubic	$a = b = c, \quad \alpha = \beta = \gamma = 90°$	
Tetragonal	$a = b \neq c, \quad \alpha = \beta = \gamma = 90°$	
Orthorhombic	$a \neq b \neq c, \quad \alpha = \beta = \gamma = 90°$	
Rhombohedral	$a = b = c, \quad \alpha = \beta = \gamma \neq 90°$	
Hexagonal	$a = b \neq c, \quad \alpha = \beta = 90°, \gamma = 120°$	
Monoclinic	$a \neq b \neq c, \quad \alpha = \gamma = 90° \neq \beta$	
Triclinic	$a \neq b \neq c, \quad \alpha \neq \beta \neq \gamma \neq 90°$	

Source: James F. Shackelford, *Introduction to Materials Science for Engineers*, 2nd ed., Macmillan, New York, 1988, 72.

THE FOURTEEN   BRAVIS LATTICES

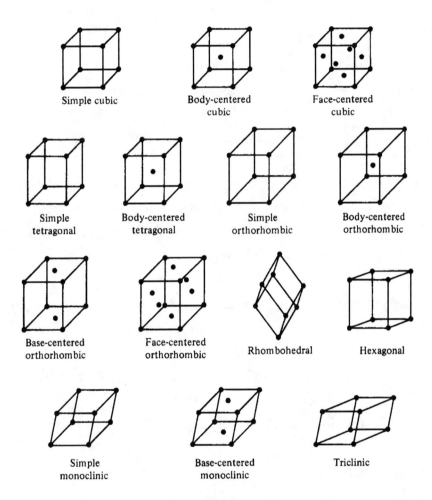

Source: James F. Shackelford, *Introduction to Materials Science for Engineers*, 2nd ed., Macmillan, New York, 1988, 73.

STRUCTURE OF SELECTED CERAMICS
Listed by Ceramic

Ceramic   Structure

BORIDES

Chromium Diboride (CrB_2)   hexagonal, AlB_2 structure (C-32 type)
                                     isomorphous with other transition metal
                                     diborides
                                     a=2.969Å; c=3.066Å; c/a=1.03

Hafnium Diboride (HfB_2)   hexagonal, AlB_2 structure (C-32 type)
                                     isomorphous with TiB_2 and ZrB_2
                                     a=3.141 ± 0.002 Å; c=3.470 ± 0.002 Å; c/
                                     a=1.105

Tantalum Diboride (TaB_2)   hexagonal, AlB_2 structure (C-32 type)
                                     isomorphous with other transition metal
                                     diborides
                                     a=3.078-3.088Å; c=3.241-3.265Å; c/
                                     a=1.06-1.074

low boron composition (64 atom % boron): a=3.097-3.099Å; c=3.244-
                                     3.277Å

high boron composition (72 atom % boron): a=3.057-3.060Å; c=3.291-
                                     3.290Å

Titanium Diboride (TiB_2)   hexagonal, AlB_2 structure (C-32 type)
                                     isomorphous with ZrB_2
                                     a=3.028-3.030Å; c=3.227-3.228Å; c/
                                     a=1.064

Structure of Selected Ceramics (Continued)
Listed by Ceramic

Ceramic	Structure
Zirconium Diboride (ZrB$_2$)	hexagonal, AlB$_2$ structure (C-32 type) isomorphous with TiB$_2$ a=3.1694-3.170Å; c=3.528-3.5365Å; c/a=1.114

CARBIDES

Boron Carbide (B$_4$C)	rhombic, C$_3$ chanins and B$_{12}$ icosahedral in a NaCl structure, extended along a body diagonal
Hafnium Monocarbide (HfC)	FCC(B$_1$), NaCl type isomorphous with HfB and HfN a=4.46-4.643Å

Silicon Carbide (SiC)	
low temperature form (β)	cubic
high temperature form (α)	hexagonal
β-SiC F43m space group	a=4.349-4.358Å
α-SiC C6MC space group	a=3.073Å; c=15.07Å; c/a=4.899

Tantalum Monocarbide (TaC)	FCC, NaCl type (B$_1$) a=4.42-4.456Å
Titanium Monocarbide (TiC)	FCC, NaCl type (B-1) isomorphous with TiO and TiN a=4.315-4.3316Å
Trichromium Dicarbide (Cr$_3$C$_2$)	orthorhombic D$^5_{10}$ type a=2.82Å, b=5.53Å, c=11.47Å

Structure of Selected Ceramics (Continued)
Listed by Ceramic

Ceramic	Structure
Tungsten Monocarbide (WC)	Hexagonal
	a=2.2897-2.90Å
Zirconium Monocarbide (ZrC)	FCC(B_1), NaCl type
	isomorphous with ZrB and ZrN
	a=4.669-4.694Å

NITRIDES

Aluminum Nitride (AlN)	hexagonal, Wurtzite structure
	a=3.10-3.114Å; c=4.96-4.981Å

Boron Nitride (BN)

hexagonal (common type)	graphite type structure
	a=2.5038±0.0001Å; c=6.60±0.01Å
B-N distance	1.45Å
cubic	zinc blende structure
	a=3.615Å
B-N distance	1.57Å

Titanium Mononitride (TiN)	cubic
	a=4.23Å
	homogeneity range: $TiN_{0.42}$-$TiN_{1.16}$ yields
	a=4.213 to 4.24Å

Trisilicon tetranitride (Si_3N_4)

α hexagonal	a=7.748-7.758Å; c=5.617-5.623Å
β hexagonal	a=7.608Å; c=2.911Å

Structure of Selected Ceramics (Continued)
Listed by Ceramic

Ceramic	Structure
Zirconium Mononitirde (ZrN)	cubic, NaCl type, B1 a=4.567-4.63Å

OXIDES

Ceramic	Structure
Aluminum Oxide (Al$_2$O$_3$)	hexagonal a=4.785Å; c=12.991Å; c/a=2.72
Beryllium Oxide (BeO)	hexagonal a=2.690-2.698Å; c=4.370-4.380Å
Calcium Oxide (CaO)	cubic, NaCl type a=4.8105Å
Cerium Dioxide (CeO$_2$)	cubic
Dichromium Trioxide (Cr$_2$O$_3$)	trigonal rhombic
Hafnium Dioxide (HfO$_2$)	monoclinic to 1700°C tetragonal above 1700°C a=5.1170Å; b=5.1754Å; c=5.2915Å $\beta = 99.216°$
Magnesium Oxide (MgO)	cubic, Fm3m space group a=4.313Å
Nickel monoxide (NiO)	face centered cubic, NaCl type

Structure of Selected Ceramics (Continued)
Listed by Ceramic

Ceramic	Structure
Silicon Dioxide (SiO_2)	hexagonal
Thorium Dioxide (ThO_2)	cubic, fluorite type $a=5.59525\text{-}5.5997\text{Å}$
Titanium Oxide (TiO_2)	tetragonal (rutile) $a=4.594\text{Å}$; $c=2.958\text{Å}$ at 26 °C tetragonal (anatase) rhombic (brookite)
Uranium Dioxide (UO_2)	cubic, fluorite type $a=5.471\text{Å}$
Zircoium Oxide (ZrO_2)	
to 1050°C	monoclinic $a=5.1505\text{Å}$; $b=5.2031\text{Å}$; $c=5.3154$ $\beta=99.194°$ at room temp.
1050—2100°C	tetragonal
above 2100°C	cubic
cubic (stabilized)	$a=5.132\pm0.006\text{Å}$ (8.13 mol% Y_2O_3) $a=5.145\pm0.006\text{Å}$ (11.09 mol% Y_2O_3) $a=5.146\pm0.006\text{Å}$ (12.08 mol% Y_2O_3) $a=5.153\pm0.006\text{Å}$ (15.52 mol% Y_2O_3) $a=5.162\pm0.006\text{Å}$ (17.88 mol% Y_2O_3)
Cordierite ($2MgO\ 2Al_2O_3\ 5SiO_2$)	orthorhombic

Structure of Selected Ceramics (Continued)
Listed by Ceramic

Ceramic	Structure
Mullite ($3Al_2O_3\ 2SiO_2$)	orthorhombic a=7.54±0.03Å; b=7.693±0.03Å;c=2.890±0.01
Sillimanite ($Al_2O_3\ SiO_2$)	Orthorhombic
Spinel ($Al_2O_3\ MgO$)	cubic a=8.0844Å

SILICIDES

Molybdenum Disilicide ($MoSi_2$)

> tetragonal, D_{4h}^{17} space group
> isomorphous with WSi_2
> a=3.197-3.20Å; c=7.85-7.871

Tungsten Disilicide (WSi_2)

> tetragonal, D_{4h}^{17} space group
> isomorphous with $MoSi_2$
> a=3.212±0.005Å; c=7.880±0.005

To convert Å to nm, multiply by 10.

Source: Data compiled by J.S. Park from *No. 1 Materials Index*, Peter T.B. Shaffer, Plenum Press, New York, (1964); *Smithells Metals Reference Book*, Eric A. Brandes, ed., in association with Fulmer Research Institute Ltd. 6th ed. London, Butterworths, Boston, (1983); and *Ceramic Abstracts*, American Ceramic Society (1986-1991).

DENSITY OF SELECTED TOOL STEELS

Type	Density Mg/m^3
W1	7.84
S1	7.88
S5	7.76
S7	7.76
A2	7.86
H11	7.75
H13	7.76
H21	8.28
H26	8.67
T1	8.67
T15	8.19
M2	8.16
L2	7.86
L6	7.86

Source: Data from *ASM Metals Reference Book, Second Edition*, American Society for Metals, Metals Park, Ohio 44073, p242, (1984).

Density Range of Tool Steels

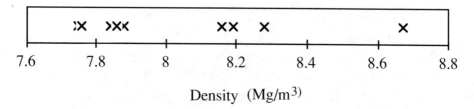

Density (Mg/m^3)

DENSITY OF SELECTED ALLOY CAST IRONS

Description	Density (Mg/m^2)
Abrasion–Resistant White Irons	
Low–C white iron	7.6 to 7.8
Martensitic nickel–chromium iron	7.6 to 7.8
Corrosion–Resistant Irons	
High– Silicon iron	7.0 to 7.05
High–chromium iron	7.3 to 7.5
High–nickel gray iron	7.4 to 7.6
High–nickel ductile iron	7.4
Heat–Resistant Gray Irons	
Medium–silicon iron	6.8 to 7.1
High–chromium iron	7.3 to 7.5
High–nickel iron	7.3 to 7.5
Nickel–chromium–silicon iron	7.33 to 7.45
High–aluminum iron	5.5 to 6.4
Heat–Resistant Ductile Irons	
Medium–silicon ductile iron	7.1
High–nickel ductile (20 Ni)	7.4
High–nickel ductile (23 Ni)	7.4

Source: Data from *ASM Metals Reference Book, Second Edition*, American Society for Metals, Metals Park, Ohio 44073, p172, (1984).

Density Range of Selected Alloy Cast Irons

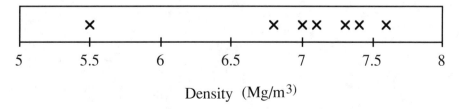

Density (Mg/m^3)

DENSITY OF SELECTED CERAMICS
Listed by Ceramic

Ceramic	Density (g/cm^3)
BORIDES	
Chromium Diboride (CrB_2)	5.6
Hafnium Diboride (HfB_2)	11.2
Tantalum Diboride (TaB_2)	12.60
Titanium Diboride (TiB_2)	4.5-4.62
Zirconium Diboride (ZrB_2)	6.09-6.102
CARBIDES	
Boron Carbide (B_4C)	2.51
Hafnium Monocarbide (HfC)	12.52-12.70
Silicon Carbide (SiC)	
(hexagonal)	3.217
(cubic)	3.210
Tantalum Monocarbide (TaC)	14.48-14.65
Titanium Monocarbide (TiC)	4.92-4.938
Trichromium Dicarbide (Cr_3C_2)	6.70
Tungsten Monocarbide (WC)	15.8
Zirconium Monocarbide (ZrC)	6.44-6.73
NITRIDES	
Aluminum Nitride (AlN)	3.26-3.30
Boron Nitride (BN)	
(cubic)	3.49
(hexagonal)	2.27

Density of Selected Ceramics (Continued)
Listed by Ceramic

Ceramic	Density (g/cm^3)
Titanium Mononitride (TiN)	5.43
Trisilicon tetranitride (Si$_3$N$_4$)	
(α)	3.184
(β)	3.187
Zirconium Mononitride (TiN)	7.349
OXIDES	
Aluminum Oxide (Al$_2$O$_3$)	3.97-3.986
Beryllium Oxide (BeO)	3.01-3.03
Calcium Oxide (CaO)	3.32
Cerium Dioxide (CeO$_2$)	7.28
Dichromium Trioxide (Cr$_2$O$_3$)	5.21
Hafnium Dioxide (HfO$_2$)	9.68
Magnesium Oxide (MgO)	3.581
Nickel monoxide (NiO)	6.8-7.45
Thorium Dioxide (ThO$_2$)	9.821
Titanium Oxide (TiO$_2$)	
(anatase)	3.84
(brookite)	4.17
(rutile)	4.25
Uranium Dioxide (UO$_2$)	10.949-10.97

Density of Selected Ceramics (Continued)
Listed by Ceramic

Ceramic	Density (g/cm^3)
Zircoium Oxide (ZrO_2)	
(monoclinic)	5.56
(CaO stabilized)	5.5
(MgO stabilized)	5.43
(plasma sprayed)	5.6-5.7
Cordierite ($2MgO\ 2Al_2O_3\ 5SiO_2$)	1.61-2.51
Mullite ($3Al_2O_3\ 2SiO_2$)	2.6-3.26
(theoretical)	3.16-3.22
Sillimanite ($Al_2O_3\ SiO_2$)	3.23-3.24
Spinel ($Al_2O_3\ MgO$)	3.580
Zircon ($SiO_2\ ZrO_2$)	4.6
SILICIDES	
Molybdenum Disilicide ($MoSi_2$)	6.24-6.29
Tungsten Disilicide (WSi_2)	9.25-9.3

Source: Data compiled by J.S. Park from *No. 1 Materials Index*, Peter T.B. Shaffer, Plenum Press, New York, (1964); *Smithells Metals Reference Book*, Eric A. Brandes, ed., in association with Fulmer Research Institute Ltd. 6th ed. London, Butterworths, Boston, (1983); and *Ceramic Abstracts*, American Ceramic Society (1986-1991).

SPECIFIC GRAVITY OF SELECTED POLYMERS
Listed by Polymer

Polymer	Specific Gravity (ASTM D792)
ABS Resins; Molded, Extruded	
Medium impact	1.05—1.07
High impact	1.02—1.04
Very high impact	1.01—1.06
Low temperature impact	1.02—1.04
Heat resistant	1.06—1.08
Acrylics; Cast, Molded, Extruded	
Cast Resin Sheets, Rods:	
General purpose, type I	1.17—1.19
General purpose, type II	1.18—1.20
Moldings:	
Grades 5, 6, 8	1.18—1.19
High impact grade	1.12—1.16
Thermoset Carbonate	
Allyl diglycol carbonate	1.32
Alkyds; Molded	
Putty (encapsulating)	2.05—2.15
Rope (general purpose)	2.20—2.22
Granular (high speed molding)	2.21—2.24
Glass reinforced (heavy duty parts)	2.02—2.10

Specific Gravity of Selected Polymers (Continued)
Listed by Polymer

Polymer	Specific Gravity (ASTM D792)
Cellulose Acetate; Molded, Extruded	
ASTM Grade:	
H6—1	
H4—1	1.29—1.31
H2—1	1.25—1.31
MH—1, MH—2	1.24—1.31
MS—1, MS—2	1.23—1.30
S2—1	1.22—1.30
Cellulose Acetate Butyrate; Molded, Extruded	
ASTM Grade:	
H4	1.22
MH	1.18—1.20
S2	1.15—1.18
Cellusose Acetate Propionate; Molded, Extruded	
ASTM Grade:	
1	1.22
3	1.20—1.21
6	1.19
Chlorinated Polymers:	
Chlorinated polyether	1.4
Chlorinated polyvinyl chloride	1.54

Specific Gravity of Selected Polymers (Continued)
Listed by Polymer

Polymer	Specific Gravity (ASTM D792)
Polycarbonates:	
Polycarbonate	1.2
Polycarbonate (40% glass fiber reinforced)	1.51
Chlorinated Polymers	
Chlorinated polyether	1.4
Chlorinated polyvinyl chloride	1.54
Polycarbonates	
Polycarbonate	1.2
Polycarbonate (40% glass fiber reinforced)	1.51
Diallyl Phthalates; Molded	
Orlon filled	1.31—1.35
Dacron filled	1.40—1.65
Asbestos filled	1.50—1.96
Glass fiber filled	1.55—1.85
Fluorocarbons; Molded,Extruded	
Polytrifluoro chloroethylene (PTFCE)	2.10—2.15
Polytetrafluoroethylene (PTFE)	2.1—2.3
Ceramic reinforced (PTFE)	2.2—2.4
Fluorinated ethylene propylene(FEP)	2.12—2.17
Polyvinylidene— fluoride (PVDF)	1.77

Specific Gravity of Selected Polymers (Continued)
Listed by Polymer

Polymer	Specific Gravity (ASTM D792)
Epoxies; Cast, Molded, Reinforced	
Standard epoxies (diglycidyl ethers of bisphenol A)	
Cast rigid	1.15
Cast flexible	1.14-1.18
Molded	1.80-2.0
General purpose glass cloth laminate	1.8
High strength laminate	1.84
Filament wound composite	2.18-2.17
Epoxies—Molded, Extruded	
High performance resins	
(cycloaliphatic diepoxides)	
Cast, rigid	1.24
Molded	1.7
Glass cloth laminate	1.97
Epoxy novolacs	
Cast, rigid	1.22
Glass cloth laminate	1.97
Melamines; Molded	
Filler & type	
Unfilled	1.48
Cellulose electrical	1.43—1.50
Glass fiber	1.8—2.0
Alpha cellulose and mineral	1.5(a), 1.72(mineral)

Specific Gravity of Selected Polymers (Continued)
Listed by Polymer

Polymer	Specific Gravity (ASTM D792)
Nylons; Molded, Extruded	
Type 6	
General purpose	1.12—1.14
Glass fiber (30%) reinforced	1.35—1.42
Cast	1.15
Flexible copolymers	1.12—1.14
Type 8	1.09
Type 11	1.04
Type 12	1.01
Nylons; Molded, Extruded	
6/6 Nylon	
General purpose molding	1.13—1.15
Glass fiber reinforced	1.37–1.47
Glass fiber Molybdenum disulfide filled	1.37—1.41
General purpose extrusion	1.13–1.15
6/10 Nylon	
General purpose	1.07—1.09
Glass fiber (30%) reinforced	1.3
Phenolics; Molded	
Type and filler	
General: woodflour and flock	1.32—1.46
Shock: paper, flock, or pulp	1.34—1.46
High shock: chopped fabric or cord	1.36—1.43
Very high shock: glass fiber	1.75—1.90

Specific Gravity of Selected Polymers (Continued)
Listed by Polymer

Polymer	Specific Gravity (ASTM D792)
Polyacetals	
Homopolymer:	
Standard	1.425
20% glass reinforced	1.56
22% TFE reinforced	1.54
Copolymer:	
Standard	1.41
25% glass reinforced	1.61
High flow	1.41
Phenolics: Molded	
Arc resistant—mineral	1.5—3.0
Rubber phenolic—woodflour or flock	1.24—1.35
Rubber phenolic—chopped fabric	1.30—1.35
Rubber phenolic—asbestos	1.60—1.65
ABS–Polycarbonate Alloy	1.14
PVC–Acrylic Alloy	
PVC–acrylic sheet	1.35
PVC–acrylic injection molded	1.3
Polymides	
Unreinforced	1.19—1.47
Glass reinforced	1.60—1.95

Specific Gravity of Selected Polymers (Continued)
Listed by Polymer

Polymer	Specific Gravity (ASTM D792)
Polyester; Thermoplastic	
Injection Moldings:	
General purpose grade	1.31
Glass reinforced grades	1.52
Glass reinforced self extinguishing	1.58
General purpose grade	1.31
Glass reinforced grade	1.45
Asbestos—filled grade	1.46
Polyesters: Thermosets	
Cast polyyester	
Rigid	1.12—1.46
Flexible	1.06—1.25
Reinforced polyester moldings	
High strength (glass fibers)	1.8—2.0
Heat and chemical resistant (asbestos)	1.5—1.75
Sheet molding compounds, general purpose	1.65—1.80
Phenylene Oxides	
SE—100	1.1
SE—1	1.06
Glass fiber reinforced	1.21,1.27
Phenylene oxides (Noryl)	
Standard	1.24
Glass fiber reinforced	1.41,1.55
Polyarylsulfone	1.36

Specific Gravity of Selected Polymers (Continued)
Listed by Polymer

Polymer	Specific Gravity (ASTM D792)
Polypropylene:	
General purpose	0.900—0.910
High impact	0.900—0.910
Asbestos filled	1.11—1.36
Glass reinforced	1.04—1.22
Flame retardant	1.2
Polyphenylene sulfide:	
Standard	1.34—1.35
40% glass reinforced	1.6—1.64
Polyethylenes; Molded, Extruded	
Type I—lower density (0.910—0.925)	
Melt index 0.3—3.6	0.910—0.925
Melt index 6—26	0.918—0.925
Melt index 200	0.91
Type II—medium density (0.926—0.940)	
Melt index 20	0.93
Melt index 1.0—1.9	0.930—0.940
Type III—higher density (0.941—0.965)	
Melt index 0.2—0.9	0.96
Melt index 0.1—12.0	0.950—0.955
Melt index 1.5—15	0.96
High molecular weight	0.94

Specific Gravity of Selected Polymers (Continued)
Listed by Polymer

Polymer	Specific Gravity (ASTM D792)
Olefin Copolymers; Molded	
EEA (ethylene ethyl acrylate)	0.93
EVA (ethylene vinyl acetate)	0.94
Ethylene butene	0.95
Propylene—ethylene	0.91
Ionomer	0.94
Polyallomer	0.898—0.904
Polystyrenes; Molded	
Polystyrenes	
General purpose	1.04
Medium impact	1.04—1.07
High impact	1.04—1.07
Glass fiber -30% reinforced	1.29
Styrene acrylonitrile (SAN)	1.04—1.07
Glass fiber (30%) reinforced SAN	1.35
Polyvinyl Chloride And Copolymers; Molded, Extruded	
Nonrigid—general	1.20—1.55
Nonrigid—electrical	1.16—1.40
Rigid—normal impact	1.32—1.44
Vinylidene chloride	1.68—1.75

Specific Gravity of Selected Polymers (Continued)
Listed by Polymer

Polymer	Specific Gravity (ASTM D792)
Silicones; Molded, Laminated	
Fibrous (glass) reinforced silicones	1.88
Granular (silica) reinforced silicones	1.86—2.00
Woven glass fabric/ silicone laminate	1.75—1.8
Ureas; Molded	
Alpha—cellulose filled (ASTM Type 1)	1.45—1.55
Cellulose filled (ASTM Type 2)	1.52
Woodflour filled	1.45—1.49

Source: data compiled by J.S. Park from Charles T. Lynch, *CRC Handbook of Materials Science*, Vol. 3, CRC Press, Boca Raton, Florida, 1975 and *Engineered Materials Handbook*, Vol.2, Engineering Plastics, ASM International, Metals Park, Ohio, 1988.

COMPOSITION LIMITS OF

| Designations | | | Composition (%) | | |
AISI	SAE	UNS	C	Mn	Si

Molybdenum high speed steels

M2	M2	T11302.	0.78-0.88; 0.95-1.05	0.15-0.40	0.20-0.45

Tungsten high speed steels

T1	T1	T12001	0.65-0.80	0.10-0.40	0.20-0.40
T15		T12015	1.50-1.60	0.15-0.40	0.15-0.40

Chromium hot work steels

H11	H11	T20811	0.33-0.43	0.20-0.50	0.80-1.20
H13	H13	T20813	0.32-0.45	0.20-0.50	0.80-1.20

Tungsten hot work steels

H21	H21	T20821	0.26-0.36	0.15-0.40	0.15-0.50
H26		T20826	0.45-0.55(b)	0.15-0.40	0.15-0.40

Air-hardening medium-alloy cold work steels

A2	A2	T30102	0.95-1.05	1.00 max	0.50 max
A3		T30103	1.20-1.30	0.40-0.60	0.50 max

Shock-resisting steels

S1	S1	T41901	0.40-0.55	0.10-0.40	0.15-1.20
S5	S5	T41905	0.50-0.65	0.60-1.00	1.75-2.25
S7		T41907	0.45-0.55	0.20-0.80	0.20-1.00

Low-alloy special-purpose tool steels

L2		T61202	0.45-1.00(b)	0.10-0.90	0.50 max
L6	L6	T61206	0.65-0.75	0.25-0.80	0.50 max

SELECTED TOOL STEELS

		Composition Continued (%)			
Cr	Ni	Mo	W	V	Co
3.75-4.50	0.30 max	4.50-5.50	5.50-6.75	1.75-2.20	
3.75-4.00	0.30 max		17.25-18.75	0.90-1.30	
3.75-5.00	0.30 max	1.00 max	11.75-13.00	4.50-5.25	4.75-5.25
4.75-5.50	0.30 max	1.10-1.60		0.30-0.60	
4.75-5.50	0.30 max	1.10-1.75		0.80-1.20	
3.00-3.75	0.30 max		8.50-10.00	0.30-0.60	
3.75-4.50	0.30 max		17.25-19.00	0.75-1.25	
4.75-5.50	0.30 max	0.90-1.40		0.15-0.50	
4.75-5.50	0.30 max	0.90-1.40		0.80-1.40	
1.00-1.80	0.30 max	0.50 max	1.50-3.00	0.15-0.30	
0.35 max		0.20-1.35		0.35 max	
3.00-3.50		1.30-1.80		0.20-0.30(d)	
0.70-1.20		0.25 max		0.10-0.30	
0.60-1.20	1.25-2.00	0.50 max		0.20-0.30(d)	

Composition Limits of

Designations			Composition (%)		
AISI	SAE	UNS	C	Mn	Si

Water-hardening tool steels

| W1 | W108, W109, W110, W112 | T72301 | 0.70-1.50(e) | 0.10-0.40 | 0.10-0.40 |

Source: Data from ASM Metals Reference Book, Second Edition, American Society for Metals, Metals Park, Ohio 44073, p.239, (1984).

M2 Tool Steel Composition

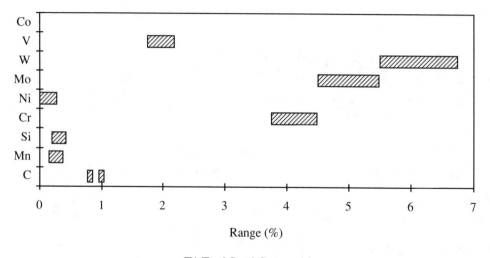

Range (%)

T1 Tool Steel Composition

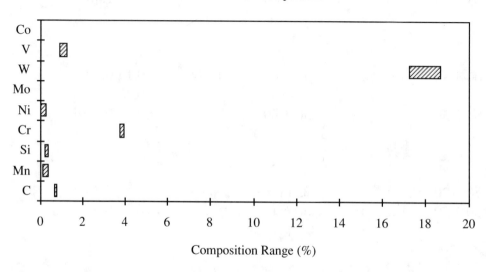

Composition Range (%)

Selected Tool Steels (Continued)

Composition Continued (%)

Cr	Ni	Mo	W	V	Co
0.15 max	0.20 max	0.10 max	0.15 max	0.10 max	

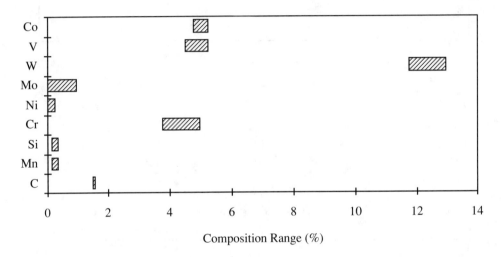

T15 Tool Steel Composition

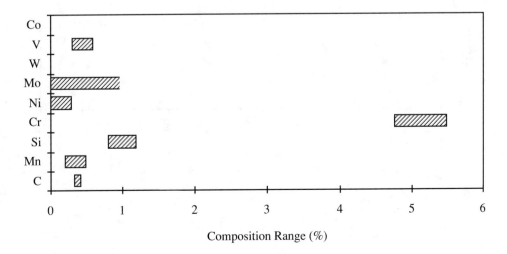

H11 Tool Steel Composition

Composition Limits of Selected Tool Steels (Continued)

H21 Tool Steel Composition

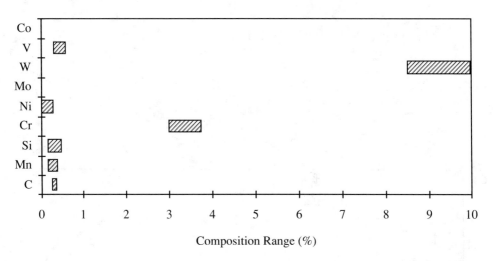

H26 Tool Steel Composition

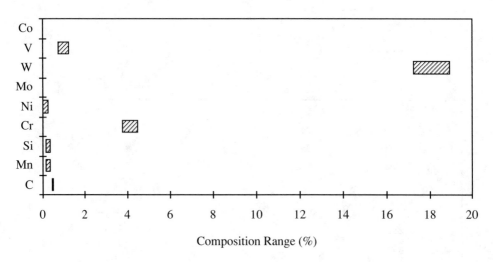

Composition Limits of Selected Tool Steels (Continued)

A2 Tool Steel Composition

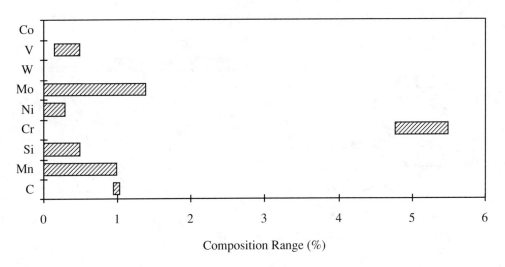

Composition Range (%)

A3 Tool Steel Composition

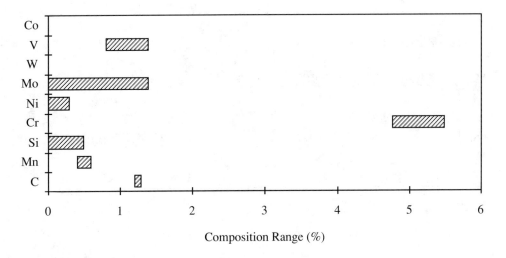

Composition Range (%)

Composition Limits of Selected Tool Steels (Continued)

S1 Tool Steel Composition

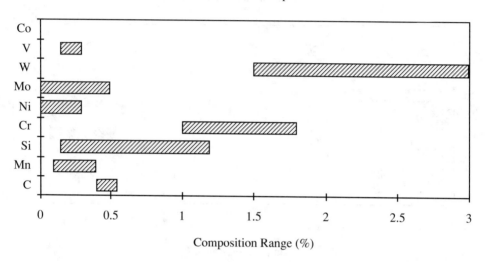

Composition Range (%)

S5 Tool Steel Composition

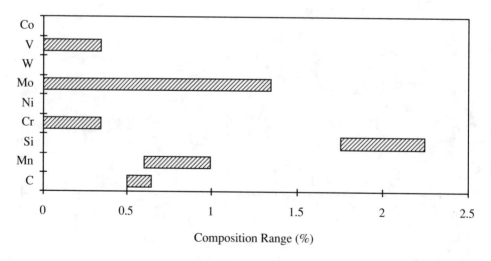

Composition Range (%)

Composition Limits of Selected Tool Steels (Continued)

S7 Tool Steel Composition

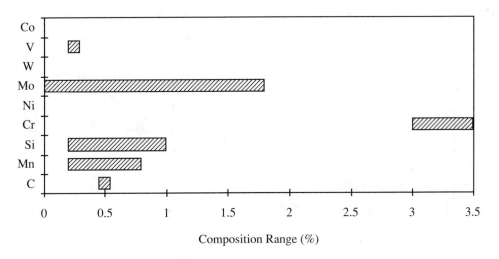

Composition Range (%)

L2 Tool Steel Composition

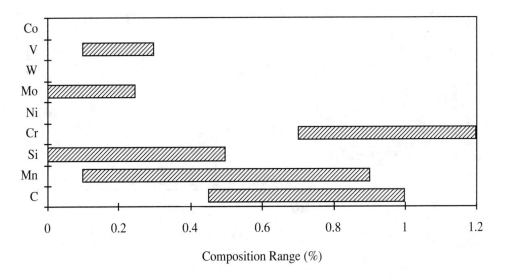

Composition Range (%)

Composition Limits of Selected Tool Steels (Continued)

L6 Tool Steel Composition

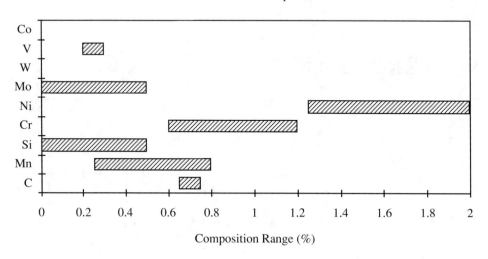

W1 Tool Steel Composition

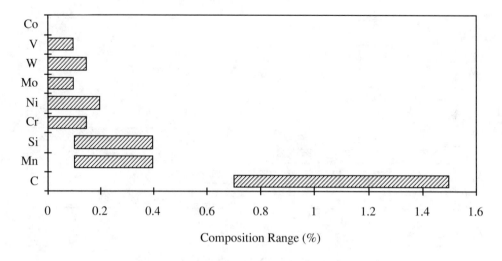

Source: Data from *ASM Metals Reference Book, Second Edition,* American Society for Metals, Metals Park, Ohio 44073, p.239, (1984).

COMPOSITION LIMITS OF SELECTED GRAY CAST IRONS

UNS	SAE grade	TC	Mn	Si	P	S
			Composition Limits (%)			
F10004	G1800	3.40 to 3.70	0.50 to 0.80	2.80 to 2.30	0.15	0.15
F10005	G2500	3.20 to 3.50	0.60 to 0.90	2.40 to 2.00	0.12	0.15
F10009	G2500a	3.40 min	0.60 to 0.90	1.60 to 2.10	0.12	0.12
F10006	G3000	3.10 to 3.40	0.60 to 0.90	2.30 to 1.90	0.10	0.16
F10007	G3500	3.00 to 3.30	0.60 to 0.90	2.20 to 1.80	0.08	0.16
F10010	G3500b	3.40 min	0.60 to 0.90	1.30 to 1.80	0.08	0.12
F10011	G3500c	3.50 min	0.60 to 0.90	1.30 to 1.80	0.08	0.12
F10008	G4000	3.00 to 3.30	0.70 to 1.00	2.10 to 1.80	0.07	0.16
F10012	G4000d	3.10 to 3.60	0.60 to 0.90	1.95 to 2.40	0.07	0.12

F10004 Gray Cast Iron Composition

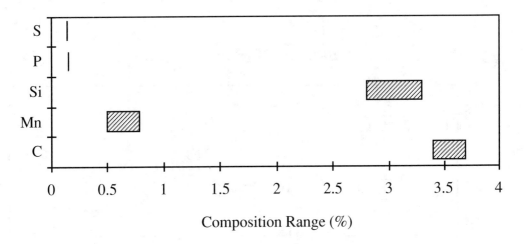

Composition Range (%)

Source: Data from *ASM Metals Reference Book, Second Edition,* American Society for Metals, Metals Park, Ohio 44073, p.166, (1984).

Composition Limits of Selected Gray Cast Irons (Continued)

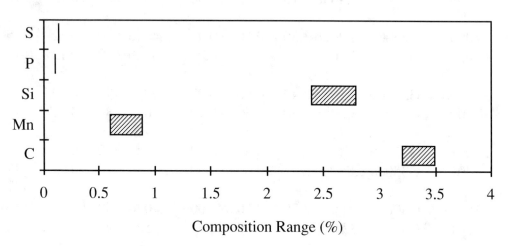

F10005 Gray Cast Iron Composition

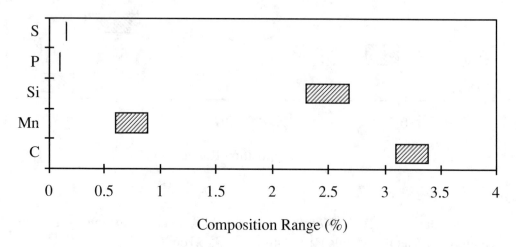

F10006 Gray Cast Iron Composition

Composition Limits of Selected Gray Cast Irons (Continued)

F10007 Gray Cast Iron Composition

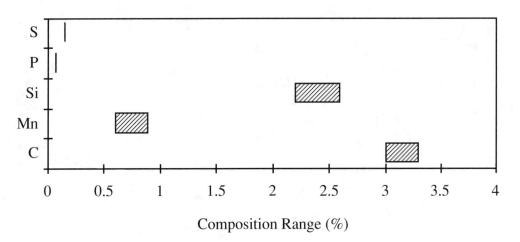

Composition Range (%)

F10008 Gray Cast Iron Composition

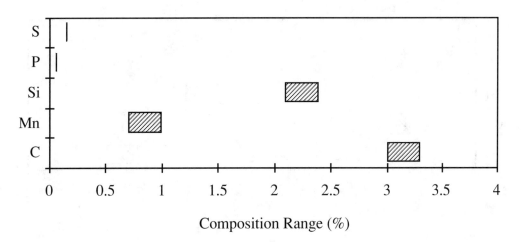

Composition Range (%)

Composition Limits of Selected Gray Cast Irons (Continued)

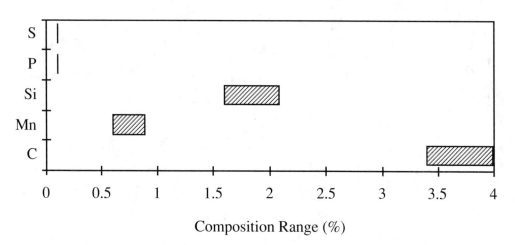

F10009 Gray Cast Iron Composition

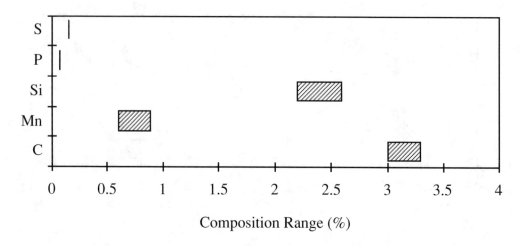

F10010 Gray Cast Iron Composition

Composition Limits of Selected Gray Cast Irons (Continued)

F10011 Gray Cast Iron Composition

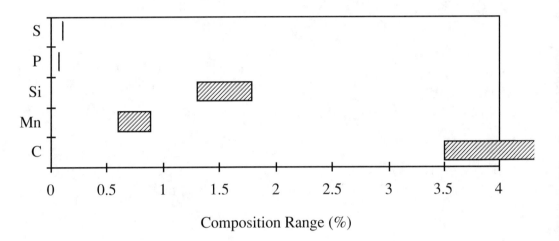

Composition Range (%)

F10012 Gray Cast Iron Composition

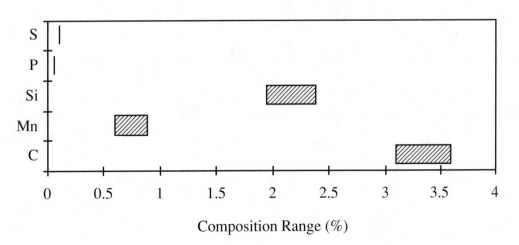

Composition Range (%)

COMPOSITION LIMITS OF SELECTED DUCTILE IRONS

Specification No. Grade or Class	UNS	TC	Si	Composition (%) Mn	P	S
ASTM A395 ASME SA395 60-40-18	F32800	3.00 min	2.50 max	0.08 max		
ASTM A476 SAE AM55316 80-60-03	F34100	3.00 min	3.0 max	0.08 max		
SAE J434c D4018	F32800	3.20 –4.10	1.80-3.00	0.10-1.00	0.015-0.10	0.005–0.035
MIL-1-24137 (Ships) Class A	F33101	3.0 min	2.50 max	0.08 max		
Class B	F43020	2.40-3.00	1.80-3.20	0.80-1.50	0.20 max	
Class C	F43021	2.70-3.10	2.00-3.00	1.90-2.50	0.15 max	

Source: Data from *ASM Metals Reference Book, Second Edition*, American Society for Metals, Metals Park, Ohio 44073, p.168-169, (1984).

Composition Limits of Selected Ductile Irons (Continued)

ASTM A395 Ductile Iron Composition

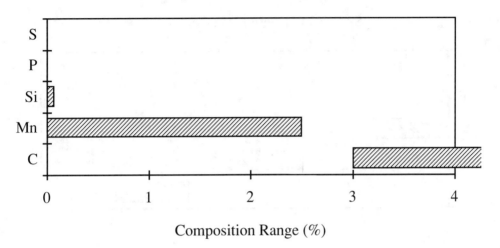

ASTM A476 Ductile Iron Composition

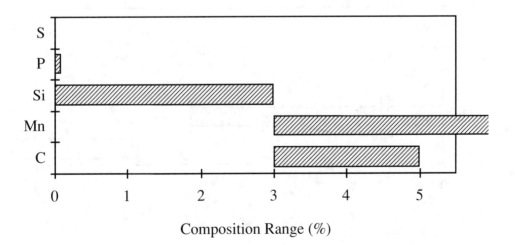

Composition Limits of Selected Ductile Irons (Continued)

SAE J434c Ductile Iron Composition

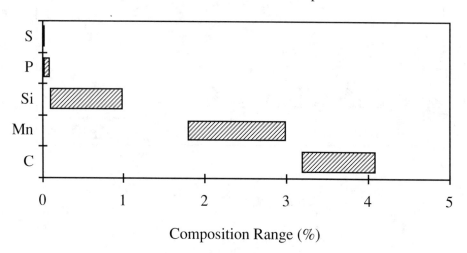

Composition Range (%)

MlL-1-24137 Class A Ductile Iron Composition

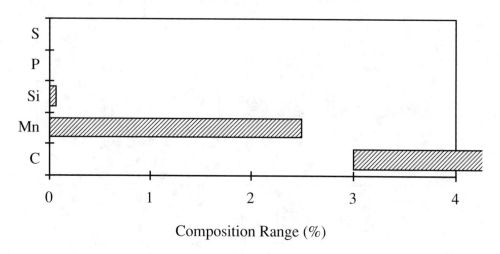

Composition Range (%)

Composition Limits of Selected Ductile Irons (Continued)

MlL-1-24137 Class B Ductile Iron Composition

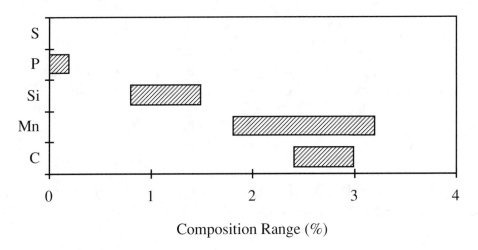

Composition Range (%)

MlL-1-24137 Class C Ductile Iron Composition

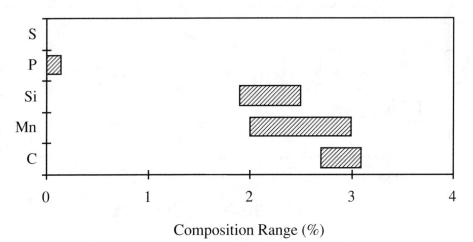

Composition Range (%)

COMPOSITION RANGES FOR SELECTED MALLEABLE IRONS

Type	Composition (%)				
	TC	Mn	Si	P	S
Ferritic					
Grade 32510	2.30-2.70	0.25-0.55	1.00-1.75	0.05 max	0.03-0.18
Grade 35018	2.00-2.45	0.25-0.55	1.00-1.35	0.05 max	0.03-0.18
Pearlitic	2.00-2.70	0.25-1.25	1.00-1.75	0.05 max	0.03-0.18

Source: Data from *ASM Metals Reference Book, Second Edition*, American Society for Metals, Metals Park, Ohio 44073, p170, (1984).

Grade 32510 Ferritic Iron Composition

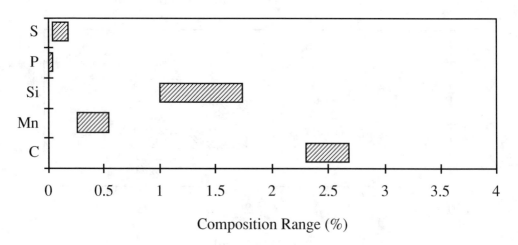

Composition Range (%)

Composition Ranges for Selected Malleable Irons (Continued)

Grade 35018 Ferritic Iron Composition

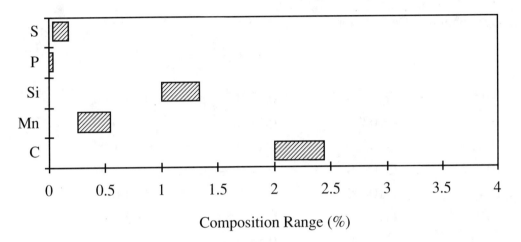

Composition Range (%)

Pearlitic Iron Composition

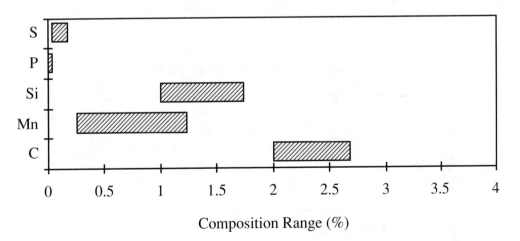

Composition Range (%)

COMPOSITION RANGES FOR SELECTED CARBON STEELS

Composition Range (%)

AISI–SAE Designation	UNS Designation	C	Mn
1015	G10150	0.12-0.18	0.30-0.60
1020	G10200	0.17-0.23	0.30-0.60
1022	G10220	0.17-0.23	0.70-1.00
1030	G10900	0.27-0.34	0.60-0.90
1040	G10400	0.36-0.44	0.60-0.90
1050	G10500	0.47-0.55	0.60-0.90
1060	G10600	0.55-0.66	0.60-0.90
1080	G10800	0.74-0.88	0.60-0.90
1095	G10950	0.90-1.04	0.30-0.50

Source: Data from *ASM Metals Reference Book, Second Edition*, American Society for Metals, Metals Park, Ohio 44073, p184, (1984).

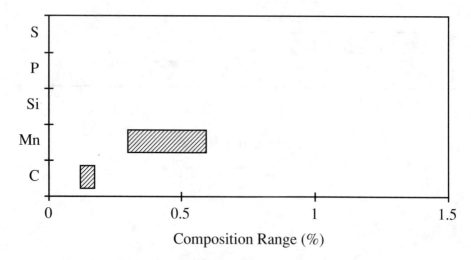

1015 Carbon Steel Composition

Composition Ranges for Selected Carbon Steels (Continued)

1020 Carbon Steel Composition

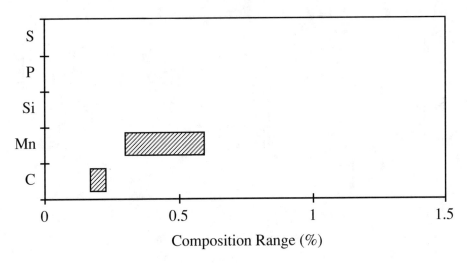

1022 Carbon Steel Composition

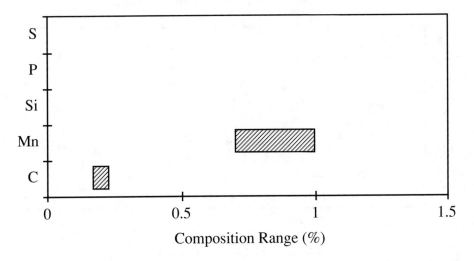

Composition Ranges for Selected Carbon Steels (Continued)

1030 Carbon Steel Composition

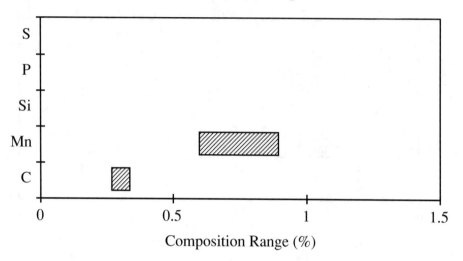

Composition Range (%)

1040 Carbon Steel Composition

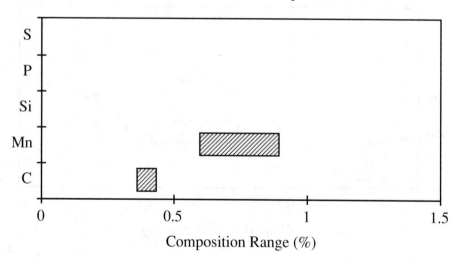

Composition Range (%)

Composition Ranges for Selected Carbon Steels (Continued)

1050 Carbon Steel Composition

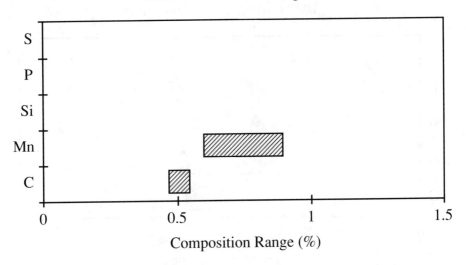

Composition Range (%)

1060 Carbon Steel Composition

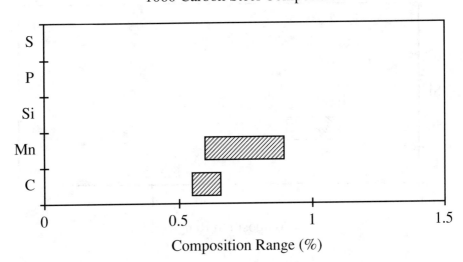

Composition Range (%)

Composition Ranges for Selected Carbon Steels (Continued)

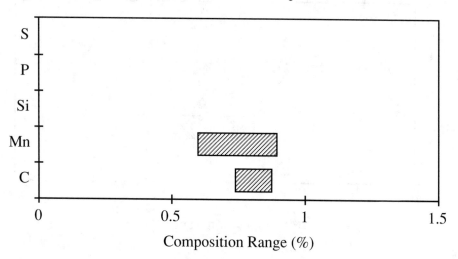

1080 Carbon Steel Composition

Composition Range (%)

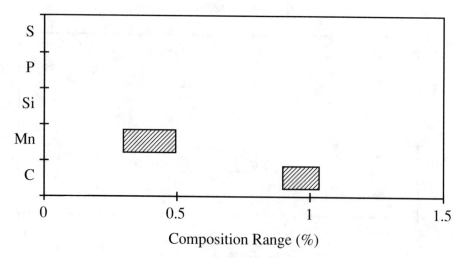

1095 Carbon Steel Composition

Composition Range (%)

COMPOSITION RANGES FOR SELECTED RESULFURIZED CARBON STEELS

AISI–SAE Designation	UNS Designation	Composition Range (%)		
		C	Mn	S
1118	G11180	0.14-0.20	1.30-1.60	0.08-0.13
1137	G11370	0.32-0.39	1.35-1.65	0.08-0.13
1141	G11410	0.37-0.45	1.35-1.65	0.08-0.13
1144	G11440	0.40-0.48	1.35-1.65	0.24-0.33

Source: Data from *ASM Metals Reference Book, Second Edition*, American Society for Metals, Metals Park, Ohio 44073, p185, (1984).

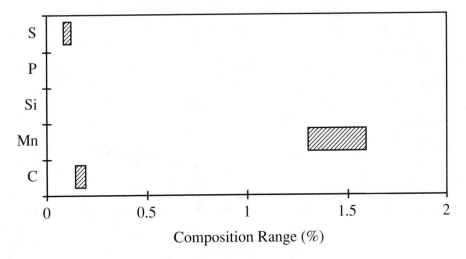

1118 Carbon Steel Composition

Composition Ranges for Selected Resulfurized Carbon Steels
(Continued)

1137 Carbon Steel Composition

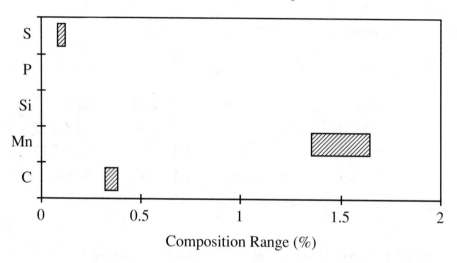

Composition Range (%)

1141 Carbon Steel Composition

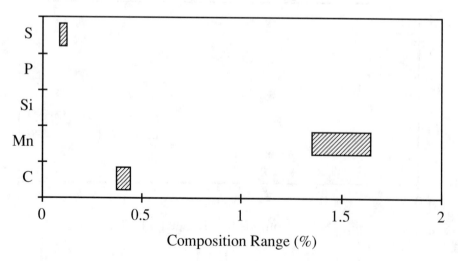

Composition Range (%)

Composition Ranges for Selected Resulfurized Carbon Steels
(Continued)

1144 Carbon Steel Composition

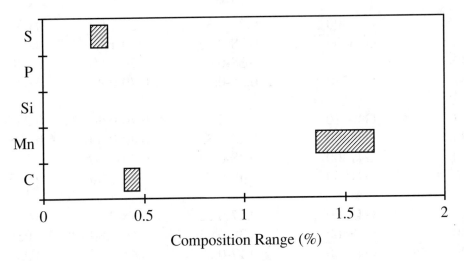

COMPOSITION RANGES FOR

Composition Range (%)

AISI–SAE Designation	UNS Designation	C	Mn	P (max)
1330	G13300	0.28-0.33	1.60-1.90	0.035
1340	G13400	0.38-0.43	1.60-1.90	0.035
3140		0.38-0.43	0.70-0.90	0.040
4037	G40370	0.35-0.40	0.70-0.90	0.035
4042	G40420	0.40-0.45	0.70-0.90	0.035
4130	G41300	0.28-0.33	0.40-0.60	0.035
4140	G41400	0.38-0.43	0.75-1.00	0.035
4150	G41500	0.48-0.53	0.75-1.00	0.035
4320	G43200	0.17-0.22	0.45-0.65	0.035
4340	G43400	0.38-0.43	0.60-0.80	0.035
4620	G46200	0.17-0.22	0.45-0.65	0.035
4820	G48200	0.18-0.23	0.50-0.70	0.035
5046	G50460	0.43-0.48	0.75-1.00	0.035
50B46	G50461	0.44-0.49	0.75-1.00	0.035
5060	G50600	0.56-0.64	0.75-1.00	0.035
50B60	G50461	0.56-0.64	0.75-1.00	0.035
5130	G51300	0.28-0.33	0.70-0.90	0.035
5140	G51400	0.38-0.43	0.70-0.90	0.035
5150	G51500	0.48-0.53	0.70-0.90	0.035
5160	G51600	0.56-0.64	0.75-1.00	0.035
51B60	G51601	0.56-0.64	0.75-1.00	0.035
6150(a)	G61500	0.48-0.53	0.70-0.90	0.035
81B45	G81451	0.43-0.48	0.75-1.00	0.035
8620	G86200	0.18-0.23	0.70-0.90	0.035
8630	G86300	0.28-0.33	0.70-0.90	0.035
8640	G86400	0.38-0.43	0.75-1.00	0.035
86B45	G86451	0.43-0.48	0.75-1.00	0.035
8650	G86500	0.48-0.53	0.75-1.00	0.035

SELECTED ALLOY STEELS

S (max)	Si	Cr	Ni	Mo
0.040	0.15-0.30			
0.040	0.15-0.30			
0.040	0.20-0.35	0.55-0.75	1.10-1.40	
0.040	0.15-0.30	0.20-0.30		
0.040	0.15-0.30	0.20-0.30		
0.040	0.15-0.30	0.80-1.10		0.15-0.25
0.040	0.15-0.30	0.80-1.10		0.15-0.25
0.040	0.15-0.30	0.80-1.10		0.15-0.25
0.040	0.15-0.30	0.40-0.60	1.65-2.00	0.20-0.30
0.040	0.15-0.30	0.70-0.90	1.65-2.00	0.20-0.30
0.040	0.15-0.30	1.65-2.00		0.20-0.30
0.040	0.15-0.30	3.25-3.75		0.20-0.30
0.040	0.15-0.30	0.20-0.35		
0.040	0.15-0.30	0.20-0.35		
0.040	0.15-0.30	0.40-0.60		
0.040	0.15-0.30	0.40-0.60		
0.040	0.15-0.30	0.80-1.10		
0.040	0.15-0.30	0.70-0.90		
0.040	0.15-0.30	0.70-0.90		
0.040	0.15-0.30	0.70-0.90		
0.040	0.15-0.30	0.70-0.90		
0.040	0.15-0.30	0.80-1.10		
0.040	0.15-0.30	0.35-0.55	0.20-0.40	0.08-0.15
0.040	0.15-0.30	0.40-0.60	0.40-0.70	0.15-0.25
0.040	0.15-0.30	0.40-0.60	0.40-0.70	0.15-0.25
0.040	0.15-0.30	0.40-0.60	0.40-0.70	0.15-0.25
0.040	0.15-0.30	0.40-0.60	0.40-0.70	0.15-0.25
0.040	0.15-0.30	0.40-0.60	0.40-0.70	0.15-0.25

Composition Ranges for

Composition Range (%)

AISI–SAE Designation	UNS Designation	C	Mn	P (max)
8660	G86600	0.56-0.64	0.75-1.00	0.035
8740	G87400	0.38-0.43	0.75-1.00	0.035
9255	G92550	0.51-0.59	0.70-0.95	0.035
9260	G92600	0.56-0.64	0.75-1.00	0.035
9310	G93106	0.08-0.13	0.45-0.65	0.025
94B30	G94301	0.28-0.33	0.75-1.00	0.035

(a) Contains 0.15% min vanadium.

1330 Alloy Steel Composition

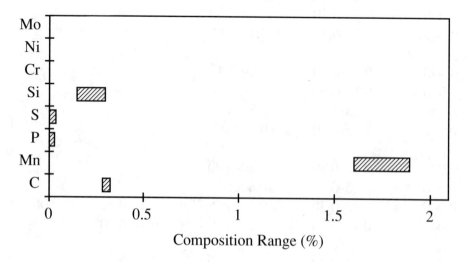

Composition Range (%)

Selected Alloy Steels (Continued)

S (max)	Si	Cr	Ni	Mo
0.040	0.15-0.30	0.40-0.60	0.40-0.70	0.15-0.25
0.040	0.15-0.30	0.40-0.60	0.40-0.70	0.20-0.30
0.040	1.80-2.20			
0.040	1.80-2.20			
0.025	0.15-0.30	1.00-1.40	3.00-3.50	0.08-0.15
0.040	0.15-0.30	0.30-0.50	0.30-0.60	0.08-0.15

Source: Data from *ASM Metals Reference Book, Second Edition*, American
Society for Metals, Metals Park, Ohio 44073, p186-193, (1984).

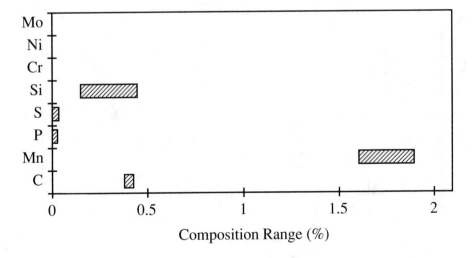

1340 Alloy Steel Composition

Composition Ranges for

3140 Alloy Steel Composition

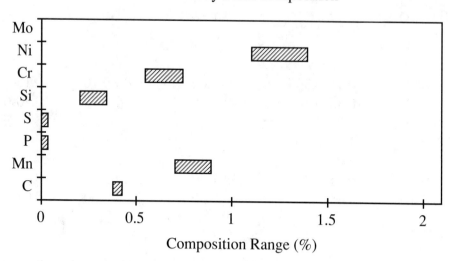

4037 Alloy Steel Composition

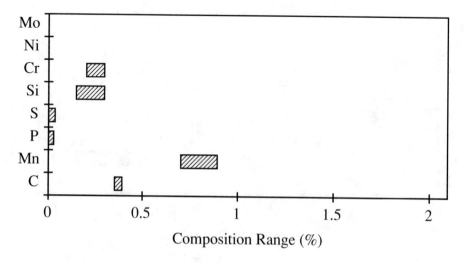

Selected Alloy Steels (Continued)

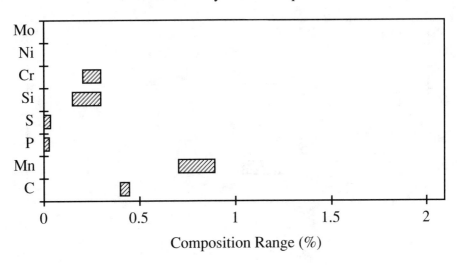

4042 Alloy Steel Composition

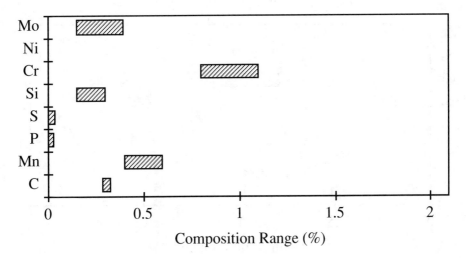

4130 Alloy Steel Composition

Composition Ranges for

4140 Alloy Steel Composition

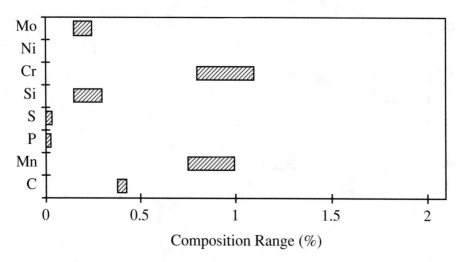

4150 Alloy Steel Composition

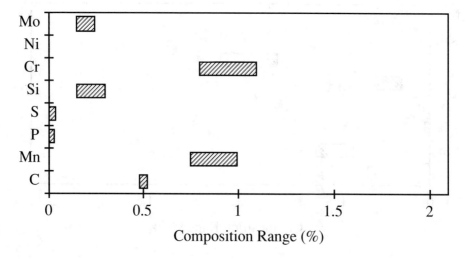

Selected Alloy Steels (Continued)

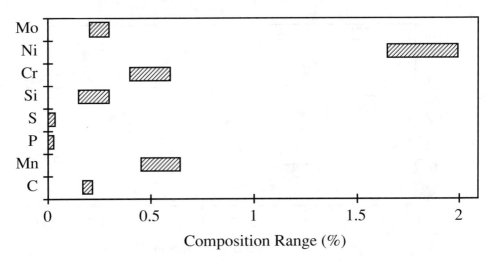

4320 Alloy Steel Composition

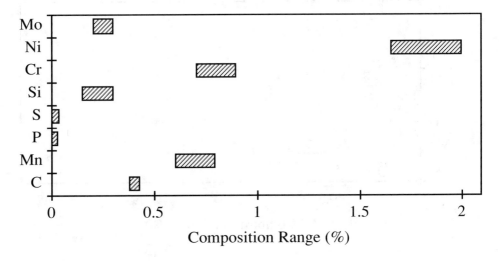

4340 Alloy Steel Composition

Composition Ranges for

4620 Alloy Steel Composition

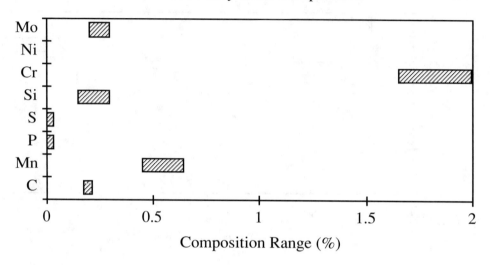

4820 Alloy Steel Composition

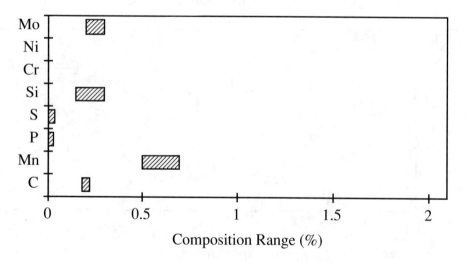

Selected Alloy Steels (Continued)

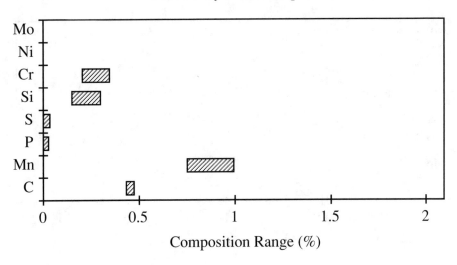

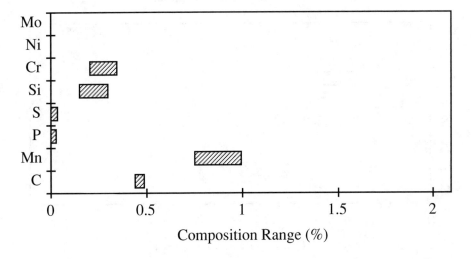

Composition Ranges for

5060 Alloy Steel Composition

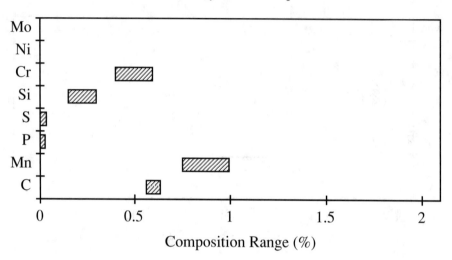

50B60 Alloy Steel Composition

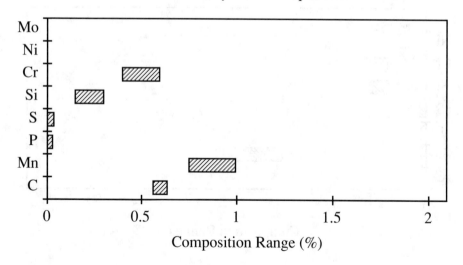

Selected Alloy Steels (Continued)

5130 Alloy Steel Composition

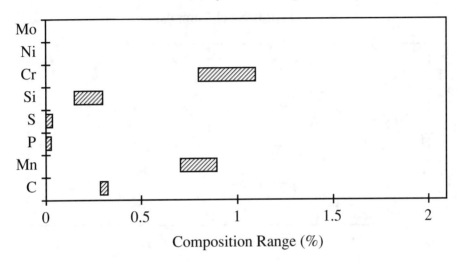

5140 Alloy Steel Composition

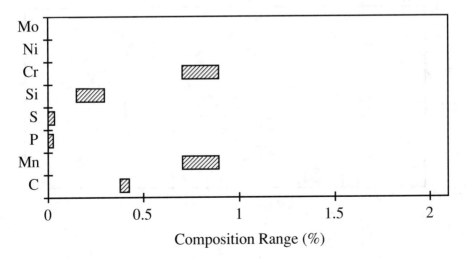

5150 Alloy Steel Composition

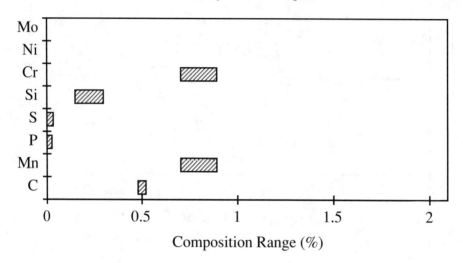

Composition Range (%)

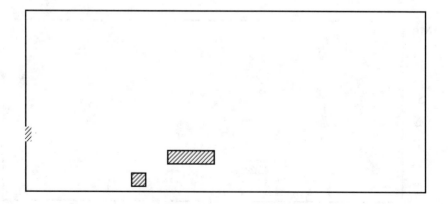

Selected Alloy Steels (Continued)

51B60 Alloy Steel Composition

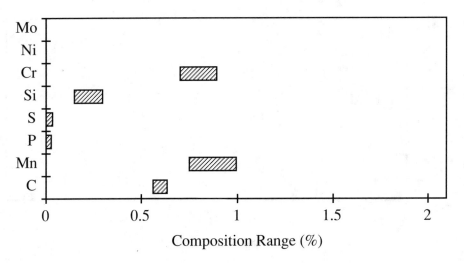

Composition Range (%)

6150 Alloy Steel Composition

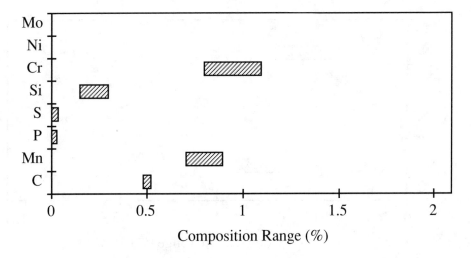

Composition Range (%)

Composition Ranges for

81B45 Alloy Steel Composition

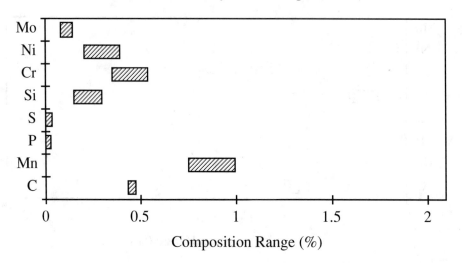

Composition Range (%)

8620 Alloy Steel Composition

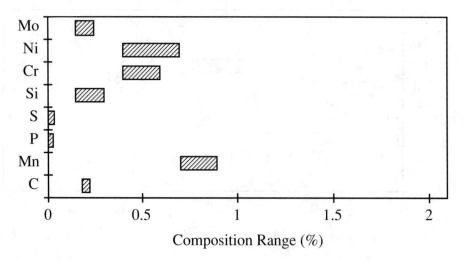

Composition Range (%)

Selected Alloy Steels (Continued)

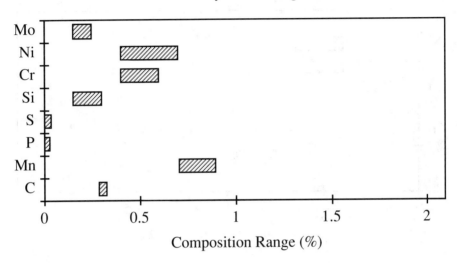

8630 Alloy Steel Composition

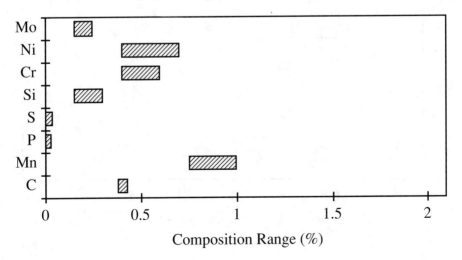

8640 Alloy Steel Composition

Composition Ranges for

8650 Alloy Steel Composition

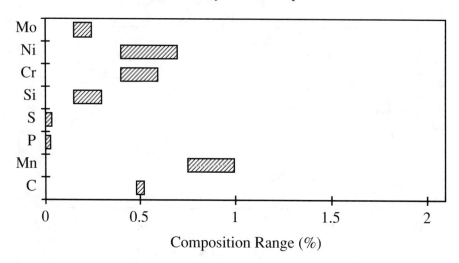

8660 Alloy Steel Composition

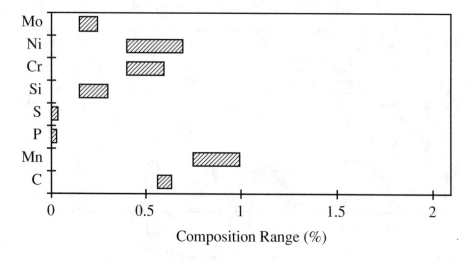

Selected Alloy Steels (Continued)

86B45 Alloy Steel Composition

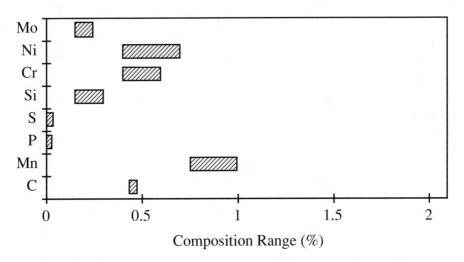

8740 Alloy Steel Composition

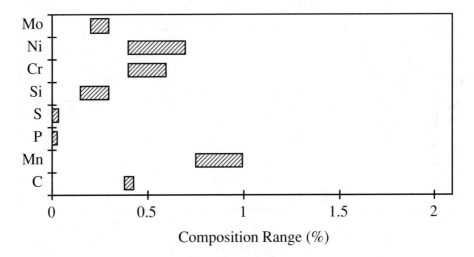

Composition Ranges for

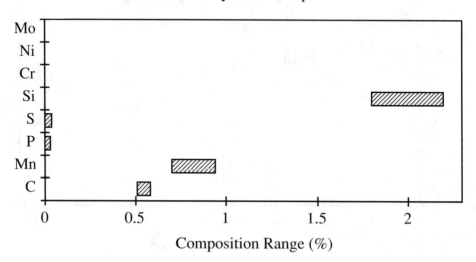

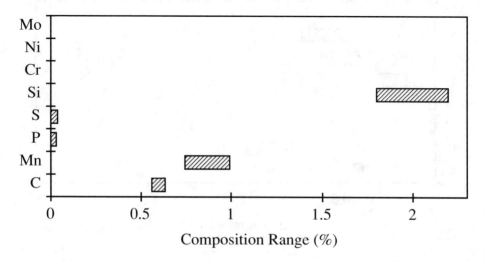

Selected Alloy Steels (Continued)

9310 Alloy Steel Composition

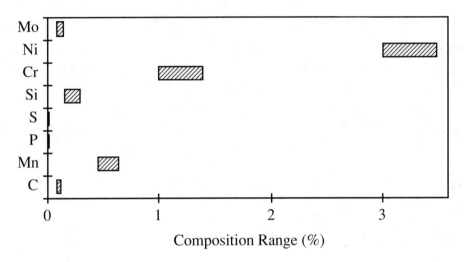

94B30 Alloy Steel Composition

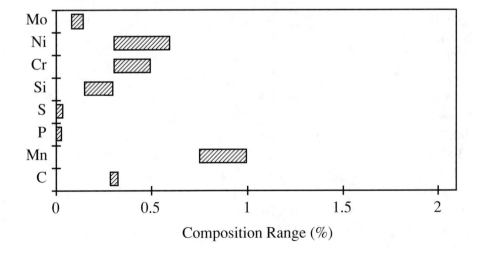

COMPOSITION RANGES FOR

AA Number	Composition, (%) Cu	Mg
201.0	4.6	0.35
206.0	4.6	0.25
A206.0	4.6	0.25
208.0	4.0	—
242.0	4.0	1.5
295.0	4.5	—
296.0	4.5	—
308.0	4.5	—
319.0	3.5	—
336.0	1.0	1.0
354.0	1.8	0.50
355.0	1.2	0.50
C355.0	1.2	0.50
356.0	0.25 (max)	0.32
A356.0	0.20 (max)	0.35
357.0	—	0.50
A357.0	—	0.6
359.0	—	0.6
360.0	—	0.50
A360.0	—	0.50
380.0	3.5	—
A380.0	3.5	—
383.0	2.5	—
384.0	3.8	
A384.0	3.8	—
390.0	4.5	0.6
A390.0	4.5	0.6
413.0	—	—
A413.0		—
4430	0.6 (max)	—
A443.0	0.30 (max)	—

SELECTED CAST ALUMINUM ALLOYS

Mn	Si	Others
0.35	—	0.7 Ag, 0.25 Ti
0.35	0.10 (max)	0.22 Ti, 0.15 Fe (max)
0.35	0.05 (max)	0.22 Ti, 0.10 Fe (max)
—	3.0	—
—		2.0 Ni
—	0.8	—
—	2.5	
—	5.5	—
—	6.0	—
	12.0	2.5 Ni
—	9.0	—
0.50 (max)	5.0	0.6 Fe (max), 0.35 Zn (max)
0.10 (max)	5.0	0.20 Fe (max), 0.10 Zn (max)
0.35 (max)	7.0	0.6 Fe (max), 0.35 Zn (max)
0.10 (max)	7.0	0.20 Fe (max), 0.10 Zn (max)
—	7.0	—
—	7.0	0.15 Ti, 0.005 Be
—	9.0	—
—	9.5	2.0 Fe (max)
	9.5	1.3 Fe (max)
—	8.5	2.0 Fe (max)
—	8.5	1.3 Fe (max)
—	10.5	—
—	11.2	3.0 Zn (max)
—	11.2	1.0 Zn (max)
—	17.0	1.3 Zn (max)
—	17.0	0.5 Zn (max)
—	12.0	2.0 Fe (max)
—	12.0	1.3 Fe (max)
—	5.2	—
—	5.2	—

Composition Ranges for

AA Number	Composition, (%) Cu	Mg
B443.0	0.15 (max)	—
C443.0	0.6 (max)	—
514.0	—	4.0
518.0	—	8.0
520.0	—	10.0
535.0	—	6.8
A535.0		7.0
B535.0	—	7.0
712.0	—	0.6
713.0	0.7	0.35
771.0	—	0.9
850.0	1.0	—

Selected Cast Aluminum Alloys (Continued)

Mn	Si	Others
—	5.2	—
—	5.2	2.0 Fe (max)
—	—	—
—		
—	—	—
0.18	—	0.18 Ti
0.18	—	—
—	—	0.18 Ti
—	—	5.8 Zn, 0.5 Cr, 0.20 Ti
—	—	7.5 Zn, 0.7 Cu
—	—	7.0 Zn, 0.13 Cr, 0.15 Ti
—	—	6.2 Sn, 1.0 Ni

Source: Data from *ASM Metals Reference Book, Second Edition*, American Society for Metals, Metals Park, Ohio 44073, p.303 (1984).

COMPOSITION RANGES FOR

Composition (%)

AA Number	Al		Si	Cu	Mn
1060	99.60	min	—	—	—
1100	99.00	min	—	0.12	—
2011	93.7		—	5.5	—
2014	93.5		0.8	4.4	0.8
2024	93.5		—	4.4	0.6
2219	93.0		—	6.3	0.3
2319	93.0		—	6.3	0.3
2618	93.7		0.18	2.3	—
3003	98.6		—	—	0.12
3004	97.8		—	—	1.2
3105	99.0			—	0.55
4032	85.0		12.2	0.9	
4043	94.8		5.2	—	—
5005	99.2		—	—	—
5050	98.6		—	—	—
5052	97.2		—	—	—
5056	95.0		—	—	0.12
5083	94.7		—	—	0.7
5086	95.4		—	—	0.4
5154	96.2		—	—	—
5182	95.2		—	—	0.35
5252	97.5		—	—	—
5254	96.2		—	—	—
5356	94.6		—	—	0.12
5454	96.3			—	0.8
5456	93.9		—	—	0.8

SELECTED WROUGHT ALUMINUM ALLOYS

Mg	Cr	Zn	Other
—	—	—	—
—	—	—	—
—	—	—	0.4Bi; 0.4Pb
0.5	—	—	—
1.5	—	—	—
—	—	—	0.06Ti; 0.10V; 0.18Zr
—	—	—	0.18Zn, 0.15Ti; 0.10V
1.6	—	—	1.1Fe; 1.0Ni; 0.07Ti
1.2	—	—	—
1.0	—	—	—
0.50	—	—	—
1.0	—	—	0.9Ni
—	—	—	—
0.8	—	—	—
1.4	—	—	—
2.5	0.25	—	—
5.0	0.12	—	—
4.4	0.15	—	—
4.0	0.15	—	—
3.5	0.25	—	—
4.5	—	—	—
2.5	—	—	—
3.5	0.25	—	—
5.0	0.12	—	0.13Ti
2.7	0.12	—	—
5.1	0.12	—	—

Composition Ranges for

AA Number	Composition (%)			
	Al	Si	Cu	Mn
5457	98.7	—	—	0.3
5652	97.2	—	—	—
5657	99.2	—	—	—
6005	98.7	0.8	—	—
6009	97.7	0.8	0.35	0.5
6010	97.3	1.0	0.35	0.5
6061	97.9	0.6	0.28	—
6063	98.9	0.4	—	—
6066	95.7	1.4	1.0	0.8
6070	96.8	1.4	0.28	0.7
6101	98.9	0.5	—	—
6151	98.2	0.9	—	—
6201	98.5	0.7	—	—
6205	98.4	0.8	—	0.1
7049	88.2	—	1.5	—
7075	90.0	—	1.6	—

Selected Wrought Aluminum Alloys (Continued)

Mg	Cr	Zn	Other
1.0	—	—	—
2.5	0.25	—	—
0.8	—	—	—
0.5	—	—	—
0.6	—	—	—
0.8	—	—	—
1.0	0.2	—	—
0.7	—	—	—
1.1	—	—	—
0.8	—	—	—
0.6	—	—	—
0.6	0.25	—	—
0.8	—	—	—
0.5	0.1	—	0.1Zr
2.5	0.15	7.6	—
2.5	0.23	5.6	—

Source: Data from *ASM Metals Reference Book, Second Edition*, American Society for Metals, Metals Park, Ohio 44073, p.292, (1984).

TYPICAL COMPOSITION OF

Glass–Ceramic	Typical composition (wt %)		
	SiO_2	Li_2O	Al_2O_3
Li_2O-Al_2O_3-SiO_2 system	74	4	16
MgO-Al_2O_3-SiO_2 system	65	—	19
Li_2O-MgO-SiO_2 system	73	11	—
Li_2O-ZnO-SiO_2 system	58	23	—

Li_2O-Al_2O_3-SiO_2 System
Typical Compositon

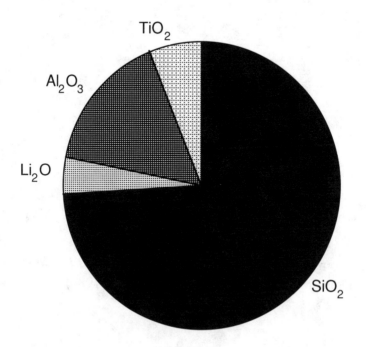

SELECTED GLASS-CERAMICS

MgO	ZnO	B_2O_3	TiO_2 (nucleating agent)	P_2O_5 (nucleating agent)
—	—	—	6	—
9	—	—	7	—
7	—	6	—	3
—	16	—	—	3

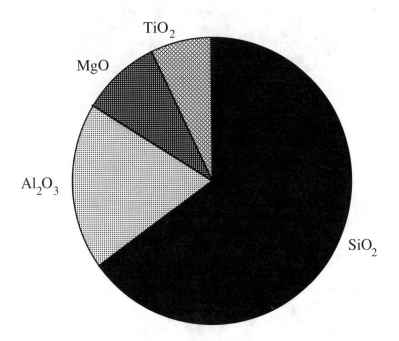

MgO-Al_2O_3-SiO_2 System
Typical Compositon

Typical Composition of Selected Glass-Ceramics (Continued)

Li_2O-MgO-SiO_2 System
Typical Compositon

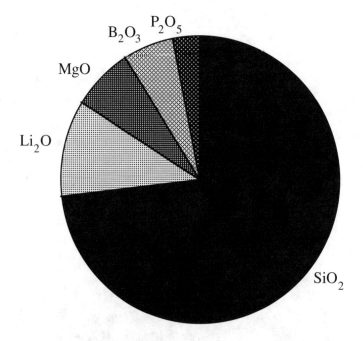

Typical Composition of Selected Glass-Ceramics (Continued)

Li_2O-ZnO-SiO_2 System
Typical Compositon

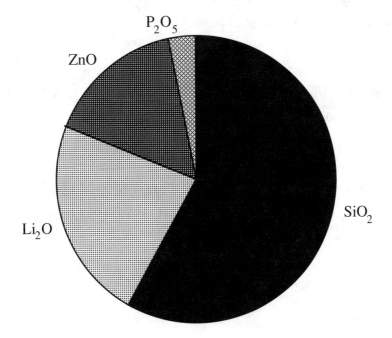

Source: Data compiled by J.S. Park from P.C. McMillan, *Glass-Ceramics*, 2nd edition, Academic Press, 1979

PHASE DIAGRAM SOURCES

Phase Diagrams are especially useful tools for the field of materials science and engineering. In the last decade, a substantial effort has been made within the materials community to provide a comprehensive set of accurate phase equilibria information. Cooperative efforts involving academia, industry, and government have been coordinated through the professional societies, ASM International and the American Ceramic Society. As a result, the following references are available and new updates will become available on a regular basis.

Binary Alloy Phase Diagrams, Vols. 1 and 2, T.B. Massalski, ed., American Society for Metals, Metals Park, Ohio, 1986.

Phase Diagrams for Ceramists, Vols. 1-8, American Ceramic Society, Columbus, Ohio, 1964, 1969, 1975, 1981, 1983, 1987, 1989, and 1989.

Mechanical Properties

MECHANICAL PROPERTIES OF SELECTED TOOL STEELS

Type	Condition	Tensile Strength (MPa)	0.2% yield Strength (MPa)	Elong- ation (%)	Area Reduction (%)	Hard– ness (HRC)	Impact Energy (J)
L2	Annealed	710	510	25	50	96 HRB	
	Oil quenched from 855 °C and single tempered at:						
	205 °C	2000	1790	5	15	54	28(b)
	315 °C	1790	1655	10	30	52	19(b)
	425 °C	1550	1380	12	35	47	26(b)
	540 °C	1275	1170	15	45	41	39(b)
	650 °C	930	760	25	55	30	125(b)
L6	Annealed	655	380	25	55	93 HRB	
	Oil quenched from 845 °C and single tempered at:						
	315 °C	2000	1790	4	9	54	12(b)
	425 °C	1585	1380	8	20	46	18(b)
	540 °C	1345	1100	12	30	42	23(b)
	650 °C	965	830	20	48	32	81(b)
S1	Annealed	690	415	24	52	96 HRB	
	Oil quenched from 930 °C and single tempered at:						
	205 °C	2070	1895			57.5	249(c)
	315 °C	2030	1860	4	12	54	233(c)
	425 °C	1790	1690	5	17	50.5	203(c)
	540 °C	1680	1525	9	23	47.5	230(c)
	650 °C	1345	1240	12	37	42	

Mechanical Properties of Selected Tool Steels (Continued)

Type	Condition	Tensile Strength (MPa)	0.2% yield Strength (MPa)	Elong- ation (%)	Area Reduction (%)	Hard- ness (HRC)	Impact Energy (J)
S5	Annealed	725	440	25	50	96 HRB	
	Oil quenched from 870 °C and single tempered at:						
	205 °C	2345	1930	5	20	59	206(c)
	315 °C	2240	1860	7	24	58	232(c)
	425 °C	1895	1690	9	28	52	243(c)
	540 °C	1520	1380	10	30	48	188(c)
	650 °C	1035	1170	15	40	37	
S7	Annealed	640	380	25	55	95 HRB	
	Fan cooled from 940 °C and single tempered at:						
	205 °C	2170	1450	7	20	58	244(c)
	315 °C	1965	1585	9	25	55	309(c)
	425 °C	1895	1410	10	29	53	243(c)
	540 °C	1820	1380	10	33	51	324(c)
	650 °C	1240	1035	14	45	39	358(c)

Area Reduction in 50 mm or 2 in.
(b) Charpy V-notch.
(c) Charpy unnotched.

Source: Data from ASM Metals Reference Book, Second Edition, American Society for Metals, Metals Park, Ohio 44073, p241, (1984).

Mechanical Properties of Selected Tool Steels (Continued)

L2 Tool Steel Strength vs. Quenching Temperature

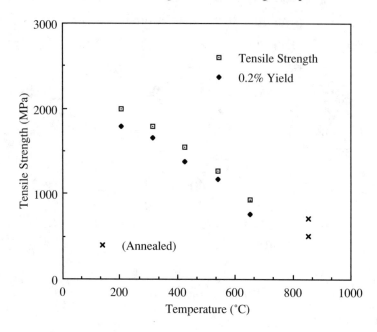

L2 Tool Steel Ductility vs. Quenching Temperature

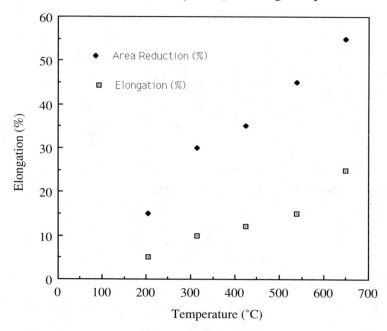

Mechanical Properties of Selected Tool Steels (Continued)

Tool Steel Strength vs. Quenching Temperature

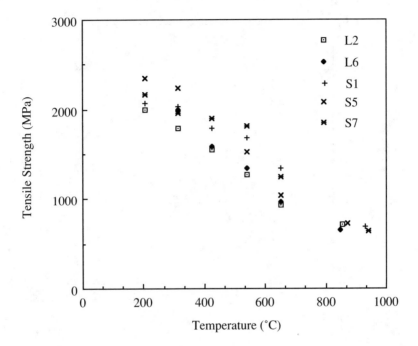

Tool Steel Yield Strength vs. Quenching Temperature

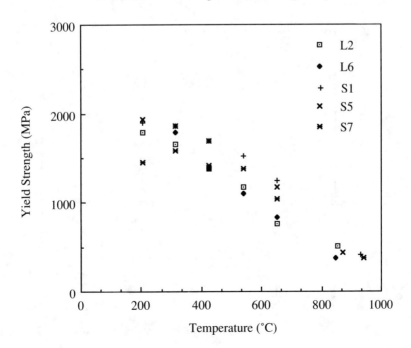

Mechanical Properties of Selected Tool Steels (Continued)

Tool Steel Ductility vs. Quenching Temperature

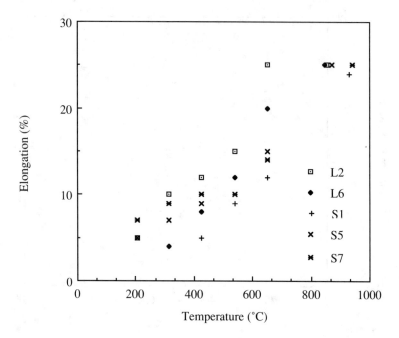

Tool Steel Ductility vs. Quenching Temperature

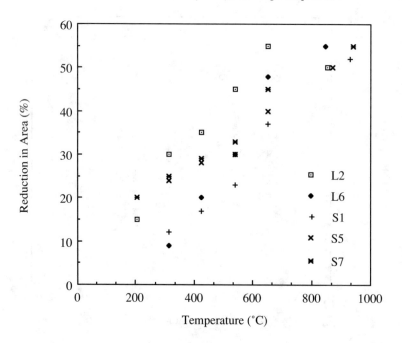

Mechanical Properties of Selected Tool Steels (Continued)

Tool Steel Hardness vs. Quenching Temperature

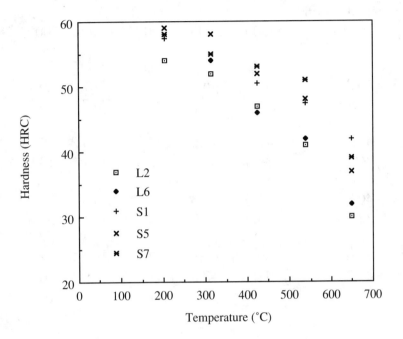

Tool Steel Impact Energy vs. Quenching Temperature

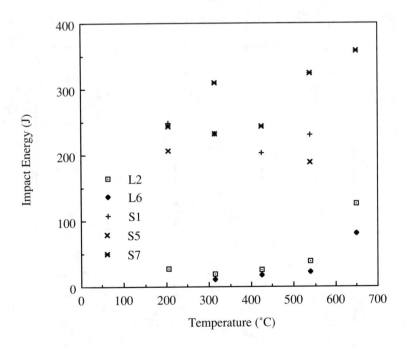

TOOL STEEL SOFTENING AFTER 100 HOURS FOR VARIOUS TEMPERATURES

Type	Original Hardness (HRC)	Hardness (HRC) after 100 h at					
		480°C	540°C	600°C	650°C	700°C	760°C
H13	60.2	48.7	46.3	29.0	22.7	20.1	13.9
	41.7	38.6	39.3	27.7	23.7	20.2	13.2
H21	49.2	48.7	47.6	37.2	27.4	19.8	16.2
	36.7	34.8	34.9	32.6	27.1	19.8	14.9
H23	40.8	40.0	40.6	40.8	38.6	33.2	26.8
	38.9	38.9	38.0	38.0	37.1	32.6	26.6
H26	61.0	60.6	60.3	47.1	38.4	26.9	21.3
	42.9	42.4	42.3	41.3	34.9	26.4	21.1

Source: Data from *ASM Metals Reference Book, Second Edition*, American Society for Metals, Metals Park, Ohio 44073, p.426, (1984).

Tool Steel Softening vs. Temperature (100 hours)

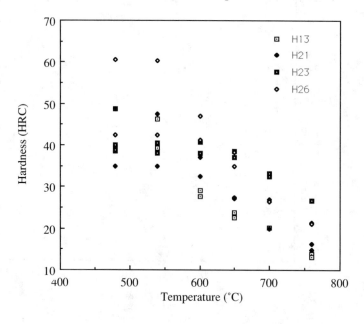

MECHANICAL PROPERTIES OF SELECTED GRAY CAST IRONS

SAE grade	Hardness (HB)	Maximum Tensile Strength (MPa)
G1800	187 max	118
G2500	170 to 229	173
G2500a	170 to 229	173
G3000	187 to 241	207
C3500	207 to 255	241
G3500b	207 to 255	1241
G3500c	207 to 255	1241
G4000	217 to 269	276
G4000d	241 to 321	1276

Hardness Range of Selected Gray Cast Irons

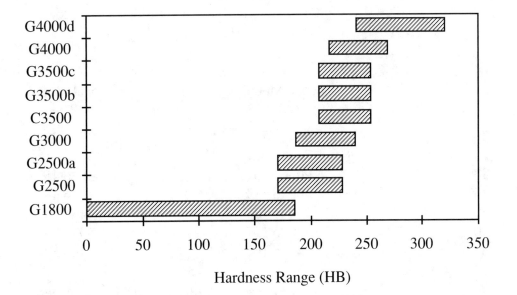

Hardness Range (HB)

Mechanical Properties of Selected
Gray Cast Irons (Continued)

Maximum Tensile Strength Values for Selected Gray Cast Irons

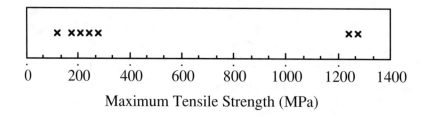

Maximum Tensile Strength (MPa)

ASTM Class	Tensile Strength (MPa)	Torsional Shear Strength (MPa)	Compressive Strength (MPa)	Reversed Bending Fatigue Limit (MPa)	Hardness (HB)
20	152	179	572	69	156
25	179	220	669	79	174
30	214	276	752	97	210
35	252	334	855	110	212
40	293	393	965	128	235
50	362	503	1130	148	262
60	431	610	1293	169	302

ASTM Class	Tensile Modulus (GPa)	Torsional Modulus (GPa)
20	66 to 97	27 to 39
25	79 to 102	32 to 41
30	90 to 113	36 to 45
35	100 to 119	40 to 48
40	110 to 138	44 to 54
50	130 to 157	50 to 55
60	141 to 162	54 to 59

Source: Data from *ASM Metals Reference Book, Second Edition*, American Society for Metals, Metals Park, Ohio 44073, p166-167, (1984).

MECHANICAL PROPERTIES OF SELECTED DUCTILE IRONS

Specification Number	Grade or Class	Hardness (HB)	Tensile Strength (MPa)	Yield Strength (MPa)	Elong–ation (%)
ASTM A395-76 ASME SA395	60-40-18	143-187	414	276	18
ASTM A476-70(d); SAE AMS5316	80-60-03	201 min	552	414	3
ASTM A536-72, MIL-1-11466B(MR)	60-40-18		414	276	18
	65-45-12		448	310	12
	80-55-06		552	379	6
	100-70-03		689	483	3
	120-90-02		827	621	2
SAE J434c	D4018	170 max	414	276	18
	D4512	156-217	448	310	12
	D5506	187-255	552	379	6
	D7003	241-302	689	483	3
MIL-I-24137(Ships)	Class A	190 max	414	310	15
	Class B	190 max	379	207	7
	Class C	175 max	345	172	20

Hardness Range of Ductile Irons

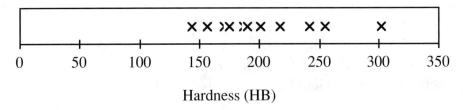

Hardness (HB)

Mechanical Properties of Selected Ductile Irons (Continued)

Tensile Strength Range of Ductile Irons

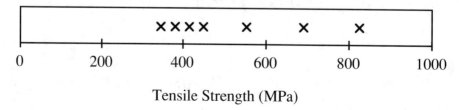

Yield Strength of Ductile Irons

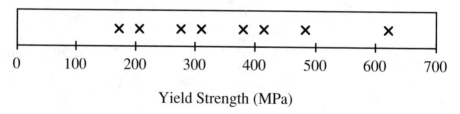

Elongation of Ductile Irons

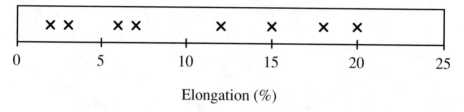

Source: Data from *ASM Metals Reference Book, Second Edition*, American Society for Metals, Metals Park, Ohio 44073, p169, (1984).

AVERAGE MECHANICAL PROPERTIES OF TREATED DUCTILE IRONS

Treatment	Tension Modulus (MPa)	Compression Modulus (MPa)	Comp. Poisson's Ratio	Torsion Modulus (MPa)	Torsion Poisson's Ratio
60-40-18	169	164	0.26	63	0.29
65-45-12	168	163	0.31	64	0.29
80-55-06	168	165	0.31	62	0.31
120 90-02	164	164	0.27	63.4	0.28

Source: Data from *ASM Metals Reference Book, Second Edition*, American Society for Metals, Metals Park, Ohio 44073, p169-170, (1984).

Poisson's Ratio vs. Treatment for Ductile Irons

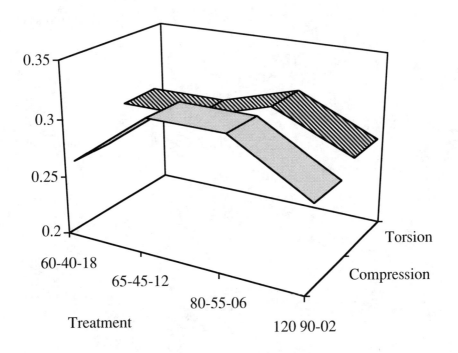

Average Mechanical Properties of Treated Ductile Irons
(Continued)

Modulus vs. Treatment for Ductile Irons

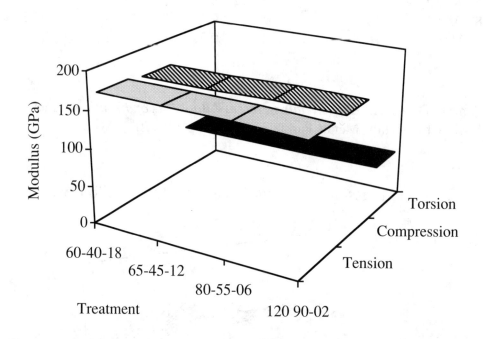

MECHANICAL PROPERTIES OF SELECTED MALLEABLE IRON CASTINGS

Specification Number	grade or class	Tensile Strength (MPa)	Yield Strength (MPa)	Hardness (HB)	Elong-ation (%)
Ferritic					
ASTM A47, A338; ANSI G48.1; FED QQ-I-666c	32510	345	224	156 max	10
	35018	365	241	156 max	18
ASTM A197		276	207	156 max	5
Pearlitic and Martensitic					
ASTM A220; ANSI C48.2; MIL-I-11444B	40010	414	276	149-197	10
	45008	448	310	156-197	8
	45006	448	310	156-207	6
	50005	483	345	179-229	5
	60004	552	414	197-241	4
	70003	586	483	217-269	3
	80002	655	552	241-285	2
	90001	724	621	269-321	1
Automotive					
ASTM A602; SAE J158	M3210(c)	345	224	156 max	10
	M4504(d)	448	310	163-217	4
	M5003(d)	517	345	187-241	3
	M5503(e)	517	379	187-241	3
	M7002(e)	621	483	229-269	2
	M8501(e)	724	586	269-302	1

[d] Air quenched and tempered

[e] Liquid quenched and tempered

Mechanical Properties of Selected Malleable Iron Castings
(Continued)

Tensile Strength of Malleable Irons

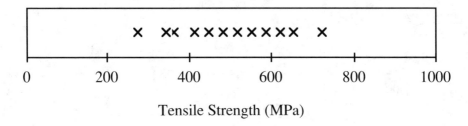

Tensile Strength (MPa)

Yield Strength of Malleable Irons

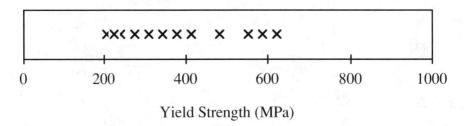

Yield Strength (MPa)

Mechanical Properties of Selected Malleable Iron Castings
(Continued)

Hardness of Malleable Irons

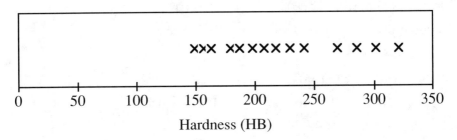

Hardness (HB)

Ductility of Malleable Irons

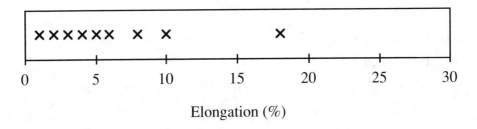

Elongation (%)

Source: Data from *ASM Metals Reference Book, Second Edition*, American Society for Metals, Metals Park, Ohio 44073, p171, (1984).

YOUNG'S MODULUS OF SELECTED CERAMICS
Listed by Ceramic

Ceramic	Young's Modulus
Borides	
Chromium Diboride (CrB_2)	30.6×10^6 psi
Tantalum Diboride (TaB_2)	37×10^6 psi
Titanium Diboride (TiB_2)	53.2×10^6 psi
(6.0 μm grain size, $\rho = 4.46 g/cm^3$)	81.6×10^6 psi
(3.5 μm grain size, $\rho = 4.37 g/cm^3$, 0.8wt% Ni)	
	75.0×10^6 psi
(6.0 μm grain size, $\rho = 4.56 g/cm^3$, 0.16wt% Ni)	
	77.9×10^6 psi
(12.0 μm grain size, $\rho = 4.66 g/cm^3$, 9.6wt% Ni)	
	6.29×10^6 psi
Zirconium Diboride (ZrB_2)	$49.8\text{-}63.8 \times 10^6$ psi
(22.4% density, foam)	3.305×10^6 psi
Carbides	
Boron Carbide (B_4C)	$42\text{-}65.2 \times 10^6$ psi at room temp.
Hafnium Monocarbide (HfC)	
($\rho = 11.94 g/cm^3$)	61.55×10^6 psi at room temp.

To convert psi to MPa, multiply by 145.

Young's Modulus of Selected Ceramics (Continued)
Listed by Ceramic

Ceramic	Young's Modulus
Silicon Carbide (SiC)	
(pressureless sintered)	43.9×10^6 psi at room temp.
(hot pressed)	63.8×10^6 psi at room temp.
(self bonded)	59.5×10^6 psi at room temp.
(cubic, CVD)	$60.2\text{-}63.9 \times 10^6$ psi at room temp.
($\rho = 3.128$ g/cm^3)	58.2×10^6 psi at room temp.
($\rho = 3.120$ g/cm^3)	59.52×10^6 psi at room temp.
(hot pressed)	$62.4\text{-}65.3 \times 10^6$ psi at 20°C
(sintered)	$54.38\text{-}60.9 \times 10^6$ psi at 20°C
(reaction sintered)	$50.75\text{-}54.38 \times 10^6$ psi at 20°C
	55×10^6 psi at 400°C
	53×10^6 psi at 800°C
	51×10^6 psi at 1200°C
(hot presses)	55.1×10^6 psi at 1400°C
(sintered)	$43.5\text{-}58.0 \times 10^6$ psi at 1400°C
(reaction sintered)	$29\text{-}46.4 \times 10^6$ psi at 1400°C
Tantalum Monocarbide (TaC)	$41.3\text{-}91.3 \times 10^6$ psi at room temp.
Titanium Monocarbide (TiC)	63.715×10^6 psi at room temp.
	$45\text{-}55 \times 10^6$ psi at 1000 ^{o}C
Trichromium Dicarbide (Cr$_3$C$_2$)	54.1×10^6 psi
Tungsten Monocarbide (WC)	$96.91\text{-}103.5 \times 10^6$ psi at room temp.
Zirconium Monocarbide (ZrC)	$28.3\text{-}69.6 \times 10^6$ psi at room temp.

To convert psi to MPa, multiply by 145.

Young's Modulus of Selected Ceramics (Continued)
Listed by Ceramic

Ceramic	Young's Modulus
Nitrides	
Aluminum Nitride (AlN)	50×10^6 psi at 25°C
	46×10^6 psi at 1000°C
	40×10^6 psi at 1400°C
Boron Nitride (BN)	
parallel to c axis	4.91×10^6 psi at 23°C
	3.47×10^6 psi at 300°C
	0.51×10^6 psi at 700°C
parallel to a axis	12.46×10^6 psi at 23°C
	8.79×10^6 psi at 300°C
	1.54×10^6 psi at 700°C
	1.65×10^6 psi at 1000°C
Titanium Mononitride (TiN)	$11.47\text{-}36.3 \times 10^6$ psi
Trisilicon tetranitride (Si_3N_4)	
(hot pressed)	$36.25\text{-}47.13 \times 10^6$ psi at 20°C
(sintered)	$28.28\text{-}45.68 \times 10^6$ psi at 20°C
(reaction sintered)	$14.5\text{-}31.9 \times 10^6$ psi at 20°C
(hot pressed)	$25.38\text{-}36.25 \times 10^6$ psi at 1400°C
(reaction sintered)	$17.4\text{-}29.0 \times 10^6$ psi at 1400°C

To convert psi to MPa, multiply by 145.

Young's Modulus of Selected Ceramics (Continued)
Listed by Ceramic

Ceramic	Young's Modulus
Oxides	
Aluminum Oxide (Al_2O_3)	$50\text{-}59.3 \times 10^6$ psi at room temp.
	$50\text{-}57.275 \times 10^6$ psi at 500°C
	51.2×10^6 psi at 800°C
	$45.5\text{-}50 \times 10^6$ psi at 1000°C
	$39.8\text{-}53.65 \times 10^6$ psi at 1200°C
	32×10^6 psi at 1250°C
	32.7×10^6 psi at 1400°C
	25.6×10^6 psi at 1500°C
Beryllium Oxide (BeO)	$42.8\text{-}45.5 \times 10^6$ psi at room temp.
	40×10^6 psi at 800°C
	33×10^6 psi at 1000°C
	20×10^6 psi at 1145°C
Cerium Dioxide (CeO_2)	24.9×10^6 psi
Dichromium Trioxide (Cr_2O_3)	$>14.9 \times 10^6$ psi
Hafnium Dioxide (HfO_2)	8.2×10^6 psi
Magnesium Oxide (MgO)	$30.5\text{-}36.3 \times 10^6$ psi at room temp.
	29.5×10^6 psi at 600°C
	21×10^6 psi at 1000°C
	10×10^6 psi at 1200°C
	4×10^6 psi at 1300°C
($\rho = 3.506$ g/cm^3)	42.74×10^6 psi at room temp.

Young's Modulus of Selected Ceramics (Continued)
Listed by Ceramic

Ceramic	Young's Modulus
Thorium Dioxide (ThO_2)	$17.9\text{-}34.87 \times 10^6$ psi at room temp.
	$18\text{-}18.5 \times 10^6$ psi at 800°C
	17.1×10^6 psi at 1000°C
	12.8×10^6 psi at 1200°C
Titanium Oxide (TiO_2)	41×10^6 psi
Uranium Dioxide (UO_2)	21×10^6 psi at 0-1000°C
	25×10^6 psi at 20°C
(ρ=10.37 g/cm^3)	27.98×10^6 psi at room temp.
Zirconium Oxide (ZrO_2)	
(partially stabilized)	29.7×10^6 psi at room temp.
(fully stabilized)	$14.1\text{-}30.0 \times 10^6$ psi at room temp.
(plasma sprayed)	6.96×10^6 psi at room temp.
	3.05×10^6 psi at 1100°C
	$24.8\text{-}27 \times 10^6$ psi at room temp.
	2×10^6 psi at 500°C
	18.9×10^6 psi at 800°C
	$18.5\text{-}25 \times 10^6$ psi at 1000°C
	$17.1\text{-}18.0 \times 10^6$ psi at 1200°C
	14.2×10^6 psi at 1400°C
	12.8×10^6 psi at 1500°C
	36×10^6 psi at 20°C
(stabilized, ρ=5.634 g/cm^3)	19.96×10^6 psi at room temp.
Cordierite ($2MgO \cdot 2Al_2O_3 \cdot 5SiO_2$)	20.16×10^6 psi
(glass)	13.92×10^6 psi

Young's Modulus of Selected Ceramics (Continued)
Listed by Ceramic

Ceramic   Young's Modulus

Mullite ($3Al_2O_3\ 2SiO_2$)

(ρ=2.779 g/cm^3) 20.75×10^6 psi at room temp.

(ρ=2.77 g/cm^3) 18.42×10^6 psi at 25°C.

(ρ=2.77 g/cm^3) 18.89×10^6 psi at 400°C.

(ρ=2.77 g/cm^3) 14.79×10^6 psi at 800°C.

(ρ=2.77 g/cm^3) 4.00×10^6 psi at 1200°C.

(full density) 33.35×10^6 psi at room temp.

Spinel ($Al_2O_3\ MgO$) 34.5×10^6 psi at room temp.

                                                 34.4×10^6 psi at 200°C

                                                 34.5×10^6 psi at 400°C

                                                 34×10^6 psi at 600°C

                                                 32.9×10^6 psi at 800°C

                                                 30.4×10^6 psi at 1000°C

                                                 25.0×10^6 psi at 1200°C

                                                 20.1×10^6 psi at 1300°C

(ρ=3.510 g/cm^3) 38.23×10^6 psi at room temp.

Zircon ($SiO_2\ ZrO_2$) 24×10^6 psi at room temp.

Silicide

Molybdenum Disilicide ($MoSi_2$) $39.3\text{-}56.36 \times 10^6$ psi at room temp.

To convert psi to MPa, multiply by 145.

Source: Data compiled by J.S. Park from *No. 1 Materials Index*, Peter T.B. Shaffer, Plenum Press, New York, (1964); *Smithells Metals Reference Book*, Eric A. Brandes, ed., in association with Fulmer Research Institute Ltd. 6th ed. London, Butterworths, Boston, (1983); and *Ceramic Abstracts*, American Ceramic Society (1986-1991)

MODULUS OF ELASTICITY IN TENSION FOR SELECTED POLYMERS
Listed by Polymer

Polymer	Modulus of Elasticity in Tension, (ASTM D638) (10^5 psi)
ABS Resins; Molded, Extruded	
Medium impact	3.3—4.0
High impact	2.6—3.2
Very high impact	2.0—3.1
Low temperature impact	2.0—3.1
Heat resistant	3.5—4.2
Acrylics; Cast, Molded, Extruded	
Cast Resin Sheets, Rods:	
General purpose, type I	3.5—4.5
General purpose, type II	4.0—5.0
Moldings:	
Grades 5, 6, 8	3.5—5.0
High impact grade	2.3—3.3
Chlorinated Polymers	
Chlorinated polyether	1.5
Chlorinated polyvinyl chloride	3.7
Polycarbonates	
Polycarbonate	3.45
Polycarbonate (40% glass fiber reinforced)	17
Diallyl Phthalates; Molded	
Orlon filled	6
Asbestos filled	12

Modulus of Elasticity in Tension for
Selected Polymers (Continued)
Listed by Polymer

Polymer	Modulus of Elasticity in Tension, (ASTM D638) (10^5 psi)
Fluorocarbons; Molded,Extruded	
Polytrifluoro chloroethylene (PTFCE)	1.9—3.0
Polytetrafluoroethylene (PTFE)	0.38—0.65
Ceramic reinforced (PTFE)	1.5—2.0
Fluorinated ethylene propylene(FEP)	0.5—0.7
Polyvinylidene— fluoride (PVDF)	1.7—2
Epoxies; Cast, Molded, Reinforced	
Standard epoxies (diglycidyl ethers of bisphenol A)	
Cast rigid	4.5
Cast flexible	0.5—2.5
Molded:	
General purpose glass cloth laminate	33—36
High strength laminate	57—58
Filament wound composite	72—64
Epoxies—Molded, Extruded	
High performance resins	
(cycloaliphatic diepoxides)	
Cast, rigid	4—5
Molded	
Glass cloth laminate	32—33
Epoxy novolacs	
Cast, rigid	4.8—5.0
Glass cloth laminate	27.5

Modulus of Elasticity in Tension for
Selected Polymers (Continued)
Listed by Polymer

Polymer	Modulus of Elasticity in Tension, (ASTM D638) (10^5 psi)
Melamines; Molded	
Unfilled	
Cellulose electrical	10—11
Phenolics; Molded	
Type and filler	
General: woodflour and flock	8—13
Shock: paper, flock, or pulp	8—12
High shock: chopped fabric or cord	9—14
Very high shock: glass fiber	30—33
Phenolics: Molded	
Arc resistant—mineral	10—30
Rubber phenolic—woodflour or flock	4—6
Rubber phenolic—chopped fabric	3.5—6
Rubber phenolic—asbestos	5—9
Polyesters: Thermosets	
Cast polyyester	
Rigid	1.5—6.5
Flexible	0.001—0.10
Reinforced polyester moldings	
High strength (glass fibers)	16—20
Heat and chemical resistsnt (asbestos)	12—15
Sheet molding compounds, general purpose	15—20

Modulus of Elasticity in Tension for
Selected Polymers (Continued)
Listed by Polymer

Polymer	Modulus of Elasticity in Tension, (ASTM D638) (10^5 psi)
Polyethylenes; Molded, Extruded	
Type I—lower density (0.910—0.925)	
Melt index 0.3—3.6	0.21—0.27
Melt index 6—26	0.20—0.24
Polyvinyl Chloride And Copolymers;	ASTM D412
Molded, Extruded	
Nonrigid—general	0.004—0.03
Nonrigid—electrical	0.01—0.03
Rigid—normal impact	3 5—4.0
Vinylidene chloride	0.7—2.0
Silicones; Molded, Laminated	ASTM D651
Woven glass fabric/ silicone laminate	28
Ureas; Molded	
Alpha—cellulose filled (ASTM Type l)	13—16
Woodflour filled	11—14

To convert psi to MPa, multiply by 145.

Source: data compiled by J.S. Park from Charles T. Lynch, *CRC Handbook of Materials Science*, Vol. 3, CRC Press, Boca Raton, Florida, 1975 and *Engineered Materials Handbook*, Vol.2, Engineering Plastics, ASM International, Metals Park, Ohio, 1988.

POISSON'S RATIO FOR SELECTED CERAMICS
Listed by Ceramic

Ceramic	Poisson's Ratio
BORIDES	
Titanium Diboride (TiB_2)	0.09-0.28
(6.0 μm grain size, ρ=4.46g/cm^3)	0.10
(3.5 μm grain size, ρ=4.37g/cm^3, 0.8wt% Ni)	0.12
(6.0 μm grain size, ρ=4.56g/cm^3, 0.16wt% Ni)	0.11
(12.0 μm grain size, ρ=4.66g/cm^3, 9.6wt% Ni)	0.15
Zirconium Diboride (ZrB_2)	0.144
CARBIDES	
Boron Carbide (B_4C)	0.207
Hafnium Monocarbide (HfC)	0.166
Silicon Carbide (SiC)	
(ρ = 3.128 g/cm^3)	0.183-0.192 at room temp.
Tantalum Monocarbide (TaC)	0.1719 -0.24
Titanium Monocarbide (TiC)	0.187-189
Tungsten Monocarbide (WC)	0.24
Zirconium Monocarbide (ZrC)	
(ρ = 6.118 g/cm^3)	0.257
NITRIDES	
Trisilicon tetranitride (Si_3N_4)	0.24
(presureless sintered)	0.22-0.27

Poisson's Ratio for Selected Ceramics (Continued)
Listed by Ceramic

Ceramic	Poisson's Ratio
OXIDES	
Aluminum Oxide (Al_2O_3)	0.21-0.27
Beryllium Oxide (BeO)	0.26-0.34
Cerium Dioxide (CeO_2)	0.27-0.31
Magnesium Oxide (MgO) ($\rho = 3.506$ g/cm^3)	0.163 at room temp.
Thorium Dioxide (ThO_2) ($\rho=9.722$ g/cm^3)	0.275
Titanium Oxide (TiO_2)	0.28
Uranium Dioxide (UO_2) ($\rho=10.37$ g/cm^3)	0.302
Zircoium Oxide (ZrO_2)	0.324-0.337 at room temp.
(partially stabilized)	0.23
(fully stabilized)	0.23-0.32
(plasma sprayed)	0.25
Cordierite (2MgO 2Al_2O_3 5SiO_2)	
($\rho=2.3$g/cm^3)	0.21
($\rho=2.1$g/cm^3)	0.17
(glass)	0.26

Poisson's Ratio for Selected Ceramics (Continued)
Listed by Ceramic

Ceramic	Poisson's Ratio
Mullite ($3Al_2O_3 \ 2SiO_2$) ($\rho=2.779$ g/cm^3)	0.238
Spinel ($Al_2O_3 \ MgO$) ($\rho=3.510$ g/cm^3)	0.294
SILICIDE	
Molybdenum Disilicide ($MoSi_2$)	0.158-0.172

Source: Data compiled by J.S. Park from *No. 1 Materials Index*, Peter T.B. Shaffer, Plenum Press, New York, (1964); *Smithells Metals Reference Book*, Eric A. Brandes, ed., in association with Fulmer Research Institute Ltd. 6th ed. London, Butterworths, Boston, (1983); and *Ceramic Abstracts*, American Ceramic Society (1986-1991)

YIELD STRENGTH OF SELECTED
CAST ALUMINUM ALLOYS
Listed by Alloy

Alloy AA No.	Temper	Yield Strength (MPa)
201.0	T4	215
	T6	435
	T7	415
206.0, A206.0	T7	345
208.0	F	97
242.0	T21	125
	T571	205
	T77	160
	T571	235
	T61	290
295.0	T4	110
	T6	165
	T62	220
296.0	T4	130
	T6	180
	T7	140
308.0	F	110
319.0	F	125
	T6	165
	F	130
	T6	185
336.0	T551	195
	T65	295
354.0	T61	285
355.0	T51	160
	T6	175

Yield Strength of Selected Cast Aluminum Alloys (Continued)
Listed by Alloy

Alloy AA No.	Temper	Yield Strength (MPa)
	T61	240
	T7	250
	T71	200
	T51	165
	T6	190
	T62	280
	T7	210
	T71	215
356.0	T51	140
	T6	165
	T7	210
	T71	145
	T6	185
	T7	165
357.0, A357.0	T62	290
359.0	T61	255
	T62	290
360.0	F	170
A360.0	F	165
380.0	F	165
383.0	F	150
384.0, A384.0	F	165
390.0	F	240
	T5	260
A390.0	F,T5	180
	T6	280

Yield Strength of Selected Cast Aluminum Alloys (Continued)
Listed by Alloy

Alloy AA No.	Temper	Yield Strength (MPa)
	T7	250
	F,T5	200
	T6	310
	T7	260
413.0	F	140
A413.0	F	130
443.0	F	55
B443.0	F	62
C443.0	F	110
514.0	F	85
518.0	F	190
520.0	T4	180
535.0	F	140
712.0	F	170
713.0	T5	150
	T5	150
771.0	T6	275
850.0	T5	75

Source: Data from *ASM Metals Reference Book, Second Edition*, American Society for Metals, Metals Park, Ohio 44073, (1984).

YIELD STRENGTH OF SELECTED
CAST ALUMINUM ALLOYS
Listed by Value

Alloy AA No.	Temper	Yield Strength (MPa)
443.0	F	55
B443.0	F	62
850.0	T5	75
514.0	F	85
208.0	F	97
295.0	T4	110
308.0	F	110
C443.0	F	110
242.0	T21	125
319.0	F	125
296.0	T4	130
319.0	F	130
A413.0	F	130
296.0	T7	140
356.0	T51	140
413.0	F	140
535.0	F	140
356.0	T71	145
383.0	F	150
713.0	T5	150
713.0	T5	150
242.0	T77	160
355.0	T51	160
295.0	T6	165
319.0	T6	165

Yield Strength of Selected Cast Aluminum Alloys (Continued)
Listed by Value

Alloy AA No.	Temper	Yield Strength (MPa)
355.0	T51	165
356.0	T6	165
356.0	T7	165
A360.0	F	165
380.0	F	165
384.0, A384.0	F	165
360.0	F	170
712.0	F	170
355.0	T6	175
296.0	T6	180
A390.0	F,T5	180
520.0	T4	180
319.0	T6	185
356.0	T6	185
355.0	T6	190
518.0	F	190
336.0	T551	195
355.0	T71	200
A390.0	F,T5	200
242.0	T571	205
355.0	T7	210
356.0	T7	210
201.0	T4	215
355.0	T71	215
295.0	T62	220
242.0	T571	235

Yield Strength of Selected Cast Aluminum Alloys (Continued)
Listed by Value

Alloy AA No.	Temper	Yield Strength (MPa)
355.0	T61	240
390.0	F	240
355.0	T7	250
A390.0	T7	250
359.0	T61	255
390.0	T5	260
A390.0	T7	260
771.0	T6	275
355.0	T62	280
A390.0	T6	280
354.0	T61	285
242.0	T61	290
357.0, A357.0	T62	290
359.0	T62	290
336.0	T65	295
A390.0	T6	310
206.0, A206.0	T7	345
201.0	T7	415
201.0	T6	435

Source: Data from *ASM Metals Reference Book, Second Edition*, American Society for Metals, Metals Park, Ohio 44073, (1984).

YIELD STRENGTH OF SELECTED WROUGHT ALUMINUM ALLOYS
Listed by Alloy

Alloy	Temper	Yield Strength (MPa)
1050	O	28
	H14	105
	H16	125
	H18	145
1060	O	28
	H12	76
	H14	90
	H16	105
	H18	125
1100	O	34
	H12	105
	H14	115
	H16	140
	H18	150
1350	O	28
	H12	83
	H14	97
	H16	110
	H19	165
2011	T3	295
	T8	310
2014	O	97
	T4	290
	T6	415

Yield Strength of Selected Wrought Aluminum Alloys
(Continued)
Listed by Alloy

Alloy	Temper	Yield Strength (MPa)
Alclad 2014	0	69
	T3	275
	T4	255
	T6	415
2024	0	76
	T3	345
	T4, T351	325
	T361	395
Alclad 2024	0	76
	T	310
	T4, T351	290
	T361	365
	T81, T851	415
	T861	455
2036	T4	195
2048		415
2124	T851	440
2218	T61	305
	T71	275
	T72	255
2219	0	76
	T42	185
	T31, T351	250
	T37	315
	T62	290
	T81, T851	350
	T87	395

Yield Strength of Selected Wrought Aluminum Alloys
(Continued)
Listed by Alloy

Alloy	Temper	Yield Strength (MPa)
2618	All	370
3003	0	42
Alclad	H12	125
3003	H14	145
	H16	170
	H18	185
3004	0	69
Alclad	H32	170
3004	H34	200
	H36	230
	H38	250
3105	0	55
	H12	130
	H14	150
	H16	170
	H18	195
	H25	160
4032	T6	315
4043	0	69
	H18	270
5005	0	41
	H12	130
	H14	150
	H16	170
	H18	195
	H32	115

Yield Strength of Selected Wrought Aluminum Alloys
(Continued)
Listed by Alloy

Alloy	Temper	Yield Strength (MPa)
	H34	140
	H36	165
	H38	185
5050	0	55
	H32	145
	H34	165
	H36	180
	H38	200
5052	0	90
	H32	195
	H34	215
	H36	240
	H38	255
5056	0	150
	H18	405
	H38	345
5083	0	145
	H112	195
	H113	230
	H321	230
	H323, H32	250
	H343, H34	285
5086	0	115
	H32, H116, H117	205
	H34	255
	H112	130

Yield Strength of Selected Wrought Aluminum Alloys
(Continued)
Listed by Alloy

Alloy	Temper	Yield Strength (MPa)
5154	0	115
	H32	205
	H34	230
	H36	250
	H38	270
	H112	115
5182	0	140
	H32	235
	H34	285
	H19(n)	395
5252	H25	170
	H28, H38	240
5254	0	115
5254	H32	205
	H34	230
	H36	250
	H38	270
	H112	115
5454	0	115
	H32	205
	H34	240
	H36	275
	H38	310
	H111	180
	H112	125
	H311	180

Yield Strength of Selected Wrought Aluminum Alloys
(Continued)
Listed by Alloy

Alloy	Temper	Yield Strength (MPa)
5456	0	160
	H111	230
	H112	165
	H321, H116	255
5457	0	48
	H25	160
	H28, H38	185
5652	0	90
	H32	195
	H34	215
	H36	240
	H38	255
5657	H25	140
	H28, H38	165
6005	T1	105
	T5	240
6009	T4	130
	T6	325
6010	T4	170
6061	0	55
	T4, T451	145
	T6, T651	275
Alclad 6061	0	48
	T4, T451	130
	T6, T651	255

Yield Strength of Selected Wrought Aluminum Alloys
(Continued)
Listed by Alloy

Alloy	Temper	Yield Strength (MPa)
6063	0	48
	TI	90
	T4	90
	T5	145
	T6	215
	T83	240
	T831	185
	T832	270
6066	0	83
	T4, T451	205
	T6, T651	360
6070	0	69
	T4	170
	T6	350
6101	Hlll	76
6151	T6	195
6201	T6	300
	T81	310
6205	Tl	140
	T5	290
6262	T9	380
6351	T4	150
	T6	285
6463	Tl	90
	T5	145

Yield Strength of Selected Wrought Aluminum Alloys
(Continued)
Listed by Alloy

Alloy	Temper	Yield Strength (MPa)
	T6	215
7005	0	83
	T53	345
	T6,T63,T6351	315
7049	T73	
7050	T736	455
7072	0	
	H12	
	H14	
7075	0	105
	T6,T651	505
	T73	435
Alclad 7075	0	95
	T6,T651	460
7175	T66	525
	T736	455
7475	T61	460

Source: Data from *ASM Metals Reference Book, Second Edition*, American Society for Metals, Metals Park, Ohio 44073, p.299—302, (1984).

YIELD STRENGTH OF SELECTED WROUGHT ALUMINUM ALLOYS
Listed by Yield Strength

Alloy	Temper	Yield Strength (MPa)
1050	0	28
1060	0	28
1350	0	28
1100	0	34
5005	0	41
3003	0	42
5457	0	48
Alclad 6061	0	48
6063	0	48
3105	0	55
5050	0	55
6061	0	55
Alclad 2014	0	69
3004	0	69
4043	0	69
6070	0	69
1060	H12	76
2024	0	76
Alclad 2024	0	76
2219	0	76
6101	Hlll	76
1350	H12	83
6066	0	83
7005	0	83
1060	H14	90
5052	0	90

Yield Strength of Selected Wrought Aluminum Alloys
(Continued)
Listed by Yield Strength

Alloy	Temper	Yield Strength (MPa)
5652	0	90
6063	TI	90
6063	T4	90
6463	Tl	90
Alclad 7075	0	95
1350	H14	97
2014	0	97
1050	H14	105
1060	H16	105
1100	H12	105
6005	T1	105
7075	0	105
1350	H16	110
1100	H14	115
5005	H32	115
5086	0	115
5154	0	115
5154	H112	115
5254	0	115
5254	H112	115
5454	0	115
1050	H16	125
1060	H18	125
Alclad	H12	125
5454	H112	125
3105	H12	130

Yield Strength of Selected Wrought Aluminum Alloys
(Continued)
Listed by Yield Strength

Alloy	Temper	Yield Strength (MPa)
5005	H12	130
5086	H112	130
6009	T4	130
Alclad 6061	T4, T451	130
1100	H16	140
5005	H34	140
5182	0	140
5657	H25	140
6205	Tl	140
1050	H18	145
3003	H14	145
5050	H32	145
5083	0	145
6061	T4, T451	145
6063	T5	145
6463	T5	145
1100	H18	150
3105	H14	150
5005	H14	150
5056	0	150
6351	T4	150
3105	H25	160
5456	0	160
5457	H25	160
1350	H19	165
5005	H36	165

Yield Strength of Selected Wrought Aluminum Alloys
(Continued)
Listed by Yield Strength

Alloy	Temper	Yield Strength (MPa)
5050	H34	165
5456	H112	165
5657	H28, H38	165
3003	H16	170
Alclad	H32	170
3105	H16	170
5005	H16	170
5252	H25	170
6010	T4	170
6070	T4	170
5050	H36	180
5454	H111	180
5454	H311	180
2219	T42	185
3003	H18	185
5005	H38	185
5457	H28, H38	185
6063	T831	185
2036	T4	195
3105	H18	195
5005	H18	195
5052	H32	195
5083	H112	195
5652	H32	195
6151	T6	195
3004	H34	200

Yield Strength of Selected Wrought Aluminum Alloys
(Continued)
Listed by Yield Strength

Alloy	Temper	Yield Strength (MPa)
5050	H38	200
5086	H32, H116, H117	205
5154	H32	205
5254	H32	205
5454	H32	205
6066	T4, T451	205
5052	H34	215
5652	H34	215
6063	T6	215
6463	T6	215
3004	H36	230
5083	H113	230
5083	H321	230
5154	H34	230
5254	H34	230
5456	H111	230
5182	H32	235
5052	H36	240
5252	H28, H38	240
5454	H34	240
5652	H36	240
6005	T5	240
6063	T83	240
2219	T31, T351	250
3004	H38	250
5083	H323, H32	250

Yield Strength of Selected Wrought Aluminum Alloys
(Continued)
Listed by Yield Strength

Alloy	Temper	Yield Strength (MPa)
5154	H36	250
5254	H36	250
Alclad 2014	T4	255
2218	T72	255
5052	H38	255
5086	H34	255
5456	H321, H116	255
5652	H38	255
Alclad 6061	T6, T651	255
4043	H18	270
5154	H38	270
5254	H38	270
6063	T832	270
Alclad 2014	T3	275
2218	T71	275
5454	H36	275
6061	T6, T651	275
5083	H343, H34	285
5182	H34	285
6351	T6	285
2014	T4	290
Alclad 2024	T4, T351	290
2219	T62	290
6205	T5	290
2011	T3	295
6201	T6	300

Yield Strength of Selected Wrought Aluminum Alloys
(Continued)
Listed by Yield Strength

Alloy	Temper	Yield Strength (MPa)
2218	T61	305
2011	T8	310
Alclad 2024	T	310
5454	H38	310
6201	T81	310
2219	T37	315
4032	T6	315
7005	T6,T63,T6351	315
2024	T4, T351	325
6009	T6	325
2024	T3	345
5056	H38	345
7005	T53	345
2219	T81, T851	350
6070	T6	350
6066	T6, T651	360
Alclad 2024	T361	365
2618	All	370
6262	T9	380
2024	T361	395
2219	T87	395
5182	H19(n)	395
5056	H18	405
2014	T6	415
Alclad 2014	T6	415
Alclad 2024	T81, T851	415

Yield Strength of Selected Wrought Aluminum Alloys
(Continued)
Listed by Yield Strength

Alloy	Temper	Yield Strength (MPa)
2048		415
7075	T73	435
2124	T851	440
Alclad 2024	T861	455
7050	T736	455
7175	T736	455
Alclad 7075	T6,T651	460
7475	T61	460
7075	T6,T651	505
7175	T66	525

Source: Data from *ASM Metals Reference Book, Second Edition*, American Society for Metals, Metals Park, Ohio 44073, p.299—302, (1984).

YIELD STRENGTH OF SELECTED POLYMERS
Listed by Polymer

Polymer	Yield Strength, (ASTM D638) (10^3 psi)
Chlorinated Polyether	5.9
Polycarbonate	8.5
Nylons; Molded, Extruded	
Type 6	
General purpose	8.5—12.5
Cast	12.8
Flexible copolymers	7.5—10.0
Type 8	3.9
Type 11	8.5
Type 12	5.5—6.5
Nylons; Molded, Extruded	
6/6 Nylon	
General purpose molding	8.0—11.8
Glass fiber reinforced	25
General purpose extrusion	8.6—12.6
6/10 Nylon	
General purpose	7.1—8.5
ABS–Polycarbonate Alloy	8.2

Yield Strength of Selected Polymers (Continued)
Listed by Polymer

Polymer	Yield Strength, (ASTM D638) (10^3 psi)
PVC–Acrylic Alloy	
PVC–acrylic sheet	6.5
PVC–acrylic injection molded	5.5
Polymides	
Unreinforced	7.5
Unreinforced 2nd value	5
Glass reinforced	28
Polyacetals	
Homopolymer:	
Standard	10
Copolymer:	
Standard	8.8
25% glass reinforced	18.5
High flow	8.8
Polyester; Thermoplastic	
Injection Moldings:	
General purpose grade	7.5—8
Glass reinforced grades	17—25
Glass reinforced self extinguishing	17
General purpose grade	8.2
Glass reinforced grade	14
Asbestos—filled grade	12

Yield Strength of Selected Polymers (Continued)
Listed by Polymer

Polymer	Yield Strength, (ASTM D638) (10^3 psi)
Phenylene Oxides	
SE—100	7.8
SE—1	9.6
Glass fiber reinforced	14.5—17.0
Phenylene oxides (Noryl)	
Standard	10.2
Glass fiber reinforced	17—19
Polyarylsulfone	8—12
Polypropylene:	
General purpose	4.5—6.0
High impact	2.8—4.3
Asbestos filled	3.3—8.2
Glass reinforced	7—11
Flame retardant	3.6—4.2
Polyphenylene sulfide:	
Standard	9.511
40% glass reinforced	20—21

Yield Strength of Selected Polymers (Continued)
Listed by Polymer

Polymer	Yield Strength, (ASTM D638) (10^3 psi)
Polystyrenes; Molded	
Polystyrenes	
General purpose	5.0—10
Medium impact	3.7—6.0
High impact	2.8—5.3
Glass fiber -30% reinforced	14
Styrene acrylonitrile (SAN)	
Glass fiber (30%) reinforced SAN	18

To convert psi to MPa, multiply by 145.

Source: data compiled by J.S. Park from Charles T. Lynch, *CRC Handbook of Materials Science*, Vol. 3, CRC Press, Boca Raton, Florida, 1975 and *Engineered Materials Handbook*, Vol.2, Engineering Plastics, ASM International, Metals Park, Ohio, 1988.

TENSILE STRENGTH OF SELECTED ALUMINUM CASTING ALLOYS
Listed by Alloy

Alloy AA No.	Temper	Tensile Strength (MPa)
201.0	T4	365
	T6	485
	T7	460
206.0, A206.0	T7	435
208.0	F	145
242.0	T21	185
	T571	220
	T77	205
	T571	275
	T61	325
295.0	T4	220
	T6	250
	T62	285
296.0	T4	255
	T6	275
	T7	270
308.0	F	195
319.0	F	185
	T6	250
	F	235
	T6	280
336.0	T551	250
	T65	325
354.0	T61	380

Tensile Strength of Selected Aluminum Casting Alloys
(Continued)
Listed by Alloy

Alloy AA No.	Temper	Tensile Strength (MPa)
355.0	T51	195
	T6	240
	T61	270
	T7	265
	T71	175
	T51	210
	T6	290
	T62	310
	T7	280
	T71	250
356.0	T51	175
	T6	230
	T7	235
	T71	195
	T6	265
	T7	220
357.0, A357.0	T62	360
359.0	T61	330
	T62	345
360.0	F	325
A360.0	F	320
380.0	F	330
383.0	F	310
384.0, A384.0	F	330

Tensile Strength of Selected Aluminum Casting Alloys
(Continued)
Listed by Alloy

Alloy AA No.	Temper	Tensile Strength (MPa)
390.0	F	280
	T5	300
A390.0	F,T5	180
	T6	280
	T7	250
	F,T5	200
	T6	310
	T7	260
413.0	F	300
A413.0	F	290
443.0	F	130
B443.0	F	159
C443.0	F	228
514.0	F	170
518.0	F	310
520.0	T4	330
535.0	F	275
712.0	F	240
713.0	T5	210
	T5	220
771.0	T6	345
850.0	T5	160

Source: Data from *ASM Metals Reference Book, Second Edition*, American Society for Metals, Metals Park, Ohio 44073, (1984).

TENSILE STRENGTH OF SELECTED ALUMINUM CASTING ALLOYS
Listed by Value

Alloy AA No.	Temper	Tensile Strength (MPa)
443.0	F	130
208.0	F	145
B443.0	F	159
850.0	T5	160
514.0	F	170
355.0	T71	175
356.0	T51	175
A390.0	F,T5	180
242.0	T21	185
319.0	F	185
308.0	F	195
355.0	T51	195
356.0	T71	195
A390.0	F,T5	200
242.0	T77	205
355.0	T51	210
713.0	T5	210
242.0	T571	220
295.0	T4	220
356.0	T7	220
713.0	T5	220
C443.0	F	228
356.0	T6	230
319.0	F	235
356.0	T7	235

Tensile Strength of Selected Aluminum Casting Alloys
(Continued)
Listed by Value

Alloy AA No.	Temper	Tensile Strength (MPa)
355.0	T6	240
712.0	F	240
295.0	T6	250
319.0	T6	250
336.0	T551	250
355.0	T71	250
A390.0	T7	250
296.0	T4	255
A390.0	T7	260
355.0	T7	265
356.0	T6	265
296.0	T7	270
355.0	T61	270
242.0	T571	275
296.0	T6	275
535.0	F	275
319.0	T6	280
355.0	T7	280
390.0	F	280
A390.0	T6	280
295.0	T62	285
355.0	T6	290
A413.0	F	290
390.0	T5	300
413.0	F	300

Tensile Strength of Selected Aluminum Casting Alloys
(Continued)
Listed by Value

Alloy AA No.	Temper	Tensile Strength (MPa)
355.0	T62	310
383.0	F	310
A390.0	T6	310
518.0	F	310
A360.0	F	320
242.0	T61	325
336.0	T65	325
360.0	F	325
359.0	T61	330
380.0	F	330
384.0, A384.0	F	330
520.0	T4	330
359.0	T62	345
771.0	T6	345
357.0, A357.0	T62	360
201.0	T4	365
354.0	T61	380
206.0, A206.0	T7	435
201.0	T7	460
201.0	T6	485

Source: Data from *ASM Metals Reference Book, Second Edition*, American Society for Metals, Metals Park, Ohio 44073, (1984).

TENSILE STRENGTH OF SELECTED WROUGHT ALUMINUM ALLOYS
Listed by Alloy

Alloy	Temper	Tensile Strength (MPa)
1050	0	76
	H14	110
	H16	130
	H18	160
1060	0	69
	H12	83
	H14	97
	H16	110
	H18	130
1100	0	90
	H12	110
	H14	125
	H16	145
	H18	165
1350	0	83
	H12	97
	H14	110
	H16	125
	H19	185
2011	T3	380
	T8	405
2014	0	185
	T4	425
	T6	485

Tensile Strength of Selected Wrought Aluminum Alloys
(Continued)
Listed by Alloy

Alloy	Temper	Tensile Strength (MPa)
Alclad 2014	0	170
	T3	435
	T4	420
	T6	470
2024	0	185
	T3	485
	T4, T351	470
	T361	495
Alclad 2024	0	180
	T	450
	T4, T351	440
	T361	460
	T81, T851	450
	T861	485
2036	T4	340
2048		455
2124	T851	490
2218	T61	405
	T71	345
	T72	330
2219	0	170
	T42	360
	T31, T351	360
	T37	395
	T62	415
	T81, T851	455
	T87	475

Tensile Strength of Selected Wrought Aluminum Alloys
(Continued)
Listed by Alloy

Alloy	Temper	Tensile Strength (MPa)
2618	All	440
3003	0	110
Alclad	H12	130
3003	H14	150
	H16	180
	H18	200
3004	0	180
Alclad	H32	215
3004	H34	240
	H36	260
	H38	285
3105	0	115
	H12	150
	H14	170
	H16	195
	H18	215
	H25	180
4032	T6	380
4043	0	145
	H18	285
5005	0	125
	H12	140
	H14	160
	H16	180
	H18	200

Tensile Strength of Selected Wrought Aluminum Alloys
(Continued)
Listed by Alloy

Alloy	Temper	Tensile Strength (MPa)
	H32	140
	H34	160
	H36	180
	H38	200
5050	0	145
	H32	170
	H34	195
	H36	205
	H38	220
5052	0	195
	H32	230
	H34	260
	H36	275
	H38	290
5056	0	290
	H18	435
	H38	415
5083	0	290
	H112	305
	H113	315
	H321	315
	H323, H32	325
	H343, H34	345
5086	0	260
	H32, H116, H117	290
	H34	325

Tensile Strength of Selected Wrought Aluminum Alloys
(Continued)
Listed by Alloy

Alloy	Temper	Tensile Strength (MPa)
	H112	270
5154	0	240
	H32	270
	H34	290
	H36	310
	H38	330
	H112	240
5182	0	275
	H32	315
	H34	340
	H19(n)	420
5252	H25	235
	H28, H38	285
5254	0	240
5254	H32	270
	H34	290
	H36	310
	H38	330
	H112	240
5454	0	250
	H32	275
	H34	305
	H36	340
	H38	370
	H111	260
	H112	250
	H311	260

Tensile Strength of Selected Wrought Aluminum Alloys
(Continued)
Listed by Alloy

Alloy	Temper	Tensile Strength (MPa)
5456	0	310
	H111	325
	H112	310
	H321, H116	350
5457	0	130
	H25	180
	H28, H38	205
5652	0	195
	H32	230
	H34	260
	H36	275
	H38	290
5657	H25	160
	H28, H38	195
6005	T1	170
	T5	260
6009	T4	235
	T6	345
6010	T4	255
6061	0	125
	T4, T451	240
	T6, T651	310
Alclad 6061	0	115
	T4, T451	230
	T6, T651	290

Tensile Strength of Selected Wrought Aluminum Alloys
(Continued)
Listed by Alloy

Alloy	Temper	Tensile Strength (MPa)
6063	0	90
	TI	150
	T4	170
	T5	185
	T6	240
	T83	255
	T831	205
	T832	290
6066	0	150
	T4, T451	360
	T6, T651	395
6070	0	145
	T4	315
	T6	380
6101	Hlll	97
6151	T6	220
6201	T6	330
	T81	330
6205	Tl	260
	T5	310
6262	T9	400
6351	T4	250
	T6	310
6463	Tl	150
	T5	185
	T6	240

Tensile Strength of Selected Wrought Aluminum Alloys
(Continued)
Listed by Alloy

Alloy	Temper	Tensile Strength (MPa)
7005	0	193
	T53	393
	T6,T63,T6351	372
7050	T736	515
7075	0	230
	T6,T651	570
	T73	505
Alclad 7075	0	220
	T6,T651	525
7175	T66	595
	T736	525
7475	T61	525

Source: Data from *ASM Metals Reference Book, Second Edition*, American Society for Metals, Metals Park, Ohio 44073, p.299—302, (1984).

TENSILE STRENGTH OF SELECTED WROUGHT ALUMINUM ALLOYS
Listed by Tensile Strength

Alloy	Temper	Tensile Strength (MPa)
1060	0	69
1050	0	76
1060	H12	83
1350	0	83
1100	0	90
6063	0	90
1060	H14	97
1350	H12	97
6101	Hlll	97
1050	H14	110
1060	H16	110
1100	H12	110
1350	H14	110
3003	0	110
3105	0	115
Alclad 6061	0	115
1100	H14	125
1350	H16	125
5005	0	125
6061	0	125
1050	H16	130
1060	H18	130
Alclad	H12	130
5457	0	130
5005	H12	140
5005	H32	140

Tensile Strength of Selected Wrought Aluminum Alloys
(Continued)
Listed by Tensile Strength

Alloy	Temper	Tensile Strength (MPa)
1100	H16	145
4043	0	145
5050	0	145
6070	0	145
3003	H14	150
3105	H12	150
6063	TI	150
6066	0	150
6463	Tl	150
1050	H18	160
5005	H14	160
5005	H34	160
5657	H25	160
1100	H18	165
Alclad 2014	0	170
2219	0	170
3105	H14	170
5050	H32	170
6005	T1	170
6063	T4	170
Alclad 2024	0	180
3003	H16	180
3004	0	180
3105	H25	180
5005	H16	180
5005	H36	180

Tensile Strength of Selected Wrought Aluminum Alloys (Continued)
Listed by Tensile Strength

Alloy	Temper	Tensile Strength (MPa)
5457	H25	180
1350	H19	185
2014	O	185
2024	O	185
6063	T5	185
6463	T5	185
7005	O	193
3105	H16	195
5050	H34	195
5052	O	195
5652	O	195
5657	H28, H38	195
3003	H18	200
5005	H18	200
5005	H38	200
5050	H36	205
5457	H28, H38	205
6063	T831	205
Alclad	H32	215
3105	H18	215
5050	H38	220
6151	T6	220
Alclad 7075	O	220
5052	H32	230
5652	H32	230
Alclad 6061	T4, T451	230

Tensile Strength of Selected Wrought Aluminum Alloys
(Continued)
Listed by Tensile Strength

Alloy	Temper	Tensile Strength (MPa)
7075	0	230
5252	H25	235
6009	T4	235
3004	H34	240
5154	0	240
5154	H112	240
5254	0	240
5254	H112	240
6061	T4, T451	240
6063	T6	240
6463	T6	240
5454	0	250
5454	H112	250
6351	T4	250
6010	T4	255
6063	T83	255
3004	H36	260
5052	H34	260
5086	0	260
5454	H111	260
5454	H311	260
5652	H34	260
6005	T5	260
6205	Tl	260
5086	H112	270
5154	H32	270

Tensile Strength of Selected Wrought Aluminum Alloys (Continued)
Listed by Tensile Strength

Alloy	Temper	Tensile Strength (MPa)
5254	H32	270
5052	H36	275
5182	0	275
5454	H32	275
5652	H36	275
3004	H38	285
4043	H18	285
5252	H28, H38	285
5052	H38	290
5056	0	290
5083	0	290
5086	H32, H116, H117	290
5154	H34	290
5254	H34	290
5652	H38	290
Alclad 6061	T6, T651	290
6063	T832	290
5083	H112	305
5454	H34	305
5154	H36	310
5254	H36	310
5456	0	310
5456	H112	310
6061	T6, T651	310
6205	T5	310
6351	T6	310

Tensile Strength of Selected Wrought Aluminum Alloys
(Continued)
Listed by Tensile Strength

Alloy	Temper	Tensile Strength (MPa)
5083	H113	315
5083	H321	315
5182	H32	315
6070	T4	315
5083	H323, H32	325
5086	H34	325
5456	H111	325
2218	T72	330
5154	H38	330
5254	H38	330
6201	T6	330
6201	T81	330
2036	T4	340
5182	H34	340
5454	H36	340
2218	T71	345
5083	H343, H34	345
6009	T6	345
5456	H321, H116	350
2219	T42	360
2219	T31, T351	360
6066	T4, T451	360
5454	H38	370
7005	T6,T63,T6351	372
2011	T3	380
4032	T6	380

Tensile Strength of Selected Wrought Aluminum Alloys
(Continued)
Listed by Tensile Strength

Alloy	Temper	Tensile Strength (MPa)
6070	T6	380
7005	T53	393
2219	T37	395
6066	T6, T651	395
6262	T9	400
2011	T8	405
2218	T61	405
2219	T62	415
5056	H38	415
Alclad 2014	T4	420
5182	H19(n)	420
2014	T4	425
Alclad 2014	T3	435
5056	H18	435
Alclad 2024	T4, T351	440
2618	All	440
Alclad 2024	T	450
Alclad 2024	T81, T851	450
2048		455
2219	T81, T851	455
Alclad 2024	T361	460
Alclad 2014	T6	470
2024	T4, T351	470
2219	T87	475
2014	T6	485
2024	T3	485

Tensile Strength of Selected Wrought Aluminum Alloys
(Continued)
Listed by Tensile Strength

Alloy	Temper	Tensile Strength (MPa)
Alclad 2024	T861	485
2124	T851	490
2024	T361	495
7075	T73	505
7050	T736	515
Alclad 7075	T6,T651	525
7175	T736	525
7475	T61	525
7075	T6,T651	570
7175	T66	595

Source: Data from *ASM Metals Reference Book, Second Edition*, American Society for Metals, Metals Park, Ohio 44073, p.299—302, (1984).

TENSILE STRENGTH OF SELECTED CERAMICS
Listed by Ceramic

Ceramic	Tensile Strength

BORIDES

Ceramic	Tensile Strength
Chromium Diboride (CrB_2)	10.6×10^4 psi
Titanium Diboride (TiB_2)	18.4×10^3 psi
Zirconium Diboride (ZrB_2)	28.7×10^3 psi

CARBIDES

Ceramic	Tensile Strength
Boron Carbide (B_4C)	22.5×10^3 psi at 980 °C
Silicon Carbide (SiC)	$5\text{-}20 \times 10^3$ psi at 25°C
(hot pressed)	29×10^3 psi at 20°C
(hot pressed)	$5.75\text{-}21.75 \times 10^3$ psi at 1400°C
(reaction bonded)	11.17×10^3 psi at 20°C
Tantalum Monocarbide (TaC)	$2\text{-}42 \times 10^3$ psi
Titanium Monocarbide (TiC)	17.2×10^3 psi at 1000 °C
Tungsten Monocarbide (WC)	50×10^3 psi
Zirconium Monocarbide (ZrC)	16.0×10^3 psi at room temp.
	$11.7\text{-}14.45 \times 10^3$ psi at 980 °C
	$12.95\text{-}15.85 \times 10^3$ psi at 1250 °C

NITRIDES

Ceramic	Tensile Strength
Boron Nitride (BN)	0.35×10^3 psi at 1000 °C
	0.35×10^3 psi at 1500 °C
	1.15×10^3 psi at 1800 °C
	2.25×10^3 psi at 2000 °C
	6.80×10^3 psi at 2400 °C

Tensile Strength of Selected Ceramics (Continued)
Listed by Ceramic

Ceramic	Tensile Strength
Trisilicon tetranitride (Si$_3$N$_4$)	
(hot pressed)	54.4 x10^3 psi at 20°C
(hot pressed)	21.8 x10^3 psi at 1400°C
(reaction bonded)	24.7 x10^3 psi at 20°C
(reaction bonded)	20.3 x10^3 psi at 1400°C

OXIDES

Ceramic	Tensile Strength
Aluminum Oxide (Al$_2$O$_3$)	37-37.8 x10^3 psi at room temp.
	33.6 x10^3 psi at 300°C
	40 x10^3 psi at 500°C
	34.6 x10^3 psi at 800°C
	35 x10^3 psi at 1000°C
	33.9 x10^3 psi at 1050°C
	31.4 x10^3 psi at 1140°C
	18.5-20 x10^3 psi at 1200°C
	6.4 x10^3 psi at 1300°C
	4.3 x10^3 psi at 1400°C
	1.5 x10^3 psi at 1460°C
Beryllium Oxide (BeO)	13.5-20 x10^3 psi at room temp.
	11.1 x10^3 psi at 500°C
	7.0 x10^3 psi at 900°C
	5.0 x10^3 psi at 1000°C
	2.0 x10^3 psi at 1140°C
	0.6 x10^3 psi at 1300°C

Tensile Strength of Selected Ceramics (Continued)
Listed by Ceramic

Ceramic	Tensile Strength
Magnesium Oxide (MgO)	14×10^3 psi at room temp.
	14×10^3 psi at 200°C
	15.2×10^3 psi at 400°C
	16×10^3 psi at 800°C
	11.5×10^3 psi at 1000°C
	10×10^3 psi at 1100°C
	8×10^3 psi at 1200°C
	6×10^3 psi at 1300°C
Thorium Dioxide (ThO$_2$)	14×10^3 psi at room temp.
Zircoium Oxide (ZrO$_2$)	$17.9\text{-}20 \times 10^3$ psi at room temp.
	16.8×10^3 at 200°C
	17.5×10^3 at 400°C
	20.0×10^3 at 500°C
	17.6×10^3 at 600°C
	16.0×10^3 at 800°C
	$6.75\text{-}17.0 \times 10^3$ at 1000°C
	$13.0\text{-}13.5 \times 10^3$ at 1100°C
	12.1×10^3 at 1200°C
	10.2×10^3 at 1300°C
(MgO stabilized)	21×10^6 psi at room temp.
Cordierite (2MgO 2Al$_2$O$_3$ 5SiO$_2$)	
(ρ=2.51g/cm^3)	7.8×10^3 psi at 25°C
(ρ=2.1g/cm^3)	3.5×10^3 psi at 800°C
(ρ=1.8g/cm^3)	2.5×10^3 psi at 1200°C

Tensile Strength of Selected Ceramics (Continued)
Listed by Ceramic

Ceramic	Tensile Strength
Mullite ($3Al_2O_3\ 2SiO_2$)	16×10^3 psi at 25°C
Spinel ($Al_2O_3\ MgO$)	19.2×10^3 psi at room temp.
	13.7×10^3 psi at 550°C
	110.8×10^3 psi at 900°C
	6.1×10^3 psi at 1150°C
	1.1×10^3 psi at 1300°C
Zircon ($SiO_2\ ZrO_2$)	12.7×10^3 psi at room temp.
	8.7×10^3 psi at 1050°C
	3.6×10^3 psi at 1200°C
SILICIDE	
Molybdenum Disilicide ($MoSi_2$)	40×10^3 psi at 980°C
	42.16×10^3 psi at 1090°C
	42.8×10^3 psi at 1200°C
	41.07×10^3 psi at 1300°C

To convert psi to MPa, multiply by 145.

Source: Data compiled by J.S. Park from *No. 1 Materials Index*, Peter T.B. Shaffer, Plenum Press, New York, (1964); *Smithells Metals Reference Book*, Eric A. Brandes, ed., in association with Fulmer Research Institute Ltd. 6th ed. London, Butterworths, Boston, (1983); and *Ceramic Abstracts*, American Ceramic Society (1986-1991).

TENSILE STRENGTH OF SELECTED POLYMERS
Listed by Polymer

Polymer	Tensile Strength, (ASTM D638) (10^3 psi)
ABS Resins; Molded, Extruded	
Medium impact	6.3—8.0
High impact	5.0—6.0
Very high impact	4.5—6.0
Low temperature impact	4—6
Heat resistant	7.0—8.0
Acrylics; Cast, Molded, Extruded	
Cast Resin Sheets, Rods:	
General purpose, type I	6—9
General purpose, type II	8—10
Moldings:	
Grades 5, 6, 8	8.8—10.5
High impact grade	5.5—8.0
Thermoset Carbonate	
Allyl diglycol carbonate	5—6
Alkyds; Molded	
Putty (encapsulating)	4—5
Rope (general purpose)	7—8
Granular (high speed molding)	3—4
Glass reinforced (heavy duty parts)	5—9

Tensile Strength of Selected Polymers (Continued)
Listed by Polymer

Polymer	Tensile Strength, (ASTM D638) (10³ psi)

Polymer	Tensile Strength, (ASTM D638) (10^3 psi)
Cellulose Acetate; Molded, Extruded ASTM Grade:	(Tensile Strength at Fracture)
H4—1	7—8
H2—1	5.8—7.2
MH—1, MH—2	4.8—6.3
MS—1, MS—2	3.9—5.3
S2—1	3.0—4.4
Cellulose Acetate Butyrate; Molded, Extruded ASTM Grade:	(Tensile Strength at Fracture)
H4	6.9
MH	5.0—6.0
S2	3.0—4.0
Cellusose Acetate Propionate; Molded, Extruded ASTM Grade:	
1	5.9—6.5
3	5.1—5.9
6	4
Chlorinated Polymers	
Chlorinated polyether	6
Chlorinated polyvinyl chloride	7.3

Tensile Strength of Selected Polymers (Continued)
Listed by Polymer

Polymer	Tensile Strength, (ASTM D638) (10^3 psi)
Polycarbonates	
Polycarbonate	9.5
Polycarbonate (40% glass fiber reinforced)	18
Diallyl Phthalates; Molded	
Orlon filled	4.5—6
Dacron filled	4.6—6.2
Asbestos filled	4—6.5
Glass fiber filled	5.5—11
Fluorocarbons; Molded,Extruded	
Polytrifluoro chloroethylene (PTFCE)	4.6—5.7
Polytetrafluoroethylene (PTFE)	2.5—6.5
Ceramic reinforced (PTFE)	0.75—2.5
Fluorinated ethylene propylene(FEP)	2.5—4.0
Polyvinylidene— fluoride (PVDF)	5.2—8.6
Epoxies; Cast, Molded, Reinforced	
Standard epoxies (diglycidyl ethers of bisphenol A)	
Cast rigid	9.5-11.5
Cast flexible	1.4—7.6
Molded	8—11
General purpose glass cloth laminate	50-58
High strength laminate	160
Filament wound composite	230-240 (hoop)

Tensile Strength of Selected Polymers (Continued)
Listed by Polymer

Polymer	Tensile Strength, (ASTM D638) (10^3 psi)
Epoxies—Molded, Extruded	
High performance resins	
(cycloaliphatic diepoxides)	
Cast, rigid	8—12
Molded	5.2—5.3
Glass cloth laminate	50—52
Epoxy novolacs	
Cast, rigid	9.6—12.0
Glass cloth laminate	59.2
Melamines; Molded	
Filler & type	
Cellulose electrical	5—9
Glass fiber	6—9
Alpha cellulose and mineral	5—8
Nylons; Molded, Extruded	
Type 6	
General purpose	9.5—12.5
Glass fiber (30%) reinforced	21—24
Cast	12.8
Flexible copolymers	7.5—10.0
Type 12	7.1—8.5

Tensile Strength of Selected Polymers (Continued)
Listed by Polymer

Polymer	Tensile Strength, (ASTM D638) (10^3 psi)
6/6 Nylon	
General purpose molding	11.2—11.8
Glass fiber reinforced	25—30
Glass fiber Molybdenum disulfide filled	19—22
General purpose extrusion	1.26, 8.6
6/10 Nylon	
General purpose	7.1—8.5
Glass fiber (30%) reinforced	19
Phenolics; Molded	(ASTM D651)
Type and filler	
General: woodflour and flock	5.0—8.5
Shock: paper, flock, or pulp	5.0—8.5
High shock: chopped fabric or cord	5—9
Very high shock: glass fiber	5—10
Phenolics; Molded	(ASTM D651)
Arc resistant—mineral	6
Rubber phenolic—woodflour or flock	4.5—9
Rubber phenolic—chopped fabric	3—5
Rubber phenolic—asbestos	4
ABS–Polycarbonate Alloy	8.2

Tensile Strength of Selected Polymers (Continued)
Listed by Polymer

Polymer	Tensile Strength, (ASTM D638) (10^3 psi)
Polyacetals	
Homopolymer:	
Standard	10
20% glass reinforced	8.5
22% TFE reinforced	6.9
Copolymer:	
Standard	8.8
25% glass reinforced	18.5
High flow	8.8
Polyesters: Thermosets	
Cast polyyester	
Rigid	5—15
Flexible	1—8
Reinforced polyester moldings	
High strength (glass fibers)	5—10
Heat and chemical resistant (asbestos)	4—6
Sheet molding compounds, general purpose	15—17
Polyarylsulfone	13
Polypropylene:	
General purpose	4.5—6.0

Tensile Strength of Selected Polymers (Continued)

Listed by Polymer

Polymer	Tensile Strength, (ASTM D638) (10^3 psi)
Polyethylenes; Molded, Extruded	(ASTM D412)
Type I—lower density (0.910—0.925)	
Melt index 0.3—3.6	1.4—2.5
Melt index 6—26	1.4—2.0
Melt index 200	0.9—1.1
Type II—medium density (0.926—0.940)	
Melt index 20	2
Melt index 1.0—1.9	2.3—2.4
Type III—higher density (0.941—0.965)	
Melt index 0.2—0.9	4.4
Melt Melt index 0.1—12.0	2.9—4.0
Melt index 1.5—15	4.4
High molecular weight	5.4
Olefin Copolymers; Molded	
EEA (ethylene ethyl acrylate)	0.2
EVA (ethylene vinyl acetate)	0.36
Ethylene butene	0.35
Propylene—ethylene	0.4
Ionomer	0.4
Polyallomer	3—4.3

Tensile Strength of Selected Polymers (Continued)
Listed by Polymer

Polymer	Tensile Strength, (ASTM D638) (10³ psi)
Polystyrenes; Molded	
Polystyrenes	
General purpose	5.0—10
Medium impact	4.0—6.0
High impact	3.3—5.1
Glass fiber -30% reinforced	14
Styrene acrylonitrile (SAN)	8.3—12.0
Glass fiber (30%) reinforced SAN	18
Polyvinyl Chloride And Copolymers; Molded, Extruded D412	
Nonrigid—general	1—3.5
Nonrigid—electrical	2—3.2
Rigid—normal impact	5.5—8
Vinylidene chloride	4—8,15—40
Silicones; Molded, Laminated	(ASTM D651)
Fibrous (glass) reinforced silicones	6.5
Granular (silica) reinforced silicones	4—6
Woven glass fabric/ silicone laminate	30—35
Ureas; Molded	
Alpha—cellulose filled (ASTM Type l)	5—10

To convert psi to MPa, multiply by 145.

Source: data compiled by J.S. Park from Charles T. Lynch, *CRC Handbook of Materials Science*, Vol. 3, CRC Press, Boca Raton, Florida, 1975 and *Engineered Materials Handbook*, Vol.2, Engineering Plastics, ASM International, Metals Park, Ohio, 1988.

TOTAL ELONGATION OF SELECTED
CAST ALUMINUM ALLOYS
Listed by Alloy

Alloy AA No.	Temper	Elongation (in 2 in.) (%)
201.0	T4	20
	T6	7
	T7	4.5
206.0, A206.0	T7	11.7
208.0	F	2.5
242.0	T21	1.0
	T571	0.5
	T77	2.0
	T571	1.0
	T61	0.5
295.0	T4	8.5
	T6	5.0
	T62	2.0
296.0	T4	9.0
	T6	5.0
	T7	4.5
308.0	F	2.0
319.0	F	2.0
	T6	2.0
	F	2.5
	T6	3.0
336.0	T551	0.5
	T65	0.5
354.0	T61	6.0

Total Elongation of Selected Cast Aluminum Alloys (Continued)
Listed by Alloy

Alloy AA No.	Temper	Elongation (in 2 in.) (%)
355.0	T51	1.5
	T6	3.0
	T61	1.0
	T7	0.5
	T71	1.5
	T51	2.0
	T6	4.0
	T62	1.5
	T7	2.0
	T71	3.0
356.0	T51	2.0
	T6	3.5
	T7	2.0
	T71	3.5
	T6	5.0
	T7	6.0
357.0, A357.0	T62	8.0
359.0	T61	6.0
	T62	5.5
360.0	F	3.0
A360.0	F	5.0
380.0	F	3.0
383.0	F	3.5
384.0, A384.0	F	2.5

Total Elongation of Selected Cast Aluminum Alloys (Continued)
Listed by Alloy

Alloy AA No.	Temper	Elongation (in 2 in.) (%)
390.0	F	1.0
	T5	1.0
A390.0	F,T5	<1.0
	T6	<1.0
	T7	<1.0
	F,T5	1.0
	T6	<1.0
	T7	<1.0
413.0	F	2.5
A413.0	F	3.5
443.0	F	8.0
B443.0	F	10.0
C443.0	F	9.0
514.0	F	9.0
518.0	F	5.0—8.0
520.0	T4	16
535.0	F	13
712.0	F	5.0
713.0	T5	3.0
	T5	4.0
771.0	T6	9.0
850.0	T5	10.0

Source: Data from *ASM Metals Reference Book, Second Edition*, American Society for Metals, Metals Park, Ohio 44073, (1984).

TOTAL ELONGATION OF SELECTED
CAST ALUMINUM ALLOYS
Listed by Value

Alloy AA No.	Temper	Elongation (in 2 in.) (%)
242.0	T571	0.5
242.0	T61	0.5
336.0	T551	0.5
336.0	T65	0.5
355.0	T7	0.5
A390.0	F,T5	<1.0
A390.0	T6	<1.0
A390.0	T7	<1.0
A390.0	T6	<1.0
A390.0	T7	<1.0
242.0	T21	1.0
242.0	T571	1.0
355.0	T61	1.0
390.0	F	1.0
390.0	T5	1.0
A390.0	F,T5	1.0
355.0	T51	1.5
355.0	T71	1.5
355.0	T62	1.5
242.0	T77	2.0
295.0	T62	2.0
308.0	F	2.0
319.0	F	2.0
319.0	T6	2.0
355.0	T51	2.0

Total Elongation of Selected Cast Aluminum Alloys (Continued)
Listed by Value

Alloy AA No.	Temper	Elongation (in 2 in.) (%)
355.0	T7	2.0
356.0	T51	2.0
356.0	T7	2.0
208.0	F	2.5
319.0	F	2.5
384.0, A384.0	F	2.5
413.0	F	2.5
319.0	T6	3.0
355.0	T6	3.0
355.0	T71	3.0
360.0	F	3.0
380.0	F	3.0
713.0	T5	3.0
356.0	T6	3.5
356.0	T71	3.5
383.0	F	3.5
A413.0	F	3.5
355.0	T6	4.0
713.0	T5	4.0
201.0	T7	4.5
296.0	T7	4.5
295.0	T6	5.0
296.0	T6	5.0
356.0	T6	5.0
A360.0	F	5.0

Total Elongation of Selected Cast Aluminum Alloys (Continued)
Listed by Value

Alloy AA No.	Temper	Elongation (in 2 in.) (%)
712.0	F	5.0
518.0	F	5.0—8.0
359.0	T62	5.5
354.0	T61	6.0
356.0	T7	6.0
359.0	T61	6.0
201.0	T6	7
357.0, A357.0	T62	8.0
443.0	F	8.0
295.0	T4	8.5
296.0	T4	9.0
C443.0	F	9.0
514.0	F	9.0
771.0	T6	9.0
B443.0	F	10.0
850.0	T5	10.0
206.0, A206.0	T7	11.7
535.0	F	13
520.0	T4	16
201.0	T4	20

Source: Data from *ASM Metals Reference Book, Second Edition*, American Society for Metals, Metals Park, Ohio 44073, (1984).

TOTAL ELONGATION OF SELECTED POLYMERS
Listed by Polymer

Polymer	Elongation (in 2 in.), (ASTM D638) (%)
ABS Resins; Molded, Extruded	
Medium impact	5—20
High impact	5—50
Very high impact	20—50
Low temperature impact	30—200
Heat resistant	20
Acrylics; Cast, Molded, Extruded	
Cast Resin Sheets, Rods:	
General purpose, type I	2—7
General purpose, type II	2—7
Moldings:	
Grades 5, 6, 8	3—5
High impact grade	>25
Chlorinated Polymers	
Chlorinated polyether	130
Polycarbonates	
Polycarbonate	110
Polycarbonate (40% glass fiber reinforced)	0—5

Total Elongation of Selected Polymers (Continued)
Listed by Polymer

Polymer	Elongation (in 2 in.), (ASTM D638) (%)
Fluorocarbons; Molded,Extruded	
Polytrifluoro chloroethylene (PTFCE)	125—175
Polytetrafluoroethylene (PTFE)	250—350
Ceramic reinforced (PTFE)	10—200
Fluorinated ethylene propylene(FEP)	250—330
Polyvinylidene— fluoride (PVDF)	200—300
Epoxies; Cast, Molded, Reinforced	
Standard epoxies (diglycidyl ethers of bisphenol A)	
Cast rigid	4.4
Cast flexible	1.5-60
Epoxies—Molded, Extruded	
High performance resins	
(cycloaliphatic diepoxides)	
Cast, rigid	2—5
Epoxy novolacs	
Glass cloth laminate	2.2—4.8
Melamines; Molded	
Cellulose electrical	0.6

Total Elongation of Selected Polymers (Continued)
Listed by Polymer

Polymer	Elongation (in 2 in.), (ASTM D638) (%)
Nylons; Molded, Extruded	
Type 6	
General purpose	30—100
Glass fiber (30%) reinforced	2.2—3.6
Cast	20
Flexible copolymers	200—320
Type 8	400
Type 11	100—120
Type 12	120—350
Nylons; Molded, Extruded	
6/6 Nylon	
General purpose molding	15—60, 300
Glass fiber reinforced	1.8—2.2
Glass fiber Molybdenum disulfide filled	3
General purpose extrusion	90—240
6/10 Nylon	
General purpose	85—220
Glass fiber (30%) reinforced	1.9
Phenolics; Molded	
Type and filler	
General: woodflour and flock	0.4—0.8
High shock: chopped fabric or cord	0.37—0.57
Very high shock: glass fiber	0.2

Total Elongation of Selected Polymers (Continued)
Listed by Polymer

Polymer	Elongation (in 2 in.), (ASTM D638) (%)
Phenolics: Molded	
Rubber phenolic—woodflour or flock	0.75—2.25
ABS–Polycarbonate Alloy	110
PVC–Acrylic Alloy	
PVC–acrylic sheet	>100
PVC–acrylic injection molded	150
Polymides	
Unreinforced	<1
Unreinforced 2nd value	1.2
Glass reinforced	<1
Polyacetals	
Homopolymer:	
Standard	25
20% glass reinforced	7
22% TFE reinforced	12
Copolymer:	
Standard	60—75
25% glass reinforced	3
High flow	40

Total Elongation of Selected Polymers (Continued)
Listed by Polymer

Polymer	Elongation (in 2 in.), (ASTM D638) (%)
Polyester; Thermoplastic	
Injection Moldings:	
General purpose grade	300
Glass reinforced grades	1—5
Glass reinforced self extinguishing	5
General purpose grade	250
Glass reinforced grade	<5
Asbestos—filled grade	<5
Polyesters: Thermosets	
Cast polyyester	
Rigid	1.7—2.6
Flexible	25—300
Reinforced polyester moldings	
High strength (glass fibers)	0.3—0.5
Phenylene Oxides	
SE—100	50
SE—1	60
Glass fiber reinforced	4—6
Phenylene oxides (Noryl)	
Standard	50—100
Polyarylsulfone	15—40

Total Elongation of Selected Polymers (Continued)
Listed by Polymer

Polymer	Elongation (in 2 in.), (ASTM D638) (%)
Polypropylene:	
General purpose	100—600
High impact	30—>200
Asbestos filled	3—20
Glass reinforced	2—4
Flame retardant	3—15
Polyphenylene sulfide:	
Standard	3
40% glass reinforced	3—9
Polyethylenes; Molded, Extruded	
Type I—lower density (0.910—0.925)	(ASTM D412)
Melt index 0.3—3.6	500—725
Melt index 6—26	125—675
Melt index 200	80—100
Type II—medium density (0.926—0.940)	
Melt index 20	200
Melt index 1.0—1.9	200—425
Type III—higher density (0.941—0.965)	
Melt index 0.2—0.9	700—1,000
Melt Melt index 0.1—12.0	50—1,000
Melt index 1.5—15	100—700
High molecular weight	400

Total Elongation of Selected Polymers (Continued)
Listed by Polymer

Polymer	Elongation (in 2 in.), (ASTM D638) (%)
Olefin Copolymers; Molded	
EEA (ethylene ethyl acrylate)	650
EVA (ethylene vinyl acetate)	650
Ethylene butene	20
Ionomer	450
Polyallomer	300—400
Polystyrenes; Molded	
General purpose	1.0—2.3
Medium impact	3.0—40
Glass fiber -30% reinforced	1.1
Styrene acrylonitrile (SAN)	0.5—4.5
Glass fiber (30%) reinforced SAN	1.4—1.6
Polyvinyl Chloride And Copolymers; Molded, Extruded	
Nonrigid—general	200—450
Nonrigid—electrical	220—360
Rigid—normal impact	1—10
Vinylidene chloride	15—25, 20—30

Total Elongation of Selected Polymers (Continued)
Listed by Polymer

Polymer	Elongation (in 2 in.), (ASTM D638) (%)
Silicones; Molded, Laminated	(ASTM D651)
Fibrous (glass) reinforced silicones	<3
Granular (silica) reinforced silicones	<3
Ureas; Molded	
Alpha—cellulose filled (ASTM Type 1)	1

Source: data compiled by J.S. Park from Charles T. Lynch, *CRC Handbook of Materials Science*, Vol. 3, CRC Press, Boca Raton, Florida, 1975 and *Engineered Materials Handbook*, Vol.2, Engineering Plastics, ASM International, Metals Park, Ohio, 1988.

ELONGATION AT YIELD OF SELECTED POLYMERS
Listed by Polymer

Polymer	Elongation at Yield, (ASTM D638) (%)
Chlorinated polyether	15
Polycarbonates	
Polycarbonate	5
Nylons; Molded, Extruded	
Type 6	
Cast	5
Type 12	5.8
Nylons; Molded, Extruded	
6/6 Nylon:	
General purpose molding	5—25
General purpose extrusion	5—30
6/10 Nylon:	
General purpose	5—30
Polyacetals	
Homopolymer:	
Standard	12
Copolymer:	
Standard	12
25% glass reinforced	3
High flow	12

Elongation at Yield of Selected Polymers (Continued)
Listed by Polymer

Polymer	Elongation at Yield, (ASTM D638) (%)
Phenylene oxides (Noryl)	
Standard	5.6
Glass fiber reinforced	2—1.6
Polyarylsulfone	6.5—13
Polypropylene:	
General purpose	9—15
High impact	7—13
Asbestos filled	5
Polyphenylene sulfide:	
Standard	1.6
40% glass reinforced	1.25
Polystyrenes; Molded Polystyrenes	
General purpose	1.0—2.3
Medium impact	1.2—3.0
High impact	1.5—2.0
Glass fiber -30% reinforced	1.1
Glass fiber (30%) reinforced SAN	1.4—1.6

Source: data compiled by J.S. Park from Charles T. Lynch, *CRC Handbook of Materials Science*, Vol. 3, CRC Press, Boca Raton, Florida, 1975 and *Engineered Materials Handbook*, Vol.2, Engineering Plastics, ASM International, Metals Park, Ohio, 1988.

SHEAR STRENGTH OF SELECTED WROUGHT ALUMINUM ALLOYS

Listed by Alloy

Alloy AA No.	Temper	Shear Strength (MPa)
1050	0	62
	H14	69
	H16	76
	H18	83
1060	0	48
	H12	55
	H14	62
	H16	69
	H18	76
1100	0	62
	H12	69
	H14	76
	H16	83
	H18	90
1350	0	55
	H12	62
	H14	69
	H16	76
	H19	105
2011	T3	220
	T8	240
2014	0	125
	T4	260
	T6	290

Shear Strength of Selected Wrought Aluminum Alloys
(Continued)
Listed by Alloy

Alloy AA No.	Temper	Shear Strength (MPa)
Alclad 2014	0	125
	T3	255
	T4	255
	T6	285
2024	0	125
	T3	285
	T4, T351	285
	T361	290
Alclad 2024	0	125
	T	275
	T4, T351	275
	T361	285
	T81, T851	275
	T861	290
2218	T72	205
2618	All	260
3003	0	76
Alclad 3003	H12	83
	H14	97
	H16	105
	H18	110
3004	0	110
Alclad 3004	H32	115
	H34	125
	H36	140
	H38	145

Shear Strength of Selected Wrought Aluminum Alloys
(Continued)
Listed by Alloy

Alloy AA No.	Temper	Shear Strength (MPa)
3105	O	83
	H12	97
	H14	105
	H16	110
	H18	115
	H25	105
4032	T6	260
5005	O	76
	H12	97
	H14	97
	H16	105
	H18	110
	H32	97
	H34	97
	H36	105
	H38	110
5050	O	105
	H32	115
	H34	125
	H36	130
	H38	140
5052	O	125
	H32	140
	H34	145
	H36	160

Shear Strength of Selected Wrought Aluminum Alloys
(Continued)
Listed by Alloy

Alloy AA No.	Temper	Shear Strength (MPa)
	H38	165
5056	0	180
	H18	235
	H38	220
5083	0	170
5086	0	160
	H34	185
5154	0	150
	H32	150
	H34	165
	H36	180
	H38	195
5182	0	150
5252	H25	145
	H28, H38	160
5254	0	150
5254	H32	150
	H34	165
	H36	180
	H38	195
5454	0	160
	H32	165
	H34	180
	H111	160
	H112	160
	H311	160

Shear Strength of Selected Wrought Aluminum Alloys
(Continued)
Listed by Alloy

Alloy AA No.	Temper	Shear Strength (MPa)
5456	H321, H116	205
5457	0	83
	H25	110
	H28, H38	125
5652	0	125
	H32	140
	H34	145
	H36	160
	H38	165
5657	H25	97
	H28, H38	105
6005	T5	205
6009	T4	150
6061	0	83
	T4, T451	165
	T6, T651	205
Alclad 6061	0	76
	T4, T451	150
	T6, T651	185
6063	0	69
	TI	97
	T5	115
	T6	150
	T83	150
	T831	125
	T832	185

Shear Strength of Selected Wrought Aluminum Alloys
(Continued)
Listed by Alloy

Alloy AA No.	Temper	Shear Strength (MPa)
6066	0	97
	T4, T451	200
	T6, T651	235
6070	0	97
	T4	205
	T6	235
6151	T6	140
6205	T5	205
6262	T9	240
6351	T6	200
6463	Tl	97
	T5	115
	T6	150
7005	0	117
	T53	221
	T6,T63,T6351	214
7072	0	55
	H12	62
	H14	69
7075	0	150
	T6,T651	330
Alclad 7075	0	150
	T6,T651	315
7175	T66	325
	T736	290

Shear Strength of Selected Wrought Aluminum Alloys
(Continued)
Listed by Alloy

Alloy AA No.	Temper	Shear Strength (MPa)
7475	T651	295
	T7351	270
	T7651	270

Source: Data from *ASM Metals Reference Book, Second Edition*, American Society for Metals, Metals Park, Ohio 44073, (1984).

SHEAR STRENGTH OF SELECTED
WROUGHT ALUMINUM ALLOYS
Listed by Value

Alloy AA No.	Temper	Shear Strength (MPa)
1060	0	48
1060	H12	55
1350	0	55
7072	0	55
1050	0	62
1060	H14	62
1100	0	62
1350	H12	62
7072	H12	62
1050	H14	69
1060	H16	69
1100	H12	69
1350	H14	69
6063	0	69
7072	H14	69
1050	H16	76
1060	H18	76
1100	H14	76
1350	H16	76
3003	0	76
5005	0	76
Alclad 6061	0	76
1050	H18	83
1100	H16	83
Alclad	H12	83

Shear Strength of Selected Wrought Aluminum Alloys
(Continued)
Listed by Value

Alloy AA No.	Temper	Shear Strength (MPa)
3105	0	83
5457	0	83
6061	0	83
1100	H18	90
3003	H14	97
3105	H12	97
5005	H12	97
5005	H14	97
5005	H32	97
5005	H34	97
5657	H25	97
6063	TI	97
6066	0	97
6070	0	97
6463	Tl	97
1350	H19	105
3003	H16	105
3105	H14	105
3105	H25	105
5005	H16	105
5005	H36	105
5050	0	105
5657	H28, H38	105
3003	H18	110
3004	0	110

Shear Strength of Selected Wrought Aluminum Alloys
(Continued)
Listed by Value

Alloy AA No.	Temper	Shear Strength (MPa)
3105	H16	110
5005	H18	110
5005	H38	110
5457	H25	110
Alclad	H32	115
3105	H18	115
5050	H32	115
6063	T5	115
6463	T5	115
7005	0	117
2014	0	125
Alclad 2014	0	125
2024	0	125
Alclad 2024	0	125
3004	H34	125
5050	H34	125
5052	0	125
5457	H28, H38	125
5652	0	125
6063	T831	125
5050	H36	130
3004	H36	140
5050	H38	140
5052	H32	140
5652	H32	140

Shear Strength of Selected Wrought Aluminum Alloys
(Continued)
Listed by Value

Alloy AA No.	Temper	Shear Strength (MPa)
6151	T6	140
3004	H38	145
5052	H34	145
5252	H25	145
5652	H34	145
5154	0	150
5154	H32	150
5182	0	150
5254	0	150
5254	H32	150
6009	T4	150
Alclad 6061	T4, T451	150
6063	T6	150
6063	T83	150
6463	T6	150
7075	0	150
Alclad 7075	0	150
5052	H36	160
5086	0	160
5252	H28, H38	160
5454	0	160
5454	H111	160
5454	H112	160
5454	H311	160
5652	H36	160

Shear Strength of Selected Wrought Aluminum Alloys
(Continued)
Listed by Value

Alloy AA No.	Temper	Shear Strength (MPa)
5052	H38	165
5154	H34	165
5254	H34	165
5454	H32	165
5652	H38	165
6061	T4, T451	165
5083	0	170
5056	0	180
5154	H36	180
5254	H36	180
5454	H34	180
5086	H34	185
Alclad 6061	T6, T651	185
6063	T832	185
5154	H38	195
5254	H38	195
6066	T4, T451	200
6351	T6	200
2218	T72	205
5456	H321, H116	205
6005	T5	205
6061	T6, T651	205
6070	T4	205
6205	T5	205
7005	T6,T63,T6351	214

Shear Strength of Selected Wrought Aluminum Alloys
(Continued)
Listed by Value

Alloy AA No.	Temper	Shear Strength (MPa)
2011	T3	220
5056	H38	220
7005	T53	221
5056	H18	235
6066	T6, T651	235
6070	T6	235
2011	T8	240
6262	T9	240
Alclad 2014	T3	255
Alclad 2014	T4	255
2014	T4	260
2618	All	260
4032	T6	260
7475	T7351	270
7475	T7651	270
Alclad 2024	T	275
Alclad 2024	T4, T351	275
Alclad 2024	T81, T851	275
Alclad 2014	T6	285
2024	T3	285
2024	T4, T351	285
Alclad 2024	T361	285
2014	T6	290
2024	T361	290
Alclad 2024	T861	290

Shear Strength of Selected Wrought Aluminum Alloys
(Continued)
Listed by Value

Alloy AA No.	Temper	Shear Strength (MPa)
7175	T736	290
7475	T651	295
Alclad 7075	T6,T651	315
7175	T66	325
7075	T6,T651	330

Source: Data from *ASM Metals Reference Book, Second Edition*, American Society for Metals, Metals Park, Ohio 44073, (1984).

HARDNESS OF SELECTED
WROUGHT ALUMINUM ALLOYS
Listed by Alloy

Alloy AA No.	Temper	Hardness
1060	O	19
	H12	23
	H14	26
	H16	30
	H18	35
1100	O	23
	H12	28
	H14	32
	H16	38
	H18	44
2011	T3	95
	T8	100
2014	O	45
	T4	105
	T6	135
2024	O	47
	T3	120
	T4, T351	120
	T361	130
2218	T61	115
	T71	105
	T72	95
3003	O	28
Alclad	H12	35

Hardness of Selected Wrought Aluminum Alloys (Continued)
Listed by Alloy

Alloy AA No.	Temper	Hardness
3003	H14	40
	H16	47
	H18	55
3004	0	45
Alclad	H32	52
3004	H34	63
	H36	70
	H38	77
4032	T6	120
5005	0	28
	H32	36
	H34	41
	H36	46
	H38	51
5050	0	36
	H32	46
	H34	53
	H36	58
	H38	63
5052	0	47
	H32	60
	H34	68
	H36	73
	H38	77

Hardness of Selected Wrought Aluminum Alloys (Continued)
Listed by Alloy

Alloy AA No.	Temper	Hardness
5056	0	65
	H18	105
	H38	100
5154	0	58
	H32	67
	H34	73
	H36	78
	H38	80
	H112	63
5182	0	58
5252	H25	68
	H28, H38	75
5254	0	58
5254	H32	67
	H34	73
	H36	78
	H38	80
	H112	63
5454	0	62
	H32	73
	H34	81
	H111	70
	H112	62
	H311	70
5456	H321, H116	90

Hardness of Selected Wrought Aluminum Alloys (Continued)
Listed by Alloy

Alloy AA No.	Temper	Hardness
5457	0	32
	H25	48
	H28, H38	55
5652	0	47
	H32	60
	H34	68
	H36	73
	H38	77
5657	H25	40
	H28, H38	50
6005	T5	95
6009	T4	70
6010	T4	76
6061	0	30
	T4, T451	65
	T6, T651	95
6063	0	25
	TI	42
	T5	60
	T6	73
	T83	82
	T831	70
	T832	95
6066	0	43
	T4, T451	90
	T6, T651	120

Hardness of Selected Wrought Aluminum Alloys (Continued)
Listed by Alloy

Alloy AA No.	Temper	Hardness
6070	0	35
	T4	90
	T6	120
6151	T6	71
6201	T6	90
6205	Tl	65
	T5	95
6262	T9	120
6351	T6	95
6463	Tl	42
	T5	60
	T6	74
7049	T73	135
7072	0	20
	H12	28
	H14	32
7075	0	60
	T6,T651	150
7175	T66	150
	T736	145

Source: Data from *ASM Metals Reference Book, Second Edition*, American Society for Metals, Metals Park, Ohio 44073, (1984).

HARDNESS OF SELECTED
WROUGHT ALUMINUM ALLOYS
Listed by Value

Alloy AA No.	Temper	Hardness
1060	0	19
7072	0	20
1060	H12	23
1100	0	23
6063	0	25
1060	H14	26
1100	H12	28
3003	0	28
5005	0	28
7072	H12	28
1060	H16	30
6061	0	30
1100	H14	32
5457	0	32
7072	H14	32
1060	H18	35
Alclad	H12	35
6070	0	35
5005	H32	36
5050	0	36
1100	H16	38
3003	H14	40
5657	H25	40
5005	H34	41
6063	TI	42

Hardness of Selected Wrought Aluminum Alloys (Continued)
Listed by Value

Alloy AA No.	Temper	Hardness
6463	Tl	42
6066	0	43
1100	H18	44
2014	0	45
3004	0	45
5005	H36	46
5050	H32	46
2024	0	47
3003	H16	47
5052	0	47
5652	0	47
5457	H25	48
5657	H28, H38	50
5005	H38	51
Alclad	H32	52
5050	H34	53
3003	H18	55
5457	H28, H38	55
5050	H36	58
5154	0	58
5182	0	58
5254	0	58
5052	H32	60
5652	H32	60
6063	T5	60

Hardness of Selected Wrought Aluminum Alloys (Continued)
Listed by Value

Alloy AA No.	Temper	Hardness
6463	T5	60
7075	0	60
5454	0	62
5454	H112	62
3004	H34	63
5050	H38	63
5154	H112	63
5254	H112	63
5056	0	65
6061	T4, T451	65
6205	Tl	65
5154	H32	67
5254	H32	67
5052	H34	68
5252	H25	68
5652	H34	68
3004	H36	70
5454	H111	70
5454	H311	70
6009	T4	70
6063	T831	70
6151	T6	71
5052	H36	73
5154	H34	73
5254	H34	73

Hardness of Selected Wrought Aluminum Alloys (Continued)
Listed by Value

Alloy AA No.	Temper	Hardness
5454	H32	73
5652	H36	73
6063	T6	73
6463	T6	74
5252	H28, H38	75
6010	T4	76
3004	H38	77
5052	H38	77
5652	H38	77
5154	H36	78
5254	H36	78
5154	H38	80
5254	H38	80
5454	H34	81
6063	T83	82
5456	H321, H116	90
6066	T4, T451	90
6070	T4	90
6201	T6	90
2011	T3	95
2218	T72	95
6005	T5	95
6061	T6, T651	95
6063	T832	95
6205	T5	95

Hardness of Selected Wrought Aluminum Alloys (Continued)
Listed by Value

Alloy AA No.	Temper	Hardness
6351	T6	95
2011	T8	100
5056	H38	100
2014	T4	105
2218	T71	105
5056	H18	105
2218	T61	115
2024	T3	120
2024	T4, T351	120
4032	T6	120
6066	T6, T651	120
6070	T6	120
6262	T9	120
2024	T361	130
2014	T6	135
7049	T73	135
7175	T736	145
7075	T6,T651	150
7175	T66	150

Source: Data from *ASM Metals Reference Book, Second Edition*, American Society for Metals, Metals Park, Ohio 44073, (1984).

HARDNESS OF SELECTED CERAMICS
Listed by Ceramic

Ceramic	Hardness

BORIDES

Chromium Diboride (CrB2)

micro 100g: 1800 kg/mm^2

Vickers 50g: 1800 kg/mm^2

Knoop 100g: 1700 kg/mm^2

Hafnium Diboride (HfB2)

(polycrystalline)   Knoop 160g : 2400kg/mm at 24 °C

(single crystal)   Knoop 160g : 3800kg/mm at 24 °C

Tantalum Diboride (TaB$_2$)

micro : 1700 kg/mm^2

Knoop 30g: 2537 kg/mm^2

Knoop 100g: 2615 ± 120 kg/mm^2

Rockwell A : 89

Titanium Diboride (TiB$_2$)

Vickers 50g: 3400 kg/mm^2

Knoop 30g: 3370 kg/mm^2

Knoop 100g: 2710-3000 kg/mm^2

Knoop 160g: 3500 kg/mm^2

(single crystal)   Knoop 100g: 3250±100 kg/mm^2

Zirconium Diboride (ZrB$_2$)

Rockwell A: 87-89

Vickers 50g: 2200 kg/mm^2

Knoop 100g: 1560 kg/mm^2

Knoop 160g: 2100 kg/mm^2

(single crystal)   Knoop 160g: 2000 kg/mm^2

CARBIDES

Boron Carbide (B$_4$C)

Knoop 100g: 2800 kg/mm^2

Knoop 1000g: 2230 kg/mm^2

Vickers : 2400 kg/mm^2

Hardness of Selected Ceramics (Continued)
Listed by Ceramic

Ceramic	Hardness
Hafnium Monocarbide (HfC)	Knoop : 1790-1870 kg/mm^2
	Vickers 50g : 2533-3202 kg/mm^2
Silicon Carbide (SiC)	Moh : 9.2
	Vickers 25g : 3000-3500 kg/mm^2
	Knoop 100g : 2500-2550 kg/mm^2
	Knoop 100g : 2960 kg/mm^2 (black)
	Knoop 100g : 2745 kg/mm^2 (green)
(cubic, CVD)	Knoop or Vickers : 2853-4483 kg/mm^2
Tantalum Monocarbide (TaC)	Knoop 50g: 1800-1952 kg/mm^2
	Knoop 100g: 825 kg/mm^2
	Vickers 50g: 1800 kg/mm^2
	Rockwell A: 89
	Brinell: 840
Titanium Monocarbide (TiC)	Knoop 100g: 2470 kg/mm^2
	Knoop 1000g: 1905 kg/mm^2
	Vickers 50g: 2900-3200 kg/mm^2
	Vickers 100g: 2850-3390 kg/mm^2
	micro 20g: 3200 kg/mm^2
(98.6% density)	Rockwell A: 88-89
(99.5% density)	Rockwell A: 91-93.5
(100% density)	Rockwell A: 91-93.5
Trichromium Dicarbide (Cr$_3$C$_2$)	Knoop or Vickers : 1019-1834 kg/mm^2

Hardness of Selected Ceramics (Continued)
Listed by Ceramic

Ceramic	Hardness
Tungsten Monocarbide (WC)	Knoop 100g: 1870-1880 kg/mm^2
	Vickers 50g: 2400 kg/mm^2
	Vickers 100g: 1730 kg/mm^2
	Rockwell A: 92
(6% Co, 1-3µm grain size)	Rockwell A: 81.4 ± 0.4
(12% Co, 1-3µm grain size)	Rockwell A: 89.4 ± 0.5
(24% Co, 1-3µm grain size)	Rockwell A: 86.9 ± 0.6
(6% Co, 2-4µm grain size)	Rockwell A: 88.6 ± 0.5
(6% Co, 3-6µm grain size)	Rockwell A: 87.3 ± 0.5
Zirconium Monocarbide (ZrC)	Knoop : 2138 kg/mm^2
	Vickers 50g : 2600 kg/mm^2
	Vickers 100g : 2836-3840 kg/mm^2
	micro : 2090 kg/mm^2
	Rockwell A: 92.5
NITRIDES	
Aluminum Nitride (AlN)	Mohs: 5-5.5
	Knoop 100g: 1225-1230 kg/mm^2
(thick film)	Rockwell 15N: 94.5
(thin film)	Rockwell 15N: 94.0
Boron Nitride (BN)	Mohs: 2 (hexagonal)
Titanium Mononitirde (TiN)	Mohs: 8-10
	Knoop 30g : 2160 kg/mm^2
	Knoop 100g : 1770 kg/mm^2
Trisilicon tetranitride (Si$_3$N$_4$)	Mohs: 9+
(α)	Knoop or Vickers: 815-1936kg/mm^2
	Rockwell A: 99

Hardness of Selected Ceramics (Continued)
Listed by Ceramic

Ceramic	Hardness
Zirconium Mononitride (TiN)	Mohs: 8+
	Knoop 30g : 1983 kg/mm^2
	Knoop 100g : 1510 kg/mm^2
OXIDES	
Aluminum Oxide (Al$_2$O$_3$)	
(single crystal)	Mohs : 9
	Knoop 100g : 2000-2050 kg/mm^2
	Vickers 20g : 2600 kg/mm^2
	Vickers 50g : 2720 kg/mm^2
	R45N : 78-90
Beryllium Oxide (BeO)	Knoop 100g : 1300 kg/mm^2
	R45N : 64-67
Calcium Oxide (CaO)	Knoop 100g : 560 kg/mm^2
Dichromium Trioxide (Cr$_2$O$_3$)	Knoop or Vickers : 2955 kg/mm^2
Magnesium Oxide (MgO)	Mohs : 5.5
Silicon Dioxide (SiO$_2$)	
(parallel to optical axis)	Knoop 100g : 710 kg/mm^2
(normal to optical axis)	Knoop 100g : 790 kg/mm^2
(parallel to optical axis)	Vickers 500g : 1260 kg/mm^2
(normal to optical axis)	Vickers 500g : 1103 kg/mm^2
	Vickers 500g : 1120 kg/mm^2
(1010 face) 10 μm diagonal	Vickers 500g :1120-1230 kg/mm^2
(1011 face) 10 μm diagonal	Vickers 500g : 1040-1130 kg/mm^2
(polished 1010 face)	
10 μm diagonal	Vickers 500g : 1300 kg/mm^2

Hardness of Selected Ceramics (Continued)
Listed by Ceramic

Ceramic	Hardness
Thorium Dioxide (ThO$_2$)	Mohs : 6.5 Knoop 100g : 945 kg/mm^2
Titanium Oxide (TiO$_2$)	Knoop or Vickers : 713-1121 kg/mm^2
Uranium Dioxide (UO$_2$)	Mohs : 6-7 Knoop 100g : 600 kg/mm^2
Zirconium Oxide (ZrO$_2$)	Mohs : 6.5 Knoop 100g : 1200 kg/mm^2
(partially stabilized) mm^2	Knoop or Vickers : 1019-1121 kg/
(fully stabilized) mm^2	Knoop or Vickers : 1019-1529 kg/
Cordierite (2MgO 2Al$_2$O$_3$ 5SiO$_2$) (glass)	Vickers : 835.6 kg/mm^2 Vickers : 672.5 kg/mm^2
Mullite (3Al$_2$O$_3$ 2SiO$_2$)	Mohs: 7.5 Vickers : 1120 kg/mm^2 R45N: 71
Sillimanite (Al$_2$O$_3$ SiO$_2$)	Mohs: 6-7
Zircon (SiO$_2$ ZrO$_2$)	Mohs: 7.5

Hardness of Selected Ceramics (Continued)
Listed by Ceramic

Ceramic	Hardness
SILICIDES	
Molybdenum Disilicide ($MoSi_2$)	Knoop 100g : 1257 kg/mm^2
	Vickers 100g : 1290-1550 kg/mm^2
	Micro 50g : 1200 kg/mm^2
	Micro 100g : 1290 kg/mm^2
Tungsten Disilicide (WSi_2)	Knoop 100g : 1090 kg/mm^2
	Vickers 100g : 1090 kg/mm^2
	Vickers 10g : 1632 kg/mm^2
	Micro 50g : 1260 kg/mm^2

Source: Data compiled by J.S. Park from *No. 1 Materials Index*, Peter T.B. Shaffer, Plenum Press, New York, (1964); *Smithells Metals Reference Book*, Eric A. Brandes, ed., in association with Fulmer Research Institute Ltd. 6th ed. London, Butterworths, Boston, (1983); and *Ceramic Abstracts*, American Ceramic Society (1986-1991).

HARDNESS OF SELECTED POLYMERS
Listed by Polymer

Polymer	Hardness, (ASTM D785) (Rockwell)
ABS Resins; Molded, Extruded	
Medium impact	R108—115
High impact	R95—113
Very high impact	R85—105
Low temperature impact	R75—95
Heat resistant	R107—116
Acrylics; Cast, Molded, Extruded	
Cast Resin Sheets, Rods:	
General purpose, type I	M80—90
General purpose, type II	M96—102
Moldings:	
Grades 5, 6, 8	M80—103
High impact grade	M38—45
Thermoset Carbonate	
Allyl diglycol carbonate	M95—M100 (Barcol)
Alkyds; Molded	
Putty (encapsulating)	60—70 (Barcol)
Rope (general purpose)	70—75 (Barcol)
Granular (high speed molding)	60—70 (Barcol)
Glass reinforced (heavy duty parts)	70—80 (Barcol)

Hardness of Selected Polymers (Continued)
Listed by Polymer

Polymer	Hardness, (ASTM D785) (Rockwell)
Cellulose Acetate; Molded, Extruded ASTM Grade:	
H4—1	R103—120
H2—1	R89—112
MH—1, MH—2	R74—104
MS—1, MS—2	R54—96
S2—1	R49—88
Cellulose Acetate Butyrate; Molded, Extruded ASTM Grade:	
H4	R114
MH	R80—100
S2	R23—42
Cellusose Acetate Propionate; Molded, Extruded ASTM Grade:	
1	100—109
3	92—96
6	57
Chlorinated Polymers	
Chlorinated polyether	R100
Chlorinated polyvinyl chloride	R118
Polycarbonates	
Polycarbonate	M70
Polycarbonate (40% glass fiber reinforced)	M97

Hardness of Selected Polymers (Continued)
Listed by Polymer

Polymer	Hardness, (ASTM D785) (Rockwell)
Diallyl Phthalates; Molded	
Orlon filled	M108
Asbestos filled	M107
Glass fiber filled	M108
Fluorocarbons; Molded,Extruded	
Polytrifluoro chloroethylene (PTFCE)	R110—115
Polytetrafluoroethylene (PTFE)	52D
Ceramic reinforced (PTFE)	R35—55
Fluorinated ethylene propylene(FEP)	57—58D
Polyvinylidene— fluoride (PVDF)	109—110R
Epoxies; Cast, Molded, Reinforced	
Standard epoxies (diglycidyl ethers of bisphenol A)	
Cast rigid	106M
Cast flexible	50-100M
Molded	75-80 (Barcol)
General purpose glass cloth laminate	115—117M
High strength laminate	70—72 (Barcol)
Filament wound composite	98-120M
Epoxies—Molded, Extruded	
High performance resins (cycloaliphatic diepoxides)	
Cast, rigid	107—112
Molded	94—96D
Glass cloth laminate	75—80

Hardness of Selected Polymers (Continued)
Listed by Polymer

Polymer	Hardness, (ASTM D785) (Rockwell)
Melamines; Molded	
Filler & type	
Unfilled	E110
Cellulose electrical	M115—125
Nylons; Molded, Extruded	
Type 6	
General purpose	R118—R120
Glass fiber (30%) reinforced	R93—121
Cast	R116
Flexible copolymers	R72—Rl19
Type 11	Rl00—R108
Type 12	R106
Nylons; Molded, Extruded	
6/6 Nylon	
General purpose molding	R118—120, R108
Glass fiber reinforced	E60—E80
Glass fiber Molybdenum disulfide filled	M95—100
General purpose extrusion	R118—108
6/10 Nylon	
General purpose	R111
Glass fiber (30%) reinforced	E40—50

Hardness of Selected Polymers (Continued)
Listed by Polymer

Polymer	Hardness, (ASTM D785) (Rockwell)
Phenolics; Molded	
Type and filler	
General: woodflour and flock	E85—100
Shock: paper, flock, or pulp	E85—95
High shock: chopped fabric or cord	E80—90
Very high shock: glass fiber	E50—70
Phenolics: Molded	
Arc resistant—mineral	M105—115
Rubber phenolic—woodflour or flock	M40—90
Rubber phenolic—chopped fabric	M57
Rubber phenolic—asbestos	M50
ABS–Polycarbonate Alloy	R118
PVC–Acrylic Alloy	
PVC–acrylic sheet	R105
PVC–acrylic injection molded	R104
Polymide	
Glass reinforced	114E

Hardness of Selected Polymers (Continued)
Listed by Polymer

Polymer	Hardness, (ASTM D785) (Rockwell)
Polyacetals	
Homopolymer:	
Standard	M94
20% glass reinforced	M90
22% TFE reinforced	M78
Copolymer:	
Standard	M80
25% glass reinforced	M79
High flow	M80
Polyester; Thermoplastic	
Injection Moldings:	
General purpose grade	R117
Glass reinforced grades	R118—M90
Glass reinforced self extinguishing	R119
General purpose grade	R117
Glass reinforced grade	R117—M85
Asbestos—filled grade	M85
Polyesters: Thermosets	
Cast polyyester	
Rigid	35—50 (Barcol)
Flexible	6—40 (Barcol)
Reinforced polyester moldings	
High strength (glass fibers)	60—80 (Barcol)
Heat and chemical resistant (asbestos)	40—70 (Barcol)
Sheet molding compounds, general purpose	45—60 (Barcol)

Hardness of Selected Polymers (Continued)
Listed by Polymer

Polymer	Hardness, (ASTM D785) (Rockwell)
Phenylene Oxides	
SE—100	R115
SE—1	R119
Glass fiber reinforced	L106, L108
Phenylene oxides (Noryl)	
Standard	R120
Glass fiber reinforced	M84
Polyarylsulfone	M85—110
Polypropylene:	
General purpose	R80—R100
High impact	R28—95
Asbestos filled	R90—R110
Glass reinforced	R90—R115
Flame retardant	R60—R105
Polyphenylene sulfide:	
Standard	R120—124
40% glass reinforced	R123

Hardness of Selected Polymers (Continued)
Listed by Polymer

Polymer	Hardness, (ASTM D785) (Rockwell)
Polyethylenes; Molded, Extruded	
Type I—lower density (0.910—0.925)	
Melt index 0.3—3.6	C73, D50—52 (Shore)
Melt index 6—26	C73, D47—53 (Shore)
Melt index 200	D45 (Shore)
Type II—medium density (0.926—0.940)	
Melt index 20	D55 (Shore)
Melt index 1.0—1.9	D55—D56 (Shore)
Type III—higher density (0.941—0.965)	
Melt index 0.2—0.9	D68—70 (Shore)
Melt Melt index 0.1—12.0	D60—70 (Shore)
Melt index 1.5—15	D68—70 (Shore)
High molecular weight	60—65 (Shore)
Olefin Copolymers; Molded	
EEA (ethylene ethyl acrylate)	D35 (Shore)
EVA (ethylene vinyl acetate)	D36 (Shore)
Ethylene butene	D65 (Shore)
Propylene—ethylene ionomer	D60 (Shore)

Hardness of Selected Polymers (Continued)
Listed by Polymer

Polymer	Hardness, (ASTM D785) (Rockwell)
Polystyrenes; Molded	
Polystyrenes	
General purpose	M72
Medium impact	M47—65
High impact	M3—43
Glass fiber -30% reinforced	M85—95
Styrene acrylonitrile (SAN)	M75—85
Glass fiber (30%) reinforced SAN	M90—123
Polyvinyl Chloride And Copolymers; Molded, Extruded	
Rigid—normal impact	R110—120
Vinylidene chloride	M50—65
Polyvinyl Chloride And Copolymers; Molded, Extruded:	(ASTM D676)
Nonrigid—general	A50—100 (Shore)
Nonrigid—electrical	A78—100 (Shore)
Rigid—normal impact	D70—85 (Shore)
Vinylidene chloride	>A95 (Shore)
Silicones; Molded, Laminated	
Fibrous (glass) reinforced silicones	M87
Granular (silica) reinforced silicones	M71—95
Woven glass fabric/ silicone laminate	75 (Barcol)

Hardness of Selected Polymers (Continued)
Listed by Polymer

Polymer	Hardness, (ASTM D785) (Rockwell)
Ureas; Molded	
Alpha—cellulose filled (ASTM Type l)	E94—97, M116—120
Woodflour filled	M116—120

Source: data compiled by J.S. Park from Charles T. Lynch, *CRC Handbook of Materials Science*, Vol. 3, CRC Press, Boca Raton, Florida, 1975 and *Engineered Materials Handbook*, Vol.2, Engineering Plastics, ASM International, Metals Park, Ohio, 1988.

IMPACT STRENGTH OF SELECTED POLYMERS
Listed by Polymer

Polymer	Impact Strength (Izod notched, ASTM D256) (ft—lb / in.)
ABS Resins; Molded, Extruded	
Medium impact	2.0—4.0
High impact	3.0—5.0
Very high impact	5.0—7.5
Low temperature impact	6—10
Heat resistant	2.0—4.0
Acrylics; Cast, Molded, Extruded	
Cast Resin Sheets, Rods:	
General purpose, type I	0.4
General purpose, type II	0.4
Moldings:	
Grades 5, 6, 8	0.2—0.4
High impact grade	0.8—2.3
Thermoset Carbonate	
Allyl diglycol carbonate	0.2—0.4
Alkyds; Molded	
Putty (encapsulating)	0.25—0.35
Rope (general purpose)	2.2
Granular (high speed molding)	0.30—0.35
Glass reinforced (heavy duty parts)	8—12

To convert (ft—lb / in.) to (N•m/m), multiply by 53.38

Impact Strength of Selected Polymers (Continued)
Listed by Polymer

Polymer	Impact Strength (Izod notched, ASTM D256) (ft—lb / in.)
Cellulose Acetate Butyrate; Molded, Extruded ASTM Grade:	
H4	3
MH	4.4—6.9
S2	7.5—10.0
Cellusose Acetate Propionate; Molded, Extruded ASTM Grade:	
1	1.7—2.7
3	3.5—5.6
6	9.4
Chlorinated Polymers	
Chlorinated polyether	0.4 (D758)
Chlorinated polyvinyl chloride	6.3
Polycarbonate	12—16
Diallyl Phthalates; Molded	
Orlon filled	0.5—1.2
Dacron filled	1.7—5.0
Asbestos filled	0.30—0.50
Glass fiber filled	0.5—15.0

To convert (ft—lb / in.) to (N•m/m), multiply by 53.38

Impact Strength of Selected Polymers (Continued)
Listed by Polymer

Polymer	Impact Strength (Izod notched, ASTM D256) (ft—lb / in.)
Fluorocarbons; Molded,Extruded	
Polytrifluoro chloroethylene (PTFCE)	3.50—3.62
Polytetrafluoroethylene (PTFE)	2.0—4.0
Fluorinated ethylene propylene(FEP)	No break
Polyvinylidene— fluoride (PVDF)	3.0—10.3
Epoxies; Cast, Molded, Reinforced	
Standard epoxies (diglycidyl ethers of bisphenol A)	
Cast rigid	0.2—0.5
Cast flexible	0.3—0.2
Molded	0.4—0.5
General purpose glass cloth laminate	12—15
High strength laminate	60—61
Epoxies—Molded, Extruded	
High performance resins (cycloaliphatic diepoxides)	
Cast, rigid	0.5
Molded	0.3—0.5
Epoxy novolacs	
Cast, rigid	13—17
Melamines; Molded	
Filler & type	
Cellulose electrical	0.27—0.36
Glass fiber	0.5—12.0
Alpha cellulose and mineral	0.30—0.35(a), 0.2(mineral)

Impact Strength of Selected Polymers (Continued)
Listed by Polymer

Polymer	Impact Strength (Izod notched, ASTM D256) (ft—lb / in.)
Nylons; Molded, Extruded	
Type 6	
General purpose	0.6—1.2
Glass fiber (30%) reinforced	2.2—3.4
Cast	1.2
Flexible copolymers	1.5—19
Type 8	>16
Type 11	3.3—3.6
Type 12	1.2—4.2
Nylons; Molded, Extruded	
6/6 Nylon	(ASTM D638)
General purpose molding	0.55—1.0,2.0
Glass fiber reinforced	2.5—3.4
General purpose extrusion	1.3
6/10 Nylon	
General purpose	0.6, 1.6
Glass fiber (30%) reinforced	3.4
Phenolics; Molded	
Type and filler	
General: woodflour and flock	0.24—0.50
Shock: paper, flock, or pulp	0.4—1.0
High shock: chopped fabric or cord	0.6—8.0
Very high shock: glass fiber	10—33

Impact Strength of Selected Polymers (Continued)
Listed by Polymer

Polymer	Impact Strength (Izod notched, ASTM D256) (ft—lb / in.)
Phenolics: Molded	
Arc resistant—mineral	0.30—0.45
Rubber phenolic—woodflour or flock	0.34—1.0
Rubber phenolic—chopped fabric	2.0—2.3
Rubber phenolic—asbestos	0.3—0.4
ABS–Polycarbonate Alloy	10 (ASTM D638)
PVC–Acrylic Alloy	
PVC–acrylic sheet	15
PVC–acrylic injection molded	15
Polymides	
Unreinforced	0.5
Unreinforced 2nd value	0.5
Glass reinforced	17
Polyacetals	(ASTM D638)
Homopolymer:	
Standard	1.4
20% glass reinforced	0.8
22% TFE reinforced	0.7
Copolymer:	
Standard	1.3
25% glass reinforced	1.8
High flow	1

Impact Strength of Selected Polymers (Continued)
Listed by Polymer

Polymer	Impact Strength (Izod notched, ASTM D256) (ft—lb / in.)
Polyester; Thermoplastic	
Injection Moldings:	
General purpose grade	1.0—1.2
Glass reinforced grades	1.3—2.2
Glass reinforced self extinguishing	1.8
General purpose grade	1
Glass reinforced grade	1
Asbestos—filled grade	0.5
Polyesters: Thermosets	
Cast polyyester	
Rigid	0.18—0.40
Flexible	4
Reinforced polyester moldings	
High strength (glass fibers)	1—10
Heat and chemical resistsnt (asbestos)	0.45—1.0
Sheet molding compounds, general purpose	5—15
Phenylene Oxides	(ASTM D638)
SE—100	5
SE—1	5
Glass fiber reinforced	2.3
Phenylene oxides (Noryl)	
Standard	1.2—1.3
Glass fiber reinforced	1.8—2.0

Impact Strength of Selected Polymers (Continued)
Listed by Polymer

Polymer	Impact Strength (Izod notched, ASTM D256) (ft—lb / in.)
Polyarylsulfone	1.6—5.0
Polypropylene:	
General purpose	0.4—2.2
High impact	1.5—12
Asbestos filled	0.5—1.5
Glass reinforced	0.5—2
Flame retardant	2.2
Polyphenylene sulfide:	
Standard	0.3
40% glass reinforced	1.09
Polyethylenes; Molded, Extruded	
Type III—higher density (0.941—0.965)	
Melt index 0.2—0.9	4.0—14
Melt Melt index 0.1—12.0	0.4—6.0
Melt index 1.5—15	1.2—2.5
High molecular weight	>20
Olefin Copolymers; Molded	
Ethylene butene	0.4
Propylene—ethylene	1.1
Ionomer	9—14
Polyallomer	1.5

Impact Strength of Selected Polymers (Continued)
Listed by Polymer

Polymer	Impact Strength (Izod notched, ASTM D256) (ft—lb / in.)
Polystyrenes; Molded	(ASTM D638)
Polystyrenes	
General purpose	0.2—0.4
Medium impact	0.5—1.2
High impact	0.8—1.8
Glass fiber —30% reinforced	2.5
Styrene acrylonitrile (SAN)	0.29—0.54
Glass fiber (30%) reinforced SAN	1.35—3.0
Polyvinyl Chloride And Copolymers; Molded, Extruded	
Nonrigid—general	Variable
Nonrigid—electrical	Variable
Rigid—normal impact	0.5—10
Vinylidene chloride	2—8
Silicones; Molded, Laminated	
Fibrous (glass) reinforced silicones	10
Granular (silica) reinforced silicones	0.34
Woven glass fabric/ silicone laminate	10—25
Ureas; Molded	
Alpha—cellulose filled (ASTM Type 1)	0.20—0.35
Cellulose filled (ASTM Type 2)	0.20—0.275
Woodflour filled	0.25—0.35

To convert (ft—lb / in.) to (N•m/m), multiply by 53.38

Impact Strength of Selected Polymers (Continued)
Listed by Polymer

Impact Strength vs. Temperature for ABS Resins

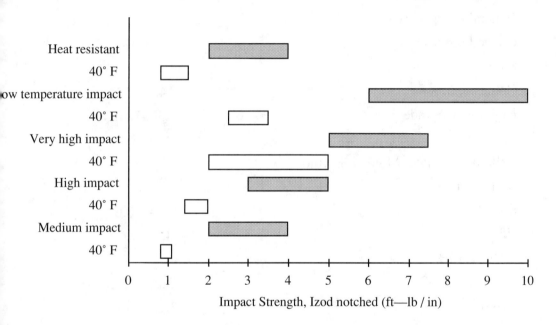

To convert (ft—lb / in.) to (N•m/m), multiply by 53.38

Source: data compiled by J.S. Park from Charles T. Lynch, *CRC Handbook of Materials Science*, Vol. 3, CRC Press, Boca Raton, Florida, 1975 and *Engineered Materials Handbook*, Vol.2, Engineering Plastics, ASM International, Metals Park, Ohio, 1988.

COMPRESSIVE YIELD STRENGTH OF SELECTED POLYMERS
Listed by Polymer

Polymer	Property, (ASTM D690 or D695) (0.1% offset, 1000 psi)
Acrylics; Cast, Molded, Extruded	
Cast Resin Sheets, Rods:	
General purpose, type I	12—14
General purpose, type II	14—18
Moldings:	
Grades 5, 6, 8	14.5—17
High impact grade	7.3—12.0
Cellulose Acetate; Molded, Extruded	
ASTM Grade:	
H4—1	6.5—10.6
H2—1	4.3—9.6
MH—1, MH—2	4.4—8.4
MS—1, MS—2	3.2—7.2
S2—1	3.15—6.1
Cellulose Acetate Butyrate; Molded, Extruded	
ASTM Grade:	
H4	8.8
MH	5.3—7.1
S2	2.6—4.3
Cellusose Acetate Propionate; Molded, Extruded	
ASTM Grade:	
1	6.2—7.3
3	4.9—5.8

Compressive Yield Strength of Selected Polymers (Continued)
Listed by Polymer

Polymer	Property, (ASTM D690 or D695) (0.1% offset, 1000 psi)
Fluorocarbons; Molded,Extruded	
Polytrifluoro chloroethylene (PTFCE)	2
Polytetrafluoroethylene (PTFE)	0.7—1.8
Ceramic reinforced (PTFE)	1.4—1.8
Fluorinated ethylene propylene(FEP)	1.6
Polyvinylidene— fluoride (PVDF)	12.8—14.2
Nylons; Molded, Extruded	
Type 6	
General purpose	9.7
Glass fiber (30%) reinforced	19—20
Cast	14
Nylons; Molded, Extruded	
6/6 Nylon	
General purpose molding	4.9
Glass fiber reinforced	20—24
General purpose extrusion	4.9
6/10 Nylon	
General purpose	3.0
Glass fiber (30%) reinforced	18
Polyacetals	
Homopolymer:	
Standard	5.2
20% glass reinforced	5.2
22% TFE reinforced	4.5

Compressive Yield Strength of Selected Polymers (Continued)
Listed by Polymer

Polymer	Property, (ASTM D690 or D695) (0.1% offset, 1000 psi)
Copolymer:	
Standard	4.5
High flow	4.5
Polypropylene:	
General purpose	5.5—6.5
High impact	4.4
Asbestos filled	7
Glass reinforced	6.5—7
Polyvinyl Chloride And Copolymers; Molded, Extruded	
Rigid—normal impact	10—11
Vinylidene chloride	75—85

To convert psi to MPa, multiply by 145.

Source: data compiled by J.S. Park from Charles T. Lynch, *CRC Handbook of Materials Science*, Vol. 3, CRC Press, Boca Raton, Florida, 1975 and *Engineered Materials Handbook*, Vol.2, Engineering Plastics, ASM International, Metals Park, Ohio, 1988.

COMPRESSIVE STRENGTH OF SELECTED POLYMERS
Listed by Polymer

Polymer	Property, (ASTM D690 or D695) (1000 psi)
ABS Resins; Molded, Extruded	(ASTM D695)
Medium impact	0.5—11.0
High impact	7.0—9.0
Heat resistant	9.3—11.0
Thermoset Carbonate	(ASTM D690)
Allyl diglycol carbonate	22.5
Alkyds; Molded	(ASTM D690)
Putty (encapsulating)	20—25
Rope (general purpose)	28
Granular (high speed molding)	16—20
Glass reinforced (heavy duty parts)	24—30
Chlorinated Polymers	(ASTM D690)
Chlorinated polyether	9
Polycarbonates	
Polycarbonate	12.5
Polycarbonate (40% glass fiber reinforced)	18.5
Diallyl Phthalates; Molded	(ASTM D695)
Orlon filled	20—25
Dacron filled	20—30
Asbestos filled	18—25
Glass fiber filled	25

Compressive Strength of Selected Polymers (Continued)
Listed by Polymer

Polymer	Property, (ASTM D690 or D695) (1000 psi)
Epoxies; Cast, Molded, Reinforced	(ASTM D695)
Standard epoxies (diglycidyl ethers of bisphenol A)	
Cast rigid	16.5—24
Molded	34-38
General purpose glass cloth laminate	50-60
High strength laminate	80-90 (edgewise)
Epoxies—Molded, Extruded	(ASTM D695)
High performance resins	
(cycloaliphatic diepoxides)	
Cast, rigid	17—19
Molded	22—26
Glass cloth laminate	67—71
Epoxy novolacs	
Cast, rigid	30—50
Glass cloth laminate	48—57
Melamines; Molded	(ASTM D695)
Filler & type	
Unfilled	40—45
Cellulose electrical	25—35
Glass fiber	20—42
Alpha cellulose and mineral	40—45(a), 26—30(mineral)

Compressive Strength of Selected Polymers (Continued)
Listed by Polymer

Polymer	Property, (ASTM D690 or D695) (1000 psi)
Phenolics; Molded	(ASTM D695)
Type and filler	
General: woodflour and flock	22—36
Shock: paper, flock, or pulp	24—35
High shock: chopped fabric or cord	15—30
Very high shock: glass fiber	17—30
Phenolics: Molded	(ASTM D695)
Arc resistant—mineral	20—30
Rubber phenolic—woodflour or flock	12—20
Rubber phenolic—chopped fabric	10—15
Rubber phenolic—asbestos	10—20
ABS	(ASTM D695)
ABS–Polycarbonate Alloy	11.1—11.8
PVC–Acrylic Alloy	
PVC–acrylic sheet	8.4
PVC–acrylic injection molded	6.2
Polymides	
Unreinforced	27.4
Unreinforced 2nd value	18.4
Glass reinforced	42

Compressive Strength of Selected Polymers (Continued)
Listed by Polymer

Polymer	Property, (ASTM D690 or D695) (1000 psi)
Polyester; Thermoplastic	(ASTM D695)
Injection Moldings:	
General purpose grade	13
Glass reinforced grades	16—18
Glass reinforced self extinguishing	18
Polyesters: Thermosets	(ASTM D690)
Cast polyyester	
Rigid	12—37
Flexible	1—17
Reinforced polyester moldings	
High strength (glass fibers)	20—26
Heat and chemical resistsnt (asbestos)	20—25
Sheet molding compounds, general purpose	22—36
Phenylene Oxides	(ASTM D695)
SE—100	12
SE—1	16.4
Glass fiber reinforced	17.6—17.9
Phenylene oxides (Noryl)	
Standard	13.9—14
Polyarylsulfone	17.8

Compressive Strength of Selected Polymers (Continued)
Listed by Polymer

Polymer	Property, (ASTM D690 or D695) (1000 psi)
Polystyrenes; Molded	(ASTM D695)
Polystyrenes	
General purpose	11.5—16.0
Medium impact	4—9
High impact	4—9
Glass fiber -30% reinforced	19
Styrene acrylonitrile (SAN)	
Glass fiber (30%) reinforced SAN	2.3
Polyvinyl Chloride And Copolymers;	(ASTM D695)
Molded, Extruded:	
Rigid—normal impact	11—12
Silicones; Molded, Laminated	(ASTM D690)
Fibrous (glass) reinforced silicones	10—12.5
Granular (silica) reinforced silicones	10.6—17
Woven glass fabric/ silicone laminate	15—24
Ureas; Molded	(ASTM D695)
Alpha—cellulose filled (ASTM Type l)	25—38
Woodflour filled	25—35

To convert psi to MPa, multiply by 145.

Source: data compiled by J.S. Park from Charles T. Lynch, *CRC Handbook of Materials Science*, Vol. 3, CRC Press, Boca Raton, Florida, 1975 and *Engineered Materials Handbook*, Vol.2, Engineering Plastics, ASM International, Metals Park, Ohio, 1988.

MODULUS OF ELASTICITY IN FLEXURE OF SELECTED POLYMERS
Listed by Polymer

Polymer	Modulus of Elasticity in Flexure (ASTM D790) (10^5 psi)
ABS Resins; Molded, Extruded	
Medium impact	3.5—4.0
High impact	2.5—3.2
Very high impact	2.0—3.2
Low temperature impact	2.0—3.2
Heat resistant	3.5—4.2
Acrylics; Cast, Molded, Extruded	
Cast Resin Sheets, Rods:	
General purpose, type I	3.5—4.5
General purpose, type II	4.0—5.0
Moldings:	
Grades 5, 6, 8	3.5—5.0
High impact grade	2.7—3.6
Thermoset Carbonate	
Allyl diglycol carbonate	2.5—3.3
Alkyds; Molded	
Rope (general purpose)	22—27
Granular (high speed molding)	22—27
Glass reinforced (heavy duty parts)	22—28

Modulus of Elasticity in Flexure of Selected Polymers (Continued)
Listed by Polymer

Polymer	Modulus of Elasticity in Flexure (ASTM D790) (10^5 psi)
Cellulose Acetate; Molded, Extruded	
ASTM Grade:	(ASTM D747)
H4—1	2.0—2.55
H2—1	1.50—2.35
MH—1, MH—2	1.50—2.15
MS—1, MS—2	1.25—1.90
S2—1	1.05—1.65
Cellulose Acetate Butyrate; Molded, Extruded	(ASTM D747)
ASTM Grade:	
H4	1.8
MH	1.20—1.40
S2	0.70—0.90
Cellusose Acetate Propionate; Molded, Extruded	
ASTM Grade:	
1	1.7—1.8
3	1.45—1.55
6	1.1
Chlorinated Polymers	
Chlorinated polyether	1.3 (0.1% offset)
Chlorinated polyvinyl chloride	3.85
Polycarbonates	
Polycarbonate	3.4
Polycarbonate (40% glass fiber reinforced)	12

Modulus of Elasticity in Flexure of Selected Polymers (Continued)
Listed by Polymer

Polymer	Modulus of Elasticity in Flexure (ASTM D790) (10^5 psi)
Fluorocarbons; Molded,Extruded	
Polytrifluoro chloroethylene (PTFCE)	2.0—2.5
Polytetrafluoroethylene (PTFE)	0.6—1.1
Ceramic reinforced (PTFE)	4.64
Fluorinated ethylene propylene(FEP)	0.8
Polyvinylidene— fluoride (PVDF)	1.75—2.0
Epoxies; Cast, Molded, Reinforced	
Standard epoxies (diglycidyl ethers of bisphenol A)	
Cast rigid	4.5—5.4
Cast flexible	0.36—3.9
Molded	15—25
General purpose glass cloth laminate	36—39
High strength laminate	53—55
Filament wound composite	69—75
Epoxies—Molded, Extruded	
High performance resins	
(cycloaliphatic diepoxides)	
Cast, rigid	4—5
Glass cloth laminate	28—31
Epoxy novolacs	
Cast, rigid	4.4—4.8
Glass cloth laminate	32—35

Modulus of Elasticity in Flexure of Selected
Polymers (Continued)
Listed by Polymer

Polymer	Modulus of Elasticity in Flexure (ASTM D790) (10^5 psi)
Melamines; Molded	
Filler & type	
Unfilled	10—13
Cellulose electrical	10—13
Glass fiber	24
Nylons; Molded, Extruded	
Type 6	
General purpose	1.4—3.9
Glass fiber (30%) reinforced	1.0—1.4
Cast	5.05
Flexible copolymers	0.92—3.2
Type 8	0.4
Type 11	1.51
Nylons; Molded, Extruded	
6/6 Nylon	
General purpose molding	4.1—4.5, 1.75
Glass fiber reinforced	10—18
Glass fiber Molybdenum disulfide filled	11—13
General purpose extrusion	1.75—4.1
6/10 Nylon	
General purpose	2.8, 1.6
Glass fiber (30%) reinforced	8.5

Modulus of Elasticity in Flexure of Selected Polymers (Continued)
Listed by Polymer

Polymer	Modulus of Elasticity in Flexure (ASTM D790) (10^5 psi)
Phenolics; Molded	
Type and filler	
General: woodflour and flock	8—12
Shock: paper, flock, or pulp	8—12
High shock: chopped fabric or cord	9—13
Very high shock: glass fiber	30—33
Phenolics: Molded	
Arc resistant—mineral	10—30
Rubber phenolic—woodflour or flock	4—6
Rubber phenolic—chopped fabric	3.5
Rubber phenolic—asbestos	5
ABS–Polycarbonate Alloy	4
PVC–Acrylic Alloy	
PVC–acrylic sheet	4
PVC–acrylic injection molded	3
Polymides	
Unreinforced	7
Unreinforced 2nd value	5
Glass reinforced	38.4

Modulus of Elasticity in Flexure of Selected Polymers (Continued)

Listed by Polymer

Polymer	Modulus of Elasticity in Flexure (ASTM D790) (10^5 psi)
Polyacetals	
Homopolymer:	
Standard	4.1
20% glass reinforced	8.8
22% TFE reinforced	4
Copolymer:	
Standard	3.75
25% glass reinforced	11
High flow	3.75
Polyester; Thermoplastic	
Injection Moldings:	
General purpose grade	3.4
Glass reinforced grades	12—15
Glass reinforced self extinguishing	12
General purpose grade	33
Glass reinforced grade	87
Asbestos—filled grade	90
Polyesters: Thermosets	
Cast polyyester	
Rigid	1—9
Flexible	0.001—0.39
Reinforced polyester moldings	
High strength (glass fibers)	15—25
Sheet molding compounds, general purpose	15—18

Modulus of Elasticity in Flexure of Selected
Polymers (Continued)
Listed by Polymer

Polymer	Modulus of Elasticity in Flexure (ASTM D790) (10^5 psi)
Phenylene Oxides	
SE—100	3.6
SE—1	3.6
Glass fiber reinforced	7.4—10.4
Phenylene oxides (Noryl)	
Standard	3.9
Glass fiber reinforced	12, 15.5
Polyarylsulfone	4
Polypropylene:	
General purpose	1.7—2.5
High impact	1.0—2.0
Asbestos filled	3.4—6.5
Glass reinforced	4—8.2
Flame retardant	1.9—6.1
Polyphenylene sulfide:	
Standard	5.5—6.0
40% glass reinforced	17—22

Modulus of Elasticity in Flexure of Selected Polymers (Continued)
Listed by Polymer

Polymer	Modulus of Elasticity in Flexure (ASTM D790) (10^5 psi)
Polyethylenes; Molded, Extruded	(ASTM D747)
Type I—lower density (0.910—0.925)	
Melt index 0.3—3.6	0.13—0.27
Melt index 6—26	0.12—0.3
Melt index 200	0.1
Type II—medium density (0.926—0.940)	
Melt index 20	0.35—0.5
Melt index 1.0—1.9	0.35—0.5
Type III—higher density (0.941—0.965)	
Melt index 0.2—0.9	1.3—1.5
Melt Melt index 0.1—12.0	0.9—0.25
Melt index 1.5—15	1.5
High molecular weight	0.75
Olefin Copolymers; Molded	
Ethylene butene	165 (psi)
Propylene—ethylene	140 (psi)
Polyallomer	0.7—1.3
Polystyrenes; Molded	
Polystyrenes:	
General purpose	4—5
Medium impact	3.5—5.0
High impact	2.3—4.0
Glass fiber -30% reinforced	12
Styrene acrylonitrile (SAN):	
Glass fiber (30%) reinforced SAN	14.5

Modulus of Elasticity in Flexure of Selected
Polymers (Continued)
Listed by Polymer

Polymer	Modulus of Elasticity in Flexure (ASTM D790) (10^5 psi)
Polyvinyl Chloride And Copolymers; Molded, Extruded	
Rigid—normal impact	3.8—5.4
Silicones; Molded, Laminated	
Fibrous (glass) reinforced silicones	25
Granular (silica) reinforced silicones	14—17
Woven glass fabric/ silicone laminate	26—32

To convert psi to MPa, multiply by 145.

Source: data compiled by J.S. Park from Charles T. Lynch, *CRC Handbook of Materials Science*, Vol. 3, CRC Press, Boca Raton, Florida, 1975 and *Engineered Materials Handbook*, Vol.2, Engineering Plastics, ASM International, Metals Park, Ohio, 1988.

FLEXURAL STRENGTH OF SELECTED POLYMERS
Listed by Polymer

Polymer	Flexural Strength, (ASTM D790) (10^3 psi)
ABS Resins; Molded, Extruded	
Medium impact	9.9—11.8
High impact	7.5—9.5
Very high impact	6.0—9.8
Low temperature impact	5—8
Heat resistant	11.0—12.0
Acrylics; Cast, Molded, Extruded	
Cast Resin Sheets, Rods:	
General purpose, type I	12—14
General purpose, type II	15—17
Moldings:	
Grades 5, 6, 8	15—16
High impact grade	8.7—12.0
Alkyds; Molded	
Putty (encapsulating)	8—11
Rope (general purpose)	19—20
Granular (high speed molding)	7—10
Glass reinforced (heavy duty parts)	12—17

Flexural Strength of Selected Polymers (Continued)
Listed by Polymer

Polymer	Flexural Strength, (ASTM D790) (10^3 psi)
Cellulose Acetate; Molded, Extruded	
ASTM Grade:	
H4—1	8.1—11.15 (yield)
H2—1	6.0—10.0 (yield)
MH—1, MH—2	4.4—8.65 (yield)
MS—1, MS—2	3.8—7.1 (yield)
S2—1	3.5—5.7 (yield)
Cellulose Acetate Butyrate; Molded, Extruded	
ASTM Grade:	
H4	9 (yield)
MH	5.6—6.7 (yield)
S2	2.5—3.95 (yield)
Cellusose Acetate Propionate; Molded, Extruded	
ASTM Grade:	
1	6.8—7.9 (yield)
3	5.6—6.2 (yield)
Chlorinated Polymers	
Chlorinated polyether	5 (0.1% offset)
Chlorinated polyvinyl chloride	14.5
Polycarbonates	
Polycarbonate	13.5
Polycarbonate (40% glass fiber reinforced)	27

Flexural Strength of Selected Polymers (Continued)
Listed by Polymer

Polymer	Flexural Strength, (ASTM D790) (10^3 psi)
Diallyl Phthalates; Molded	
Orlon filled	7.5—10.5
Dacron filled	9—11.5
Asbestos filled	8—10
Glass fiber filled	10—18
Fluorocarbons; Molded,Extruded	
Polytrifluoro chloroethylene (PTFCE)	3.5 (0.1% offset)
Fluorinated ethylene propylene(FEP)	3 (0.1% offset)
Polyvinylidene— fluoride (PVDF)	8.6—10.8 (0.1% offset)
Epoxies; Cast, Molded, Reinforced	
Standard epoxies (diglycidyl ethers of bisphenol A)	
Cast rigid	14—18
Cast flexible	1.2—12.7
Molded	19—22
General purpose glass cloth laminate	80—90
High strength laminate	165—177
Filament wound composite	180—170
Epoxies—Molded, Extruded	
High performance resins (cycloaliphatic diepoxides)	
Cast, rigid	11—16
Molded	10—12
Glass cloth laminate	70—72

Flexural Strength of Selected Polymers (Continued)
Listed by Polymer

Polymer	Flexural Strength, (ASTM D790) $(10^3$ psi)
Epoxy novolacs	
Cast, rigid	12—13
Glass cloth laminate	84—89
Melamines; Molded	
Filler & type	
Unfilled	9.5—14
Cellulose electrical	6—15
Glass fiber	14—18
Alpha cellulose and mineral	11—16(a), 8—10(mineral)
Nylons; Molded, Extruded	
Type 6	
General purpose	Unbreakable
Glass fiber (30%) reinforced	26—34
Cast	16.5
Flexible copolymers	3.4—16.4
Nylons; Molded, Extruded	
6/6 Nylon	
General purpose molding	Unbreakable
Glass fiber reinforced	26—35
Glass fiber Molybdenum disulfide filled	26—28
6/10 Nylon	
General purpose	8
Glass fiber (30%) reinforced	23

Flexural Strength of Selected Polymers (Continued)
Listed by Polymer

Polymer	Flexural Strength, (ASTM D790) (10^3 psi)
Phenolics; Molded	
Type and filler	
General: woodflour and flock	8.5—12
Shock: paper, flock, or pulp	8.0—11.5
High shock: chopped fabric or cord	8—15
Very high shock: glass fiber	10—45
Phenolics: Molded	
Arc resistant—mineral	10—13
Rubber phenolic—woodflour or flock	7—12
Rubber phenolic—chopped fabric	7
Rubber phenolic—asbestos	7
ABS–Polycarbonate Alloy	14.3
PVC–Acrylic Alloy	
PVC–acrylic sheet	10.7
PVC–acrylic injection molded	8.7
Polymides	
Unreinforced	6.6—11
Glass reinforced	56

Flexural Strength of Selected Polymers (Continued)
Listed by Polymer

Polymer	Flexural Strength, (ASTM D790) (10^3 psi)
Polyacetals	
Homopolymer:	
Standard	14.1
Copolymer:	
Standard	13
25% glass reinforced	28
High flow	13
Polyester; Thermoplastic	
Injection Moldings:	
General purpose grade	12.8
Glass reinforced grades	22—24
Glass reinforced self extinguishing	23
General purpose grade	12
Glass reinforced grade	19
Asbestos—filled grade	19
Polyesters: Thermosets	
Cast polyyester	
Rigid	8—24
Flexible	4—16
Reinforced polyester moldings	
High strength (glass fibers)	6—26
Heat and chemical resistant (asbestos)	10—13
Sheet molding compounds, general purpose	26—32

Flexural Strength of Selected Polymers (Continued)
Listed by Polymer

Polymer	Flexural Strength, (ASTM D790) (10^3 psi)
Phenylene Oxides	
SE—100	12.8
SE—1	13.5
Glass fiber reinforced	20.5—22
Phenylene oxides (Noryl)	
Standard	15.4
Glass fiber reinforced	25—28
Polyarylsulfone	16.1—17.2
Polypropylene:	
General purpose	6—7 (yield)
High impact	4.1 (yield)
Asbestos filled	7.5—9 (yield)
Glass reinforced	8—11 (yield)
Polyphenylene sulfide:	
Standard	20
40% glass reinforced	37
Polystyrenes; Molded	
Polystyrenes	
General purpose	10—15
Glass fiber —30% reinforced	17
Styrene acrylonitrile (SAN)	
Glass fiber (30%) reinforced SAN	22

Flexural Strength of Selected Polymers (Continued)
Listed by Polymer

Polymer	Flexural Strength, (ASTM D790) (10^3 psi)
Polyvinyl Chloride And Copolymers;	
Molded, Extruded:	
Rigid—normal impact	11—16
Vinylidene chloride	15—17
Silicones; Molded, Laminated	
Fibrous (glass) reinforced silicones	16—19
Granular (silica) reinforced silicones	6—10
Woven glass fabric/ silicone laminate	33—47
Ureas; Molded	
Alpha—cellulose filled (ASTM Type l)	8—18
Cellulose filled (ASTM Type 2)	7.5—13
Woodflour filled	7.5—12.0

To convert psi to MPa, multiply by 145.

Source: data compiled by J.S. Park from Charles T. Lynch, *CRC Handbook of Materials Science*, Vol. 3, CRC Press, Boca Raton, Florida, 1975 and *Engineered Materials Handbook*, Vol.2, Engineering Plastics, ASM International, Metals Park, Ohio, 1988.

FATIGUE STRENGTH OF SELECTED WROUGHT ALUMINUM ALLOYS
Listed by Alloy

Alloy AA No.	Temper	Fatigue Strength (MPa)
1060	0	21
	H12	28
	H14	34
	H16	45
	H18	45
1100	0	34
	H12	41
	H14	48
	H16	62
	H18	62
1350	H19	48
2011	T3	125
	T8	125
2014	0	90
	T4	140
	T6	125
2024	0	90
	T3	140
	T4, T351	140
	T361	125
2036	T4	125
2048		220
2219	T62	105
	T81, T851	105
	T87	105

Fatigue Strength of Selected
Wrought Aluminum Alloys (Continued)
Listed by Alloy

Alloy AA No.	Temper	Fatigue Strength (MPa)
2618	All	125
3003	0	48
Alclad	H12	55
3003	H14	62
	H16	69
	H18	69
3004	0	97
Alclad	H32	105
3004	H34	105
	H36	110
	H38	110
4032	T6	110
5050	0	83
	H32	90
	H34	90
	H36	97
	H38	97
5052	0	110
	H32	115
	H34	125
	H36	130
	H38	140
5056	0	140
	H18	150
	H38	150

Fatigue Strength of Selected
Wrought Aluminum Alloys (Continued)
Listed by Alloy

Alloy AA No.	Temper	Fatigue Strength (MPa)
5083	H321	160
5154	0	115
	H32	125
	H34	130
	H36	140
	H38	145
	H112	115
5182	0	140
5254	0	115
5254	H32	125
	H34	130
	H36	140
	H38	145
	H112	115
5652	0	110
	H32	115
	H34	125
	H36	130
	H38	140
6005	T1	97
	T5	97
6009	T4	115
6010	T4	115
6061	0	62
	T4, T451	97
	T6, T651	97

Fatigue Strength of Selected
Wrought Aluminum Alloys (Continued)
Listed by Alloy

Alloy AA No.	Temper	Fatigue Strength (MPa)
6063	0	55
	T1	62
	T5	69
	T6	69
6066	T6, T651	110
6070	0	62
	T4	90
	T6	97
6205	T5	105
6262	T9	90
6351	T6	90
6463	Tl	69
	T5	69
	T6	69
7005	T53	140
	T6,T63,T6351	125
7049	T73	295
7050	T736	240
7075	T6,T651	160
7175	T66	160
	T736	160
7475	T7351	220

Source: Data from *ASM Metals Reference Book, Second Edition*, American Society for Metals, Metals Park, Ohio 44073, (1984).

FATIGUE STRENGTH OF SELECTED WROUGHT ALUMINUM ALLOYS
Listed by Value

Alloy AA No.	Temper	Fatigue Strength (MPa)
1060	0	21
1060	H12	28
1060	H14	34
1100	0	34
1100	H12	41
1060	H16	45
1060	H18	45
1100	H14	48
1350	H19	48
3003	0	48
Alclad	H12	55
6063	0	55
1100	H16	62
1100	H18	62
3003	H14	62
6061	0	62
6063	T1	62
6070	0	62
3003	H16	69
3003	H18	69
6063	T5	69
6063	T6	69
6463	Tl	69
6463	T5	69
6463	T6	69

Fatigue Strength of Selected
Wrought Aluminum Alloys (Continued)
Listed by Value

Alloy AA No.	Temper	Fatigue Strength (MPa)
5050	0	83
2014	0	90
2024	0	90
5050	H32	90
5050	H34	90
6070	T4	90
6262	T9	90
6351	T6	90
3004	0	97
5050	H36	97
5050	H38	97
6005	T1	97
6005	T5	97
6061	T4, T451	97
6061	T6, T651	97
6070	T6	97
2219	T62	105
2219	T81, T851	105
2219	T87	105
Alclad	H32	105
3004	H34	105
6205	T5	105
3004	H36	110
3004	H38	110
4032	T6	110

Fatigue Strength of Selected
Wrought Aluminum Alloys (Continued)
Listed by Value

Alloy AA No.	Temper	Fatigue Strength (MPa)
5052	0	110
5652	0	110
6066	T6, T651	110
5052	H32	115
5154	0	115
5154	H112	115
5254	0	115
5254	H112	115
5652	H32	115
6009	T4	115
6010	T4	115
2011	T3	125
2011	T8	125
2014	T6	125
2024	T361	125
2036	T4	125
2618	All	125
5052	H34	125
5154	H32	125
5254	H32	125
5652	H34	125
7005	T6,T63,T6351	125
5052	H36	130
5154	H34	130
5254	H34	130

Fatigue Strength of Selected
Wrought Aluminum Alloys (Continued)
Listed by Value

Alloy AA No.	Temper	Fatigue Strength (MPa)
5652	H36	130
2014	T4	140
2024	T3	140
2024	T4, T351	140
5052	H38	140
5056	0	140
5154	H36	140
5182	0	140
5254	H36	140
5652	H38	140
7005	T53	140
5154	H38	145
5254	H38	145
5056	H18	150
5056	H38	150
5083	H321	160
7075	T6,T651	160
7175	T66	160
7175	T736	160
2048		220
7475	T7351	220
7050	T736	240
7049	T73	295

Source: Data from *ASM Metals Reference Book, Second Edition*, American Society for Metals, Metals Park, Ohio 44073, (1984).

COEFFICIENT OF STATIC FRICTION FOR SELECTED POLYMERS
Listed by Polymer

Polymer	Coefficient of Static Friction (Against Self) (Dimensionless)
Nylons; Molded, Extruded	
6/6 Nylon	
General purpose molding	0.04—0.13
ABS–Polycarbonate Alloy	0.2
Polycarbonates	
Polycarbonate	0.52
Nylons; Molded, Extruded	
Type 6	
Cast	0.32 (dynamic)
Polyacetals	
Homopolymer:	
Standard	0.1—0.3 (against steel)
20% glass reinforced	0.1—0.3 (against steel)
22% TFE reinforced	0.05—0.15 (against steel)
Copolymer:	
Standard	0.15 (against steel)
25% glass reinforced	0.15 (against steel)
High flow	0.15 (against steel)

Coefficient of Static Friction for Selected Polymers (Continued)
Listed by Polymer

Polymer	Coefficient of Static Friction (Against Self) (Dimensionless)
Polyester; Thermoplastic	(ASTM D1894)
Injection Moldings:	
General purpose grade	0.17
Glass reinforced grades	0.16
Glass reinforced self extinguishing	0.16
Polyester; Thermoplastic	
Injection Moldings:	
General purpose grade	0.13 (against steel)
Glass reinforced grades	0.14 (against steel)
Glass reinforced self extinguishing	0.14 (against steel)
Phenylene oxides (Noryl)	
Standard	0.67
Polyarylsulfone	0.1—0.3

Source: data compiled by J.S. Park from Charles T. Lynch, *CRC Handbook of Materials Science*, Vol. 3, CRC Press, Boca Raton, Florida, 1975 and *Engineered Materials Handbook*, Vol.2, Engineering Plastics, ASM International, Metals Park, Ohio, 1988.

ABRASION RESISTANCE OF SELECTED POLYMERS
Listed by Polymer

Polymer	Abrasion Resistance (Taber, CS—17 wheel, ASTM D1044) (mg / 1000 cycles)
Fluorocarbons; Molded,Extruded	
Polytrifluoro chloroethylene (PTFCE)	0.008 (g/cycle)
Polyvinylidene— fluoride (PVDF)	0.0006—0.0012 (g/cycle)
Polycarbonates	
Polycarbonate	10
Polycarbonate (40% glass fiber reinforced)	40
Nylons; Molded, Extruded	
Type 6	
General purpose	5
Cast	2.7
Nylons; Molded, Extruded	
6/6 Nylon	
General purpose molding	3—8
General purpose extrusion	3—5
PVC–Acrylic Alloy	
PVC–acrylic sheet	0.073 (CS—10 wheel)
PVC–acrylic injection molded	0.0058 (CS—10 wheel)
Polymides	
Unreinforced	0.08
Unreinforced 2nd value	0.004
Glass reinforced	20

Abrasion Resistance of Selected Polymers (Continued)
Listed by Polymer

Polymer	Abrasion Resistance (Taber, CS—17 wheel, ASTM D1044) (mg / 1000 cycles)
Polyacetals	
Homopolymer:	
Standard	14—20
20% glass reinforced	33
22% TFE reinforced	9
Copolymer:	
Standard	14
25% glass reinforced	40
High flow	14
Polyester; Thermoplastic	
Injection Moldings:	
General purpose grade	6.5
Glass reinforced grades	9—50
Glass reinforced self extinguishing	11
Phenylene Oxides	
SE—100	100
SE—1	20
Glass fiber reinforced	35
Phenylene oxides (Noryl)	
Standard	20
Polyarylsulfone	40

Abrasion Resistance of Selected Polymers (Continued)
Listed by Polymer

Polymer	Abrasion Resistance (Taber, CS—17 wheel, ASTM D1044) (mg / 1000 cycles)
Polystyrenes; Molded	
Glass fiber -30% reinforced	164

Source: data compiled by J.S. Park from Charles T. Lynch, *CRC Handbook of Materials Science*, Vol. 3, CRC Press, Boca Raton, Florida, 1975 and *Engineered Materials Handbook*, Vol.2, Engineering Plastics, ASM International, Metals Park, Ohio, 1988.

Electrical, Magnetic, and Optical Properties

ELECTRICAL RESISTIVITY OF SELECTED
ALLOY CAST IRONS

Description	Electrical Resistivity $(\mu\Omega \cdot m)$
Abrasion–Resistant White Irons	
Low–C white iron	0.53
Martensitic nickel–chromium iron	0.80
Corrosion–Resistant Irons	
High– Silicon iron	0.50
High–chromium iron	
High–nickel gray iron	1.0[a]
High–nickel ductile iron	1.0[a]
Heat–Resistant Gray Irons	
Medium–silicon iron	
High–chromium iron	
High–nickel iron	1.4 to 1.7
Nickel–chromium–silicon iron	1.5 to 1.7
High–aluminum iron	2.4
Heat–Resistant Ductile Irons	
Medium–silicon ductile iron	0.58 to 0.87
High–nickel ductile (20 Ni)	1.02
High–nickel ductile (23 Ni)	1.0[a]

[a] Estimated.

Source: Data from *ASM Metals Reference Book, Second Edition*, American Society for Metals, Metals Park, Ohio 44073, (1984).

RESISTIVITY OF SELECTED CERAMICS
Listed by Ceramic

Ceramic	Resistivity (Ω-cm)

BORIDES

Chromium Diboride (CrB_2)	21×10^{-6}
Hafnium Diboride (HfB_2)	$10\text{-}12 \times 10^{-6}$ at room temp.
Tantalum Diboride (TaB_2)	68×10^{-6}

Titanium Diboride (TiB_2) (polycrystalline)

(85% dense)	$26.5\text{-}28.4 \times 10^{-6}$ at room temp.
(85% dense)	9.0×10^{-6} at room temp.
(100% dense, extrapolated values)	$8.7\text{-}14.1 \times 10^{-6}$ at room temp.
	3.7×10^{-6} at liquid air temp.

Titanium Diboride (TiB_2) (monocrystalline)

(crystal length 5 cm, 39 deg. and 59 deg. orientation with respect to growth axis) $6.6 \pm 0.2 \times 10^{-6}$ at room temp.

(crystal length 1.5 cm, 16.5 deg. and 90 deg. orientation with respect to growth axis) $6.7 \pm 0.2 \times 10^{-6}$ at room temp.

Zirconium Diboride (ZrB_2)	9.2×10^{-6} at 20 $^{\circ}$C
	1.8×10^{-6} at liquid air temperature

CARBIDES

Boron Carbide (B_4C)	0.3-0.8

Resistivity of Selected Ceramics (Continued)
Listed by Ceramic

Ceramic	Resistivity (Ω-cm)
Hafnium Monocarbide (HfC)	41×10^{-6} at 4.2K
	41×10^{-6} at 80K
	45×10^{-6} at 160K
	49×10^{-6} at 240K
	60×10^{-6} at 300K
	$(30 + 0.0628T) \times10^{-6}$ from 300-2000K
Silicon Carbide (SiC)	10^2 -10^{12} at 20°C
(with 1 wt% Be additive)	3×10^{13}
(with 1 wt% B additive)	2×10^4
(with 1 wt% Al additive)	0.8
(with 1.6 wt% BeO additive)	$>10^{13}$
(with 3.2 wt% BeO additive)	4×10^{13}
(with 2.0 wt% BN additive)	1×10^{11}
Tantalum Monocarbide (TaC)	
(80% dense)	8×10^{-6} at 4.2K
(80% dense)	10×10^{-6} at 80K
(80% dense)	15×10^{-6} at 160K
(80% dense)	20×10^{-6} at 240K
(80% dense)	25×10^{-6} at 300K
Titanium Monocarbide (TiC)	0.3-0.8

Resistivity of Selected Ceramics (Continued)
Listed by Ceramic

Ceramic	Resistivity (Ω-cm)
Zirconium Monocarbide (ZrC)	41×10^{-6} at 4.2K
	45×10^{-6} at 80K
	47×10^{-6} at 160K
	53×10^{-6} at 240K
	$61\text{-}64 \times 10^{-6}$ at 300K
	97×10^{-6} at 773K
	137×10^{-6} at 1273K

NITRIDES

Aluminum Nitride (AlN)	$2 \times 10^{11}\text{-}10^{13}$ at room temp.
Boron Nitride (BN)	1.7×10^{13} at 25°C
	2.3×10^{10} at 480°C
	3.1×10^{4} at 1000°C
(20% humidity)	1.0×10^{12} at 25°C
(50% humidity)	7.0×10^{10} at 25°C
(90% humidity)	5.0×10^{9} at 25°C
Titanium Mononitirde (TiN)	$11.07\text{-}130 \times 10^{-6}$ at room temp.
	340×10^{-6} at melting temp.
	8.13×10^{-6} at liquid air
Trisilicon tetranitride (Si_3N_4)	$>10^{13}$
Zirconium Mononitirde (TiN)	$11.52\text{-}160 \times 10^{-6}$ at room temp.
	320×10^{-6} at melting temp.
	3.97×10^{-6} at liquid air

Resistivity of Selected Ceramics (Continued)
Listed by Ceramic

Ceramic	Resistivity (Ω-cm)
Aluminum Oxide (Al_2O_3)	$>10 \times 10^{14}$ at 25°C
	2×10^{13} at 100°C
	1×10^{13} at 300°C
	6.3×10^{10} at 500°C
	5.0×10^{8} at 700°C
	2×10^{6} at 1000°C
Beryllium Oxide (BeO)	$>10^{17}$ at 25°C
	$>10^{15}$ at 300°C
	$1 - 5 \times 10^{15}$ at 500°C
	$1.5 - 2 \times 10^{15}$ at 700°C
	$4 - 7 \times 10^{15}$ at 1000°C
Magnesium Oxide (MgO)	1.3×10^{15} at 27°C
	$0.2 - 1 \times 10^{8}$ at 1000°C
	4×10^{2} at 1727°C
Silicon Dioxide (SiO_2)	10^{18} at room temp.
Zircoium Oxide (ZrO_2)	
(stabilized)	2300 at 700°C
(stabilized)	77 at 1200°C
(stabilized)	9.4 at 1300°C
(stabilized)	1.6 at 1700°C
(stabilized)	0.59 at 2000°C
(stabilized)	0.37 at 2200°C

Resistivity of Selected Ceramics (Continued)
Listed by Ceramic

Ceramic	Resistivity $(\Omega\text{-cm})$
Cordierite $(2MgO\ 2Al_2O_3\ 5SiO_2)$	
$(\rho=2.3g/cm^3)$	1×10^{14} at 25°C
$(\rho=2.3g/cm^3)$	2.5×10^{11} at 100°C
$(\rho=2.3g/cm^3)$	3.3×10^{7} at 300°C
$(\rho=2.3g/cm^3)$	7.7×10^{5} at 500°C
$(\rho=2.3g/cm^3)$	8.0×10^{4} at 700°C
$(\rho=2.3g/cm^3)$	1.9×10^{4} at 900°C
$(\rho=2.1g/cm^3)$	$>1\times10^{14}$ at 25°C
$(\rho=2.1g/cm^3)$	3.0×10^{13} at 100°C
$(\rho=2.1g/cm^3)$	2.0×10^{10} at 300°C
$(\rho=2.1g/cm^3)$	9.0×10^{7} at 500°C
$(\rho=2.1g/cm^3)$	3.0×10^{6} at 700°C
$(\rho=2.1g/cm^3)$	3.5×10^{5} at 900°C
$(\rho=1.8g/cm^3)$	1.0×10^{14} at 25°C
$(\rho=1.8g/cm^3)$	1.0×10^{13} at 100°C
$(\rho=1.8g/cm^3)$	3.0×10^{9} at 300°C
$(\rho=1.8g/cm^3)$	4.9×10^{7} at 500°C
$(\rho=1.8g/cm^3)$	4.7×10^{6} at 700°C
$(\rho=1.8g/cm^3)$	7.0×10^{5} at 900°C
Mullite $(3Al_2O_3\ 2SiO_2)$	$>10^{14}$ at 25°C
	10^{10} at 300°C
	10^{8} at 500°C

Resistivity of Selected Ceramics (Continued)
Listed by Ceramic

Ceramic	Resistivity (Ω-cm)
Molybdenum Disilicide ($MoSi_2$)	21.5×10^{-6} at 22°C
	18.9×10^{-6} at -80°C
	$75\text{-}80 \times 10^{-6}$ at 1600°C
Tungsten Disilicide (WSi_2)	$33.4\text{-}54.9 \times 10^{-6}$

Source: Data compiled by J.S. Park from *No. 1 Materials Index*, Peter T.B. Shaffer, Plenum Press, New York, (1964); *Smithells Metals Reference Book*, Eric A. Brandes, ed., in association with Fulmer Research Institute Ltd. 6th ed. London, Butterworths, Boston, (1983); and *Ceramic Abstracts*, American Ceramic Society (1986-1991).

VOLUME RESISTIVITY OF SELECTED POLYMERS
Listed by Polymer

Polymer	Volume Resistivity, (ASTM D257) ($\Omega \cdot cm$)
ABS Resins; Molded, Extruded	
Medium impact	$2\text{—}4 \times 10^{15}$
High impact	$1\text{—}4 \times 10^{15}$
Very high impact	$1\text{—}4 \times 10^{15}$
Low temperature impact	$1\text{—}4 \times 10^{15}$
Heat resistant	$1\text{—}5 \times 10^{15}$
Acrylics; Cast, Molded, Extruded	
Cast Resin Sheets, Rods:	
General purpose, type I	$>10^{15}$
General purpose, type II	$>10^{15}$
Moldings:	
Grades 5, 6, 8	$>10^{14}$
High impact grade	2.0×10^{16}
Thermoset Carbonate	
Allyl diglycol carbonate	4×10^{14}
Alkyds; Molded	
Putty (encapsulating)	10^{14}
Rope (general purpose)	10^{14}
Granular (high speed molding)	$1 \times 10^{14}\text{—}1 \times 10^{15}$
Glass reinforced (heavy duty parts)	10^{14}

Volume Resistivity of Selected Polymers (Continued)
Listed by Polymer

Polymer	Volume Resistivity, (ASTM D257) $(\Omega \cdot cm)$
Cellulose Acetate; Molded, Extruded ASTM Grade:	
H6—1	10^{10}—10^{13}
H4—1	10^{10}—10^{13}
H2—1	10^{10}—10^{13}
MH—1, MH—2	10^{10}—10^{13}
MS—1, MS—2	10^{10}—10^{13}
S2—1	10^{10}—10^{13}
Cellulose Acetate Butyrate; Molded, Extruded ASTM Grade:	
H4	10^{11}—10^{14}
MH	10^{11}—10^{14}
S2	10^{11}—10^{14}
Cellusose Acetate Propionate; Molded, Extruded ASTM Grade:	
1	10^{11}—10^{14}
3	10^{11}—10^{14}
6	10^{11}—10^{14}
Chlorinated Polymers	
Chlorinated polyether	1.5×10^{16}
Chlorinated polyvinyl chloride	1×10^{15}—2×10^{16}
Polycarbonates	
Polycarbonate	2.1×10^{16}
Polycarbonate (40% glass fiber reinforced)	1.4×10^{15}

Volume Resistivity of Selected Polymers (Continued)
Listed by Polymer

Polymer	Volume Resistivity, (ASTM D257) ($\Omega \cdot cm$)
Diallyl Phthalates; Molded	
Orlon filled	6×10^4—6×10^6
Dacron filled	10^2—2.5×10^4
Asbestos filled	10^2—5×10^3
Glass fiber filled	10^4—5×10^4
Fluorocarbons; Molded,Extruded	
Polytrifluoro chloroethylene (PTFCE)	10^{18}
Polytetrafluoroethylene (PTFE)	$>10^{18}$
Ceramic reinforced (PTFE)	10^{15}
Fluorinated ethylene propylene(FEP)	$>2 \times 10^{18}$
Polyvinylidene— fluoride (PVDF)	5×10^{14}
Epoxies; Cast, Molded, Reinforced	
Standard epoxies (diglycidyl ethers of bisphenol A)	
Cast rigid	6.1×10^{15}
Cast flexible	9.1×10^5—6.7×10^9
Molded	1—5×10^{15}
High strength laminate	6.6×10^7—10^9
Epoxies—Molded, Extruded	
High performance resins	
(cycloaliphatic diepoxides)	
Cast, rigid	2.10×10^{14}
Molded	1.4—5.5×10^{14}
Epoxy novolacs	
Cast, rigid	$>10^{16}$

Volume Resistivity of Selected Polymers (Continued)
Listed by Polymer

Polymer	Volume Resistivity, (ASTM D257) $(\Omega \cdot cm)$
Melamines; Molded	
Filler & type	
Cellulose electrical	10^{12}—10^{13}
Glass fiber	1—7×10^{11}
Alpha cellulose and mineral	10^{12}
Nylons; Molded, Extruded	
Type 6	
General purpose	4.5×10^{13}
Glass fiber (30%) reinforced	2.8×10^{14}—1.5×10^{15}
Cast	2.6×10^{14}
Type 8	1.5×10^{11}
Type 11	2×10^{13}
Type 12	10^{14} —10^{15}
Nylons; Molded, Extruded	
6/6 Nylon	
General purpose molding	10^{14}—10^{15}
Glass fiber reinforced	2.6—5.5×10^{15}
General purpose extrusion	10^{15}
6/10 Nylon	
General purpose	10^{15}

Volume Resistivity of Selected Polymers (Continued)
Listed by Polymer

Polymer	Volume Resistivity, (ASTM D257) ($\Omega \cdot cm$)
Phenolics; Molded	
Type and filler	
General: woodflour and flock	10^9—10^{13}
Shock: paper, flock, or pulp	1—50×10^{11}
High shock: chopped fabric or cord	$>10^{10}$
Very high shock: glass fiber	$10^{10} \times 10^{11}$
Phenolics: Molded	
Arc resistant—mineral	$10^{10} \times 10^{12}$
Rubber phenolic—woodflour or flock	10^8—10^{11}
Rubber phenolic—chopped fabric	10^{11}
Rubber phenolic—asbestos	10^{11}
ABS–Polycarbonate Alloy	2.2×10^{16}
PVC–Acrylic Alloy	
PVC–acrylic sheet	1—5×10^{13}
PVC–acrylic injection molded	5×10^{15}
Polymides	
Unreinforced	4×10^{15}
Glass reinforced	9.2×10^{15}

Volume Resistivity of Selected Polymers (Continued)
Listed by Polymer

Polymer	Volume Resistivity, (ASTM D257) ($\Omega \cdot cm$)
Polyacetals	
Homopolymer:	
Standard	1×10^{15}
20% glass reinforced	5×10^{14}
Copolymer:	
Standard	1×10^{14}
25% glass reinforced	1.2×10^{14}
High flow	1.0×10^{14}
Polyester; Thermoplastic	
Injection Moldings:	
General purpose grade	$1—4 \times 10^{16}$
Glass reinforced grades	$3.2—3.3 \times 10^{16}$
Glass reinforced self extinguishing	3.4×10^{16}
General purpose grade	2×10^{15}
Asbestos—filled grade	3×10^{14}
Polyesters: Thermosets	
Cast polyyester	
Rigid	10^{13}
Flexible	10^{12}
Reinforced polyester moldings	
High strength (glass fibers)	$1 \times 10^{12} —1 \times 10^{13}$
Heat and chemical resistant (asbestos)	$1 \times 10^{12} —1 \times 10^{13}$
Sheet molding compounds, general purpose	$6.4 \times 10^{15} —2.2 \times 10^{16}$

Volume Resistivity of Selected Polymers (Continued)
Listed by Polymer

Polymer	Volume Resistivity, (ASTM D257) $(\Omega \cdot cm)$
Phenylene Oxides	
SE—100	10^{17}
SE—1	10^{17}
Glass fiber reinforced	10^{17}
Phenylene oxides (Noryl)	
Standard	5×10^{16}
Glass fiber reinforced	10^{17}
Polyarylsulfone	3.2—7.71×10^{16}
Polypropylene:	
General purpose	$>10^{17}$
High impact	10^{17}
Asbestos filled	1.5×10^{15}
Glass reinforced	1.7×10^{16}
Flame retardant	4×10^{16}—10^{17}
Polyphenylene sulfide:	
40% glass reinforced	4.5×10^{14}

Volume Resistivity of Selected Polymers (Continued)
Listed by Polymer

Polymer	Volume Resistivity, (ASTM D257) ($\Omega \cdot$ cm)
Polyethylenes; Molded, Extruded	
Type I—lower density (0.910—0.925)	
Melt index 0.3—3.6	10^{17}—10^{19}
Melt index 6—26	10^{17}—10^{19}
Melt index 200	10^{17}—10^{19}
Type II—medium density (0.926—0.940)	
Melt index 20	$>10^{15}$
Melt index 1.0—1.9	$>10^{15}$
Type III—higher density (0.941—0.965)	
Melt index 0.2—0.9	$>10^{15}$
Melt Melt index 0.1—12.0	$>10^{15}$
Melt index 1.5—15	$>10^{15}$
High molecular weight	$>10^{15}$.
Olefin Copolymers; Molded	
EEA (ethylene ethyl acrylate)	2.4×10^{15}
EVA (ethylene vinyl acetate)	0.15×10^{15}
Ionomer	10×10^{15}
Polyallomer	$>10^{16}$
Polystyrenes; Molded	
Polystyrenes	
General purpose	$>10^{16}$
Medium impact	$>10^{16}$
High impact	$>10^{16}$
Glass fiber -30% reinforced	3.6×10^{16}
Styrene acrylonitrile (SAN)	$>10^{16}$
Glass fiber (30%) reinforced SAN	4.4×10^{16}

Volume Resistivity of Selected Polymers (Continued)
Listed by Polymer

Polymer	Volume Resistivity, (ASTM D257) $(\Omega \cdot cm)$
Polyvinyl Chloride And Copolymers; Molded, Extruded	
Nonrigid—general	1—700×10^{12}
Nonrigid—electrical	4—300×10^{11}
Rigid—normal impact	10^{14}—10^{16}
Vinylidene chloride	10^{14}—10^{16}
Silicones; Molded, Laminated	(dry)
Fibrous (glass) reinforced silicones	9×10^{14}
Granular (silica) reinforced silicones	5×10^{14}
Woven glass fabric/ silicone laminate	2—5×10^{14}
Ureas; Molded	
Alpha—cellulose filled (ASTM Type l)	0.5—5×10^{11}
Cellulose filled (ASTM Type 2)	5—8×10^{10}

Source: data compiled by J.S. Park from Charles T. Lynch, *CRC Handbook of Materials Science*, Vol. 3, CRC Press, Boca Raton, Florida, 1975 and *Engineered Materials Handbook*, Vol.2, Engineering Plastics, ASM International, Metals Park, Ohio, 1988.

DIELECTRIC STRENGTH OF SELECTED POLYMERS
Listed by Polymer

Polymer	Dielectric Strength (Short Time, ASTM D149) (V / mil)
ABS Resins; Molded, Extruded	
Medium impact	385
High impact	350—440
Very high impact	300—375
Low temperature impact	300—415
Heat resistant	360—400
Acrylics; Cast, Molded, Extruded	
Cast Resin Sheets, Rods:	
General purpose, type I	450—530
General purpose, type II	450—500
Moldings:	
Grades 5, 6, 8	400
High impact grade	400—500
Thermoset Carbonate	(step by step)
Allyl diglycol carbonate	290 (step)
Alkyds; Molded	
Putty (encapsulating)	300—350 (step by step)
Rope (general purpose)	290 (step by step)
Granular (high speed molding)	300—350 (step by step)
Glass reinforced (heavy duty parts)	300—350 (step by step)

Dielectric Strength of Selected Polymers (Continued)
Listed by Polymer

Polymer	Dielectric Strength (Short Time, ASTM D149) (V / mil)
Cellulose Acetate; Molded, Extruded	
ASTM Grade:	
H6—1	250—600
H4—1	250—600
H2—1	250—600
MH—1, MH—2	250—600
MS—1, MS—2	250—600
S2—1	250—600
Cellulose Acetate Butyrate; Molded, Extruded	
ASTM Grade:	
H4	250—400
MH	250—400
S2	250—400
Cellusose Acetate Propionate; Molded, Extruded	
ASTM Grade:	
1	300—450
3	300—450
6	300—450
Chlorinated Polymers	
Chlorinated polyether	400
Chlorinated polyvinyl chloride	1,250—1,550

Dielectric Strength of Selected Polymers (Continued)
Listed by Polymer

Polymer	Dielectric Strength (Short Time, ASTM D149) (V / mil)
Polycarbonates	
Polycarbonate	400
Polycarbonate (40% glass fiber reinforced)	475
Diallyl Phthalates; Molded	
Orlon filled	400 (dry)
	375 (wet)
Dacron filled	376—400 (dry)
	360—391 (wet)
Asbestos filled	350—450 (dry)
	300—400 (wet)
Glass fiber filled	350—430 (dry)
	300—420 (wet)
Fluorocarbons; Molded,Extruded	
Polytrifluoro chloroethylene (PTFCE)	530—600
Polytetrafluoroethylene (PTFE)	1000—2000
Ceramic reinforced (PTFE)	300—400
Fluorinated ethylene propylene(FEP)	2100
Polyvinylidene— fluoride (PVDF)	260

Dielectric Strength of Selected Polymers (Continued)
Listed by Polymer

Polymer	Dielectric Strength (Short Time, ASTM D149) (V / mil)
Epoxies; Cast, Molded, Reinforced	(step by step)
Standard epoxies (diglycidyl ethers of bisphenol A)	
Cast rigid	>400 (step)
Cast flexible	400—410 (step)
Molded	360—400 (step)
General purpose glass cloth laminate	450—550 (step)
High strength laminate	650-750 (step)
Epoxies—Molded, Extruded	
High performance resins (cycloaliphatic diepoxides)	
Molded	280—400 (step)
Epoxy novolacs	
Cast, rigid	444
Melamines; Molded	
Filler & type	
Cellulose electrical	350—400
Glass fiber	250 —300
Alpha cellulose and mineral	375
Nylons; Molded, Extruded	
Type 6	
General purpose	385—400
Glass fiber (30%) reinforced	400—450
Cast	380
Flexible copolymers	440

Dielectric Strength of Selected Polymers (Continued)
Listed by Polymer

Polymer	Dielectric Strength (Short Time, ASTM D149) (V / mil)
Type 8	340
Type 11	425
Type 12	840
Nylons; Molded, Extruded	
6/6 Nylon	
General purpose molding	385,
Glass fiber reinforced	400, 480
Glass fiber Molybdenum disulfide filled	300—400
General purpose extrusion	
6/10 Nylon	
General purpose	470
Phenolics; Molded	
Type and filler	
General: woodflour and flock	200—425
Shock: paper, flock, or pulp	250—350
High shock: chopped fabric or cord	200—350
Very high shock: glass fiber	375—425
Phenolics: Molded	
Arc resistant—mineral	350—425
Rubber phenolic—woodflour or flock	250—375
Rubber phenolic—chopped fabric	250
Rubber phenolic—asbestos	350

Dielectric Strength of Selected Polymers (Continued)
Listed by Polymer

Polymer	Dielectric Strength (Short Time, ASTM D149) (V / mil)
ABS–Polycarbonate Alloy	500
PVC–Acrylic Alloy	
PVC–acrylic sheet	>429
PVC–acrylic injection molded	400
Polymides	
Unreinforced 2nd value	310
Glass reinforced	300
Polyacetals	
Homopolymer:	
Standard	500
20% glass reinforced	500
Copolymer:	
Standard	500
25% glass reinforced	580
High flow	500
Polyester; Thermoplastic	
Injection Moldings:	
General purpose grade	590
Glass reinforced grades	560—750
Glass reinforced self extinguishing	750
General purpose grade	420—540
Glass reinforced grade	—
Asbestos—filled grade	580

Dielectric Strength of Selected Polymers (Continued)
Listed by Polymer

Polymer	Dielectric Strength (Short Time, ASTM D149) (V / mil)
Polyesters: Thermosets	
Cast polyyester	
Rigid	300—400
Flexible	300—400
Reinforced polyester moldings	
High strength (glass fibers)	200—400
Heat and chemical resistsnt (asbestos)	350
Sheet molding compounds, general purpose	400—440
Phenylene Oxides	
SE—100	400 (1/8 in.)
SE—1	500 (1/8 in.)
Glass fiber reinforced	1,020 (1/32 in.)
Phenylene oxides (Noryl)	
Standard	425
Glass fiber reinforced	480
Polyarylsulfone	350—383
Polypropylene:	
General purpose	650 (125 mil)
High impact	450—650
Asbestos filled	450
Glass reinforced	317—475
Flame retardant	485—700

Dielectric Strength of Selected Polymers (Continued)
Listed by Polymer

Polymer	Dielectric Strength (Short Time, ASTM D149) (V / mil)
Polyphenylene sulfide:	
Standard	450—595
40% glass reinforced	490
Polyethylenes; Molded, Extruded	
Type I—lower density (0.910—0.925)	
Melt index 0.3—3.6	480
Melt index 6—26	480
Melt index 200	480
Type II—medium density (0.926—0.940)	
Melt index 20	480
Melt index 1.0—1.9	480
Type III—higher density (0.941—0.965)	
Melt index 0.2—0.9	480
Melt Melt index 0.1—12.0	480
Melt index 1.5—15	480
High molecular weight	480
Olefin Copolymers; Molded	
EEA (ethylene ethyl acrylate)	550
EVA (ethylene vinyl acetate)	525
Ionomer	1000
Polyallomer	500—650

Dielectric Strength of Selected Polymers (Continued)
Listed by Polymer

Polymer	Dielectric Strength (Short Time, ASTM D149) (V / mil)
Polystyrenes; Molded	
Polystyrenes	
General purpose	>500
Medium impact	>425
High impact	300—650
Glass fiber -30% reinforced	396
Styrene acrylonitrile (SAN)	400—500
Glass fiber (30%) reinforced SAN	515
Polyvinyl Chloride And Copolymers; Molded, Extruded	
Nonrigid—electrical	24—500
Rigid—normal impact	725—1,400
Silicones; Molded, Laminated	
Fibrous (glass) reinforced silicones	280 (in oil)
Granular (silica) reinforced silicones	380 (in oil)
Woven glass fabric/ silicone laminate	725
Ureas; Molded	
Alpha—cellulose filled (ASTM Type l)	300—400
Cellulose filled (ASTM Type 2)	340—370
Woodflour filled	300—400

Source: data compiled by J.S. Park from Charles T. Lynch, *CRC Handbook of Materials Science*, Vol. 3, CRC Press, Boca Raton, Florida, 1975 and *Engineered Materials Handbook*, Vol.2, Engineering Plastics, ASM International, Metals Park, Ohio, 1988.

DIELECTRIC CONSTANT OF SELECTED POLYMERS
Listed by Polymer

Polymer	Dielectric Constant (ASTM D150)	
	60 Hz	10^6 Hz
ABS Resins; Molded, Extruded		
Medium impact	2.8—3.2	2.75—3.0
High impact	2.8—3.2	2.7—3.0
Very high impact	2.8—3.5	2.4—3.0
Low temperature impact	2.5—3.5	2.4—3.0
Heat resistant	2.7—3.5	2.8—3.2
Acrylics; Cast, Molded, Extruded		
Cast Resin Sheets, Rods:		
General purpose, type I	3.5—4.5	2.7—3.2
General purpose, type II	3.5—4.5	2.7—3.2
Moldings:		
Grades 5, 6, 8	3.5—3.9	2.7—2.9
High impact grade	3.5—3.9	2.5—3.0
Thermoset Carbonate		
Allyl diglycol carbonate	4.4	3.5—3.8
Alkyds; Molded		
Putty (encapsulating)	5.4—5.9	4.5—4.7
Rope (general purpose)	7.4	6.8
Granular (high speed molding)	5.7—6.3	4.8—5.1
Glass reinforced (heavy duty parts)	5.2—6.0	4.5—5.0

Dielectric Constant of Selected Polymers (Continued)
Listed by Polymer

Polymer	Dielectric Constant (ASTM D150)	
	60 Hz	10^6 Hz
Cellulose Acetate; Molded, Extruded		
ASTM Grade:		
H6—1	3.5—7.5	3.2—7.0
H4—1	3.5—7.5	3.2—7.0
H2—1	3.5—7.5	3.2—7.0
MH—1, MH—2	3.5—7.5	3.2—7.0
MS—1, MS—2	3.5—7.5	3.2—7.0
S2—1	3.5—7.5	3.2—7.0
Cellulose Acetate Butyrate;		
Molded, Extruded		
ASTM Grade:		
H4	3.5—6.4	3.2—6.2
MH	3.5—6.4	3.2—6.2
S2	3.5—64	3.2—6.2
Cellusose Acetate Propionate;		
Molded, Extruded		
ASTM Grade:		
1	3.7—4.0	3.4—3.7
3	3.7—4.0	3.4—3.7
6	3.7—4.0	3.7—3.4
Chlorinated Polymers		
Chlorinated polyether	3.1	2.92
Chlorinated polyvinyl chloride	3.08	3.2—3.6

Dielectric Constant of Selected Polymers (Continued)
Listed by Polymer

Polymer	Dielectric Constant (ASTM D150)	
	60 Hz	10^6 Hz
Polycarbonates		
Polycarbonate	3.17	2.96
Polycarbonate (40% glass fiber reinforced)	3.8	3.58
Diallyl Phthalates; Molded		
Orlon filled	3.3—3.9 (Dry)	4.1—3.4 (Wet)
Dacron filled	3.5—3.8 (Dry)	3.7—3.9 (Wet)
Asbestos filled	4.5—5.2 (Dry)	4.8—6.5 (Wet)
Glass fiber filled	3.5—4.5 (Dry)	4.4—4.6 (Wet)
Fluorocarbons; Molded,Extruded		
Polytrifluoro chloroethylene (PTFCE)	2.6—2.7	
Polytetrafluoroethylene (PTFE)		
(0.01 in thickness)	2.1	
Ceramic reinforced (PTFE)	2.9—3.6	
Fluorinated ethylene propylene(FEP)		
(0.01 in thickness)	2.1	
Polyvinylidene— fluoride (PVDF)		
(0.125 in thickness)	10	
Epoxies; Cast, Molded, Reinforced		
Standard epoxies		
(diglycidyl ethers of bisphenol A)		
Cast rigid	4.02	3.42
Cast flexible	4.43-4.79	2.78-3.52
Molded	4.4-5.4	4.1-4.6
General purpose glass cloth laminate	5.3-5.4	4.7-4.8
High strength laminate	—	4.8-5.2

Dielectric Constant of Selected Polymers (Continued)
Listed by Polymer

Polymer	Dielectric Constant (ASTM D150)	
	60 Hz	10^6 Hz
Epoxies—Molded, Extruded		
High performance resins		
(cycloaliphatic diepoxides)		
Cast, rigid	3.96—4.02	3.53—3.58
Molded	4.7—5.7	4.3—4.8
Glass cloth laminate	—	5.1
Epoxy novolacs		
Cast, rigid	3.34—3.39	—
Glass cloth laminate	4.41—4.43	—
Melamines; Molded		
Filler & type		
Unfilled	7.9—11.0	6.3—7.3
Cellulose electrical	6.2—7.7	5.2—6.0
Glass fiber	7.0—11.1	6.0—7.9
Alpha cellulose	—	6.4—8.1
Alpha cellulose mineral	—	5.6
Nylons; Molded, Extruded		
Type 6		
General purpose	4.0—5.3	3.6—3.8
Glass fiber (30%) reinforced	4.6—5.6	3.9—5.4
Cast	4	3.3
Flexible copolymers	3.2—4.0	3.0—3.6
Type 8	9.3	4
Type 11	3.3 (10^3 Hz)	—
Type 12	3.6 (10^3 Hz)	—

Dielectric Constant of Selected Polymers (Continued)
Listed by Polymer

Polymer	Dielectric Constant (ASTM D150)	
	60 Hz	10^6 Hz
Nylons; Molded, Extruded		
6/6 Nylon		
General purpose molding	4	3.6
Glass fiber reinforced	40—44	3.5—4.1
6/10 Nylon		
General purpose	3.9	3.5
Phenolics; Molded		
Type and filler		
General: woodflour and flock	5.0—9.0	4.0—7.0
Shock: paper, flock, or pulp	5.6—11.0	4.5—7.0
High shock: chopped fabric or cord	6.5—15.0	4.5—7.0
Very high shock: glass fiber	7.1—7.2	4.6—6.6
Phenolics: Molded		
Arc resistant—mineral	7.4	5
Rubber phenolic—woodflour or flock	9—16	5
Rubber phenolic—chopped fabric	15	5
Rubber phenolic—asbestos	15	5
ABS–Polycarbonate Alloy	2.74	2.69
PVC–Acrylic Alloy		
PVC–acrylic sheet	3.86	3.44
PVC–acrylic injection molded	4	3.4

Dielectric Constant of Selected Polymers (Continued)
Listed by Polymer

Polymer	Dielectric Constant (ASTM D150)	
	60 Hz	10^6 Hz
Polymides		
Unreinforced	4.12	3.96
Glass reinforced	4.84	4.74
Polyacetals		
Homopolymer:		
Standard	3.7	3.7
20% glass reinforced	4	4—0
Copolymer:		
Standard	3.7 (100 Hz)	3—7
25% glass reinforced	3.9 (100 Hz)	3—9
High flow	3.7 (100 Hz)	3—7
Polyester; Thermoplastic		
Injection Moldings:		
General purpose grade	3.1—3.3	—
Glass reinforced grades	3.7—4.2	—
Glass reinforced self extinguishing	3.7—3.8	—
General purpose grade	3.16	—
Asbestos—filled grade	3.5—4.2	—

Dielectric Constant of Selected Polymers (Continued)
Listed by Polymer

Polymer	Dielectric Constant (ASTM D150)	
	60 Hz	10^6 Hz
Polyesters: Thermosets		
Cast polyyester		
Rigid	2.8—4.4	2.8—4.4
Flexible	3.18—7.0	3.7—6.1
Reinforced polyester moldings		
Sheet molding compounds, general purpose	4.62—5.0	4.55—4.75
Phenylene Oxides		
SE—100	2.65	2.64
SE—1	2.69	2.68
Glass fiber reinforced	2.93	2.92
Phenylene oxides (Noryl)		
Standard	3.06—3.15	3.03—3.10
Glass fiber reinforced	3.55	3.41
Polyarylsulfone	3.51—3.94	3.54—3.7
Polypropylene:		
General purpose	2.20—2.28	2.23—2.24
High impact	2.20—2.28	2.23—2.27
Asbestos filled	2.75	2.6—3.17
Glass reinforced	2.3—2.5	2—2.25
Flame retardant	2.46—2.79	2.45—2.70
Polyphenylene sulfide:		
Standard	—	3.22—3.8
40% glass reinforced	—	3.88

Dielectric Constant of Selected Polymers (Continued)
Listed by Polymer

Polymer	Dielectric Constant (ASTM D150)	
	60 Hz	10^6 Hz
Polyethylenes; Molded, Extruded		
Type I—lower density (0.910—0.925)		
Melt index 0.3—3.6	2.3	—
Melt index 6—26	2.3	—
Melt index 200	2.3	—
Type II—medium density (0.926—0.940)		
Melt index 20	2.3	—
Melt index 1.0—1.9	2.3	—
Type III—higher density (0.941—0.965)		
Melt index 0.2—0.9	2.3	—
Melt Melt index 0.1—12.0	2.3	—
Melt index 1.5—15	2.3	—
High molecular weight	2.3	—
Olefin Copolymers; Molded		
EEA (ethylene ethyl acrylate)	2.8	
EVA (ethylene vinyl acetate)	3.16	
Ionomer	2.4	
Polyallomer	2.3	
Polystyrenes; Molded		
Polystyrenes		
General purpose	2.45—2.65	2.45—2.65
Medium impact	2.45—4.75	2.4—3.8
High impact	2.45—4.75	2.5—4.0
Glass fiber -30% reinforced	3.1	3
Styrene acrylonitrile (SAN)	2.6—3.4	2.6—3.02
Glass fiber (30%) reinforced SAN	3.5	3.4—3.6

Dielectric Constant of Selected Polymers (Continued)
Listed by Polymer

Polymer	Dielectric Constant (ASTM D150)	
	60 Hz	10^6 Hz
Polyvinyl Chloride And Copolymers;		
Molded, Extruded		
Nonrigid—general	5.5—9.1	
Nonrigid—electrical	6.0—8.0	
Rigid—normal impact	2.3—3.7	
Vinylidene chloride	3—5	
Silicones; Molded, Laminated		
Fibrous (glass) reinforced silicones	4.34	4.28
Granular (silica) reinforced silicones	4.1—4.5	3.4 —4.3
Woven glass fabric/ silicone laminate	3.9—4.2	3.8—397
Ureas; Molded		
Alpha—cellulose filled (ASTM Type l)	7.0—9.5	6.4—6.9
Cellulose filled (ASTM Type 2)	7.2—7.3	6.4—6.5
Woodflour filled	7.0—9.5	6.4—6.9

Source: data compiled by J.S. Park from Charles T. Lynch, *CRC Handbook of Materials Science*, Vol. 3, CRC Press, Boca Raton, Florida, 1975 and *Engineered Materials Handbook*, Vol.2, Engineering Plastics, ASM International, Metals Park, Ohio, 1988.

DISSIPATION FACTOR FOR SELECTED POLYMERS
Listed by Polymer

Polymer	Dissipation Factor (ASTM D150)	
	60 Hz	10^6 Hz
ABS Resins; Molded, Extruded		
Medium impact	0.003—0.006	0.008—0.009
High impact	0.005—0.007	0.007—0.015
Very high impact	0.005—0.010	0.008—0.016
Low temperature impact	0.005—0.01	0.008—0.016
Heat resistant	0.030—0.040	0.005—0.015
Acrylics; Cast, Molded, Extruded		
Cast Resin Sheets, Rods:		
General purpose, type I	0.05—0.06	0.02—0.03
General purpose, type II	0.05—0.06	0.02—0.03
Moldings:		
Grades 5, 6, 8	0.04—0.06	0.02—0.03
High impact grade	0.03—0.04	0.01—0.02
Thermoset Carbonate		
Allyl diglycol carbonate	0.03—0.04	0.1—0.2
Alkyds; Molded		
Putty (encapsulating)	0.030—0.045	0.016—0.020
Rope (general purpose)	0.019	0.023
Granular (high speed molding)	0.030—0.040	0.017—0.020
Glass reinforced (heavy duty parts)	0.02—0.03	0.015—0.022

Dissipation Factor for Selected Polymers (Continued)
Listed by Polymer

Polymer	Dissipation Factor (ASTM D150)	
	60 Hz	10^6 Hz
Cellulose Acetate; Molded, Extruded		
ASTM Grade:		
H4—1	0.01—0.06	0.01—0.10
H2—1	0.01—0.06	0.01—0.10
MH—1, MH—2	0.01—0.06	0.01—0.10
MS—1, MS—2	0.01—0.06	0.01—0.10
S2—1	0.01—0.06	0.01—0.10
Cellulose Acetate Butyrate; Molded, Extruded		
ASTM Grade:		
H4	0.01—0.04	0.02—0.05
MH	0.01—0.04	0.02—0.05
S2	0.01—0.04	0.02—0.05
Cellusose Acetate Propionate; Molded, Extruded		
ASTM Grade:		
1	0.01—0.04	0.02—0.05
3	0.01—0.04	0.02—0.05
6	0.01—0.04	0.02—0.05
Chlorinated Polymers		
Chlorinated polyether	0.011	0.011
Chlorinated polyvinyl chloride	0.0189—0.0208	0.02

Dissipation Factor for Selected Polymers (Continued)
Listed by Polymer

Polymer	Dissipation Factor (ASTM D150)	
	60 Hz	10^6 Hz
Polycarbonates		
Polycarbonate	0.0009	0.01
Polycarbonate (40% glass fiber reinforced)	0.006	0.007
Diallyl Phthalates; Molded		
Orlon filled	0.023—0.015 (Dry)	
	0.045—0.040 (Wet)	
Dacron filled	0.004—0.016 (Dry)	
	0.009—0.017 (Wet)	
Asbestos filled	0.05—0.03 (Dry)	
	0.154—0.050 (Wet)	
Glass fiber filled	0.004—0.015 (Dry)	
	0.012—0.020 (Wet)	
Fluorocarbons; Molded,Extruded		
Polytrifluoro chloroethylene (PTFCE)	0.02	0.007—0.010
Polytetrafluoroethylene (PTFE)	0.0002	0.0002
Ceramic reinforced (PTFE)	0.0005–0.0015	0.0005–0.0015
Fluorinated ethylene propylene(FEP)	0.0003	0.0003
Polyvinylidene— fluoride (PVDF)	0.05	0.184

Dissipation Factor for Selected Polymers (Continued)
Listed by Polymer

Polymer	Dissipation Factor (ASTM D150) 60 Hz	10^6 Hz
Epoxies; Cast, Molded, Reinforced		
Standard epoxies (diglycidyl ethers of bisphenol A)		
Cast rigid	0.0074	0.032
Cast flexible	0.0048-0.0380	0.0369-0.0622
Molded	0.011-0.018	0.013—0.020
General purpose glass cloth laminate	0.004-0.006	0.024—0.026
High strength laminate	—	0.010-0.017
Epoxies—Molded, Extruded		
High performance resins		
(cycloaliphatic diepoxides)		
Cast, rigid	0.0055—0.0074	0.029—0.028
Molded	0.0071—0.025	—
Glass cloth laminate	—	0.0158
Epoxy novolacs		
Cast, rigid	0.001—0.007	—

Dissipation Factor for Selected Polymers (Continued)
Listed by Polymer

Polymer	Dissipation Factor (ASTM D150)	
	60 Hz	10^6 Hz
Melamines; Molded		
Filler & type		
Unfilled	0.048—0.162	0.031—0.040
Cellulose electrical	0.026—0.192	0.032—0.12
Glass fiber	0.14—0.23	0.020—0.03
Alpha cellulose	—	0.028
Alpha cellulose mineral	—	0.030
Nylons; Molded, Extruded		
Type 6		
General purpose	0.06—0.014	0.03—0.04
Glass fiber (30%) reinforced	0.022—0.008	0.019—0.015
Cast	0.015	0.05
Flexible copolymers	0.007—0.010	0.010—0.015
Type 8	0.19	0.08
Type 11	0.03	0.02
Type 12	0.04 (10^3 Hz)	
Nylons; Molded, Extruded		
6/6 Nylon		
General purpose molding	0.014—0.04	0.04
Glass fiber reinforced	0.018—0.009	0.017—0.018
6/10 Nylon		
General purpose	0.04	

Dissipation Factor for Selected Polymers (Continued)
Listed by Polymer

Polymer	Dissipation Factor (ASTM D150) 60 Hz	10^6 Hz
Phenolics; Molded		
Type and filler		
General: woodflour and flock	0.05—0.30	0.03—0.07
Shock: paper, flock, or pulp	0.08—0.35	0.03—0.07
High shock: chopped fabric or cord	0.08—0.45	0.03—0.09
Very high shock: glass fiber	0.02—0.03	0.02
Phenolics: Molded		
Arc resistant—mineral	0.13—0.16	0.1
Rubber phenolic—woodflour or flock	0.15—0.60	0.1—0.2
Rubber phenolic—chopped fabric	0.5	0.09
Rubber phenolic—asbestos	0.15	0.13
ABS–Polycarbonate Alloy	0.0026	0.0059
PVC–Acrylic Alloy		
PVC–acrylic sheet	0.076	0.094
PVC–acrylic injection molded	0.037	0.031
Polymides		
Unreinforced	0.003	0.011
Glass reinforced	0.0034	0.0055

Dissipation Factor for Selected Polymers (Continued)
Listed by Polymer

Polymer	Dissipation Factor (ASTM D150)	
	60 Hz	10^6 Hz
Polyacetals		
Homopolymer:		
Standard	0.0048	0.0048
20% glass reinforced	0.0047	0.0036
Copolymer:		
Standard	0.001 (100 Hz)	0.006
25% glass reinforced	0.003 (100 Hz)	0.006
High flow	0.001 (100 Hz)	0.006
Polyester; Thermoplastic		
Injection Moldings:		
General purpose grade	0.002 (10^3 Hz)	
Glass reinforced grades	0.002—0.003 (10^3 Hz)	
Glass reinforced self extinguishing	0.002 (10^3 Hz)	
General purpose grade	0.023 (10^3 Hz)	
Asbestos—filled grade	0.015 (10^3 Hz)	
Polyesters: Thermosets		
Cast polyyester		
Rigid	0.003—0.04	0.006—0.04
Flexible	0.01—0.18	0.02—0.06
Reinforced polyester moldings		
Sheet molding compounds, general purpose	0.0087—0.04	0.0086—0.022

Dissipation Factor for Selected Polymers (Continued)
Listed by Polymer

Polymer	Dissipation Factor (ASTM D150)	
	60 Hz	10^6 Hz
Phenylene Oxides		
SE—100	0.0007	0.0024
SE—1	0.0007	0.0024
Glass fiber reinforced	0.0009	0.0015
Phenylene oxides (Noryl)		
Standard	0.0008	0.0034
Glass fiber reinforced	0.0019	0.0049
Polyarylsulfone	0.0017—0.003	0.0056—0.012
Polypropylene:		
General purpose	0.0005–0.0007	0.0002–0.0003
High impact	<0.0016	0.0002—0.0003
Asbestos filled	0.007	0.002
Glass reinforced	0.002	0.003
Flame retardant	0.0007–0.017	0.0006–0.003
Polyphenylene sulfide:		
Standard	—	0.0007
40% glass reinforced	—	0.0014—0.0041

Dissipation Factor for Selected Polymers (Continued)
Listed by Polymer

Polymer	Dissipation Factor (ASTM D150) 60 Hz	10^6 Hz
Polyethylenes; Molded, Extruded		
Type I—lower density (0.910—0.925)		
Melt index 0.3—3.6	<0.0005	
Melt index 6—26	<0.0005	
Melt index 200	<0.0005	
Type II—medium density (0.926—0.940)		
Melt index 20	<0.0005	
Melt index 1.0—1.9	<0.0005	
Type III—higher density (0.941—0.965)		
Melt index 0.2—0.9	<0.0005	
Melt Melt index 0.1—12.0	<0.0005	
Melt index 1.5—15	<0.0005	
High molecular weight	<0.0005	
Olefin Copolymers; Molded		
EEA (ethylene ethyl acrylate)	0.001	
EVA (ethylene vinyl acetate)	0.003	
Ionomer	0.003	
Polyallomer	>0.0005	

Dissipation Factor for Selected Polymers (Continued)
Listed by Polymer

Polymer	Dissipation Factor (ASTM D150)	
	60 Hz	10^6 Hz
Polystyrenes; Molded		
Polystyrenes		
General purpose	0.0001–0.0003	0.0001–0.0005
Medium impact	0.0004–0.002	0.0004–0.002
High impact	0.0004–0.002	0.0004–0.002
Glass fiber -30% reinforced	0.005	0.002
Styrene acrylonitrile (SAN)	>0.006	0.007–0.010
Glass fiber (30%) reinforced SAN	0.005	0.009
Polyvinyl Chloride and Copolymers;		
Molded, Extruded		
Nonrigid—general	0.05—0.15	
Nonrigid—electrical	0.08—0.11	
Rigid—normal impact	0.020—0.03	
Vinylidene chloride	0.03—0.15	
Silicones; Molded, Laminated		
Fibrous (glass) reinforced silicones	0.01	0.004
Granular (silica) reinforced silicones	0.002—0.004	0.001—0.004
Woven glass fabric/ silicone laminate	0.02	0.002

Dissipation Factor for Selected Polymers (Continued)
Listed by Polymer

Polymer	Dissipation Factor (ASTM D150)	
	60 Hz	10^6 Hz
Ureas; Molded		
Alpha—cellulose filled (ASTM Type l)	0.035—0.043	0.028—0.032
Cellulose filled (ASTM Type 2)	0.042—0.044	0.027—0.029
Woodflour filled	0.035—0.040	0.028—0.032

Source: data compiled by J.S. Park from Charles T. Lynch, *CRC Handbook of Materials Science*, Vol. 3, CRC Press, Boca Raton, Florida, 1975 and *Engineered Materials Handbook*, Vol.2, Engineering Plastics, ASM International, Metals Park, Ohio, 1988.

ARC RESISTANCE OF SELECTED POLYMERS
Listed by Polymer

Polymer	Arc Resistance, (ASTM D495) (seconds)
Acrylics; Cast, Molded, Extruded	
Cast Resin Sheets, Rods:	
General purpose, type I	No track
General purpose, type II	No track
Moldings:	
Grades 5, 6, 8	No track
High impact grade	No track
Thermoset Carbonate	
Allyl diglycol carbonate	185
Alkyds; Molded	
Putty (encapsulating)	180
Rope (general purpose)	180
Granular (high speed molding)	180
Glass reinforced (heavy duty parts)	180
Polycarbonates	
Polycarbonate	120 (tungsten electrode)
Polycarbonate (40% glass fiber reinforced)	120 (tungsten electrode)
Diallyl Phthalates; Molded	
Orlon filled	85—115
Dacron filled	105—125
Asbestos filled	125—140
Glass fiber filled	125—140

Arc Resistance of Selected Polymers (Continued)
Listed by Polymer

Polymer	Arc Resistance, (ASTM D495) (seconds)
Fluorocarbons; Molded,Extruded	
Polytrifluoro chloroethylene (PTFCE)	>360
Polytetrafluoroethylene (PTFE)	>200
Ceramic reinforced (PTFE)	
Fluorinated ethylene propylene(FEP)	>165
Polyvinylidene— fluoride (PVDF)	50—60
Epoxies; Cast, Molded, Reinforced	
Standard epoxies (diglycidyl ethers of bisphenol A)	
Cast rigid	100
Cast flexible	75—98
Molded	135—190
General purpose glass cloth laminate	130—180
Epoxies—Molded, Extruded	
High performance resins	
(cycloaliphatic diepoxides)	
Molded	180—185
Epoxy novolacs	
Cast, rigid	120
Melamines; Molded	
Filler & type	
Unfilled	100—145
Cellulose electrical	70—135
Glass fiber	180—186
Alpha cellulose and mineral	125

Arc Resistance of Selected Polymers (Continued)
Listed by Polymer

Polymer	Arc Resistance, (ASTM D495) (seconds)
Nylons; Molded, Extruded	
Type 6	
Glass fiber (30%) reinforced	92—81
Nylons; Molded, Extruded	
6/6 Nylon	
General purpose molding	120
Glass fiber reinforced	148—100
Glass fiber Molybdenum disulfide filled	135
General purpose extrusion	120
6/10 Nylon	
General purpose	120
Phenolics; Molded	
Type and filler	
General: woodflour and flock	5—60
Shock: paper, flock, or pulp	5—60
High shock: chopped fabric or cord	5—60
Very high shock: glass fiber	60
Phenolics: Molded	
Arc resistant—mineral	180
Rubber phenolic—woodflour or flock	7—20
Rubber phenolic—chopped fabric	10—20
Rubber phenolic—asbestos	5—20
ABS–Polycarbonate Alloy	96

Arc Resistance of Selected Polymers (Continued)
Listed by Polymer

Polymer	Arc Resistance, (ASTM D495) (seconds)
PVC–Acrylic Alloy	
PVC–acrylic sheet	80
PVC–acrylic injection molded	25
Polymides	
Unreinforced	152
Glass reinforced	50—180
Polyacetals	
Homopolymer:	
Standard	129
20% glass reinforced	188
Copolymer:	
Standard	240
25% glass reinforced	136
High flow	240
Polyester; Thermoplastic	
Injection Moldings:	
General purpose grade	190
Glass reinforced grades	130
Glass reinforced self extinguishing	80
General purpose grade	125
Asbestos—filled grade	108

Arc Resistance of Selected Polymers (Continued)
Listed by Polymer

	Arc Resistance, (ASTM D495)
Polymer	(seconds)
Polyesters: Thermosets	
Cast polyyester	
Rigid	115—135
Flexible	125—145
Reinforced polyester moldings	
High strength (glass fibers)	130—170
Sheet molding compounds, general purpose	130—180
Phenylene Oxides	
SE—100	75
SE—1	75
Glass fiber reinforced	120
Phenylene oxides (Noryl)	
Standard	122
Glass fiber reinforced	114
Polyarylsulfone	67—81
Polypropylene:	
General purpose	125—136
High impact	123—140
Asbestos filled	121—125
Glass reinforced	73—77
Flame retardant	15—40

Arc Resistance of Selected Polymers (Continued)
Listed by Polymer

Polymer	Arc Resistance, (ASTM D495) (seconds)
Polyphenylene sulfide:	
40% glass reinforced	34
Polystyrenes; Molded	
Polystyrenes	
General purpose	60—135
Medium impact	20—135
High impact	20—100
Glass fiber -30% reinforced	28
Styrene acrylonitrile (SAN)	100—150
Glass fiber (30%) reinforced SAN	65
Silicones; Molded, Laminated	
Fibrous (glass) reinforced silicones	240
Granular (silica) reinforced silicones	250—310
Woven glass fabric/ silicone laminate	225—250
Ureas; Molded	
Alpha—cellulose filled (ASTM Type l)	100—135
Cellulose filled (ASTM Type 2)	85—110
Woodflour filled	80—110

Source: data compiled by J.S. Park from Charles T. Lynch, *CRC Handbook of Materials Science*, Vol. 3, CRC Press, Boca Raton, Florida, 1975 and *Engineered Materials Handbook*, Vol.2, Engineering Plastics, ASM International, Metals Park, Ohio, 1988.

DISPERSION OF OPTICAL MATERIALS AT 298 K

Material   Dispersion Equation at 298 K

Alumina (Sapphire, Single Crystal))

$$n^2-1 = \sum_{i=1}^{3} \frac{A_i\lambda^2}{\lambda^2-\lambda_i^2}$$

where

i	λ_i^2	A_i
1	0.00377588	1.023798
2	0.0122544	1.058264
3	321.3616	5.280792

(λ in mm)

ArsenicTrisulfide (Glass)

$$n^2-1 = \sum_{i=1}^{5} \frac{K_i\lambda^2}{\lambda^2-\lambda_i^2}$$

where

i	λ_i^2	K_i
1	0.0225	1.8983678
2	0.0625	1.9222979
3	0.1225	0.8765134
4	0.2025	0.1188704
5	0.705	0.9569903

(λ in μm)

Dispersion of Optical Materials at 298 K (Continued)

Material   Dispersion Equation at 298 K

Barium Fluoride (Single Crystal)

$$n^2-1 = \sum_{i=1}^{3} \frac{A_i\lambda^2}{\lambda^2-\lambda_i^2}$$

where

i	λ_i	A_i
1	0.057789	0.643356
2	0.10968	0.50676
3	46.3864	3.8261

(λ in μm)

Cadmium Sulfide (Bulk and Hexagonal Single Crystal)

$$n_0^2=5.235 + \frac{1.891\times10^7}{\lambda^2-1.651\times10^7}$$

for ordinary ray,

and

$$n_e^2=5.239+ \frac{2.076\times10^7}{\lambda^2-1.651\times10^7}$$

for extraordinary ray.

(λ in μm)

Dispersion of Optical Materials at 298 K (Continued)

Material   Dispersion Equation at 298 K

Calcium Fluoride (Single Crystal)

$$n^2-1 = \sum_{i=1}^{3} \frac{A_i\lambda^2}{\lambda^2-\lambda_i^2}$$

i	A_i	λ_ι
1	0.5675888	0.050263605
2	0.4710914	0.1003909
3	3.8484723	34.64904

(λ in μm)

Cesium Bromide (Single Crystal)

$$n^2= 5.640752 - 3.338 \times 10^{-6}\lambda^2 + \frac{0.0018612}{\lambda^2} + \frac{41110.49}{\lambda^2 -14390.4} + \frac{0.0290764}{\lambda^2 - 0.024964}$$

(λ in μm)

Dispersion of Optical Materials at 298 K (Continued)

Material   Dispersion Equation at 298 K

Cesium Iodide (Single Crystal)

$$n^2-1 = \sum_{i=1}^{5} \frac{K_i\lambda^2}{\lambda^2-\lambda_i^2}$$

where

i	λ_i^2	K_i
1	0.00052701	0.3461725
2	0.02149156	1.0080886
3	0.28551800	0.02149156
4	0.39743178	0.044944
5	3.3605359	25921

(λ in mm)

Germanium (Intrinsic Single Crystal)

$$n = A + B\lambda + C\lambda^2 + D\lambda^2 + E\lambda^4$$

where A=3.99931

B=0.391707

C=0.163492

D=-0.0000060

E=0.000000053

for $2.0\mu m \le \lambda \le 13.5\ \mu m$

Dispersion of Optical Materials at 298 K (Continued)

Material   Dispersion Equation at 298 K

Lithium Fluoride (Single Crystal)

$$n = A + BL + CL^2 + D\lambda^2 + E\lambda^4$$

where

A=1.38761

B=0.001796

C=-0.000041

D=-0.0023045

E=-0.00000557

for $0.5\mu m \leq \lambda \leq 6.0\ \mu m$

Magnesium Fluoride (Single Crystal)

$$n_o = 1.36957 + \frac{0.0035821}{\lambda - 0.14925}$$

for ordinary ray,

and

$$n_e = 1.38100 + \frac{0.0037415}{\lambda - 0.14947}$$

for extraordinary ray,

within $0.4\mu m \leq \lambda \leq 0.7\ \mu m$

Dispersion of Optical Materials at 298 K (Continued)

Material   Dispersion Equation at 298 K

Magnesium Oxide

(Single Crystal)

$$n^2 = 2.956362 - 0.1062387\,\lambda^2$$
$$-2.04968 \times 10^{-5}\lambda^4$$
$$-\frac{0.0219577}{\lambda^2 - 0.01428322}$$

Potassium Bromide

(Single Crystal)

$$n^2 = 2.3618102 - 0.00058072\,\lambda^2$$
$$+\frac{0.02305269}{\lambda^2 - 0.02425381}$$

for $0.4\mu m \le \lambda \le 0.7\ \mu m$

Potassium Chloride (Single Crystal)

$$n^2 = 2.174967 + \frac{0.08344206}{\lambda^2 - 0.0119082}$$
$$+\frac{0.00698382}{\lambda^2 - 0.025555} - 0.000513495\,\lambda^2$$
$$-0.06167587\,\lambda^4$$

for the ultraviolet

$$n^2 = 3.866619 + \frac{0.08344206}{\lambda^2 - 0.0119082}$$
$$-\frac{0.00698382}{\lambda^2 - 0.025555} - \frac{5569.715}{\lambda^2 - 3292.472}$$

for the visible

Dispersion of Optical Materials at 298 K (Continued)

Material   Dispersion Equation at 298 K

Silica (High Purity Fused)

$$n^2 = 2.978645 + \frac{0.008777808}{\lambda^2 - 0.010609} + \frac{84.06224}{\lambda^2 - 96.0000}$$

Silicon (Single Crystal)

$$n = 3.41696 + 0.138497L$$
$$+ 0.013924L^2 - 0.0000209\lambda^2$$
$$+ 0.000000148\lambda^4$$
$$\text{where } L = (\lambda^2 - 0.028)^{-1}$$

Silver Bromide (Single Crystal)

$$\frac{n^2 - 1}{n^2 + 2} = 0.48484 + \frac{0.10279\lambda^2}{\lambda^2 - 0.0900}$$
$$- 0.004796\lambda^2$$
$$\text{for } 0.54\mu m \leq \lambda \leq 0.65 \ \mu m$$

Silver Chloride (Single Crystal)

$$n = 4.00804 - 0.00085111\lambda^2 -$$
$$0.00000019762\lambda^4 + 0.079086/$$
$$(\lambda^2 - 0.04584)$$

Dispersion of Optical Materials at 298 K (Continued)

Material   Dispersion Equation at 298 K

Strontium Titanate (Single Crystal)

$$n = A + BL + CL^2 + D\lambda^2 + E\lambda^4$$

where

A=2.28355

B=0.035906

C=0.001666

D=-0.0061355

E=-0.00001502

for $1.0\ \mu m \leq \lambda \leq 5.3\ \mu m$

Thallium Bromoiodide (KRS-5, Mixed Crystal)

$$n^2 - 1 = \sum_{i=1}^{5} \frac{K_i \lambda^2}{\lambda^2 - \lambda_i^2}$$

where

i	λ_i^2	K_i
1	0.0225	1.8293958
2	0.0625	1.6675593
3	0.1225	1.1210424
4	0.2025	0.4513366
5	27089.737	12.380234

(λ in μm)

Dispersion of Optical Materials at 298 K (Continued)

Material   Dispersion Equation at 298 K

Titanium Dioxide (Rutile, Single Crystal)

$$n_o^2 = 5.913 + \frac{2.441 \times 10^7}{\lambda^2 - 0.803 \times 10^7}$$

for ordinary ray,

and

$$n_e^2 = 7.197 + \frac{3.322 \times 10^7}{\lambda^2 - 0.843 \times 10^7}$$

for extraordinary ray.

(λ in Å)

Zinc Sulfide (Single Crystal, Cubic)

$$n = 5.164 + 1.208 \times 10^7 / (\lambda^2 - 0.732 \times 10^7)$$

(λ in Å)

Source: data compiled by J.S. Park from Charles T. Lynch, *CRC Handbook of Materials Science*, Vol. 3, CRC Press, Boca Raton, Florida, 1975 and *Engineered Materials Handbook*, Vol.2, Engineering Plastics, ASM International, Metals Park, Ohio, 1988.

TRANSMISSION RANGE OF GLASS CERAMICS

Material & Crystal Structure	Transmission Region (mm, at 298 K)
Alumina (Sapphire, Single Crystal))	0.15 - 6.5
Ammonium Dihydrogen Phosphate (ADP, Single Crystal)	0.13 - 1.7
Arsenic Trisulfade (Glass)	0.6 - 13
Barium Fluoride (Single Crystal)	0.25 - 15
Cadmium Sulfide (Bulk and Hexagonal Single Crystal)	0.5 - 16
Cadmium Telluride (Hot Pressed Polycrystalline)	0.9 - 16
Calcium Carbonate (Calcite, Single Crystal)	0.2 - 5.5
Calcium Fluoride (Single Crystal)	0.13 - 12
Cesium Bromide (Single Crystal)	0.3 - 55
Cesium Iodide (Single Crystal)	0.25 - 80
Cuprous Chloride (Single Crystal)	0.4 - 19
Gallium Arsenide (Intrinsic Single Crystal)	1.0 - 15

Transmission Range of Glass Ceramics at 298 K (Continued)

Material & Crystal Structure	Transmission Region (mm, at 298 K)
Germanium (Intrinsic Single Crystal)	1.8 - 23
Indium Arsenide (Single Crystal)	3.8 - 7.0
Lead Sulfide (Single Crystal)	3.0 - 7.0
Lithium Fluoride (Single Crystal)	0.12 - 9.0
Lithium Niobate (Single Crystal)	0.33 - 5.2
Magnesium Fluoride (Film)	0.2 - 5.0
Magnesium Fluoride (Single Crystal)	0.1 - 9.7
Magnesium Oxide (Single Crystal)	0.25 - 8.5
Potassium Bromide (Single Crystal)	0.25 - 35
Potassium Iodide (Single Crystal)	0.25 - 45
Selenium (Amorphous)	1.0 - 20
Silica (High Purity Crystalline)	0.12 - 4.5

Transmission Range of Glass Ceramics at 298 K (Continued)

Material & Crystal Structure	Transmission Region (mm, at 298 K)
Silica (High Purity Fused)	0.12 - 4.5
Silicon (Single Crystal)	1.2 - 15
Silver Bromide (Single Crystal)	0.45 - 35
Silver Chloride (Single Crystal)	0.4 - 2.8
Sodium Fluoride (Single Crystal)	0.19 - 15
Strontium Titanate (Single Crystal)	0.39 - 6.8
Tellurium (Polycrystalline Film)	3.5 - 8.0
Tellurium (Single Crystal)	3.5 - 8.0
Thallium Bromoiodide (KRS-5, Mixed Crystal)	0.6 - 40
Thallium Chloribromide (KRS-6, Mixed Crystal)	0.21 - 35
Titanium Dioxide (Rutile, Single Crystal)	0.43 - 6.2

Transmission Range of Glass Ceramics at 298 K (Continued)

Material & Crystal Structure	Transmission Region (mm, at 298 K)
Zinc Selenide (Single Crystal, Cubic)	~0.5 - 22
Zinc Sulfide (Single Crystal, Cubic)	~0.6 - 15.6

External transmittance ≥ 10% with 2.0 mm thickness.

Table compiled by J.S. Park, University of California, Davis.
Data from:Jun p?

TRANSPARENCY OF SELECTED POLYMERS
Listed by Polymer

Polymer	Transparency (visible light) (ASTM D791) (%)
Acrylics; Cast, Molded, Extruded	(0.125 in.)
Cast Resin Sheets, Rods:	
General purpose, type I	91—92
General purpose, type II	91—92
Moldings:	
Grades 5, 6, 8	>92
High impact grade	90
Thermoset Carbonate	
Allyl diglycol carbonate	89—92
Alkyds; Molded	
Putty (encapsulating)	Opaque
Rope (general purpose)	Opaque
Granular (high speed molding)	Opaque
Glass reinforced (heavy duty parts)	Opaque
Cellulose Acetate; Molded, Extruded	
ASTM Grade:	
H6—1	75—90
H4—1	75—90
H2—1	80—90
MH—1, MH—2	80—90
MS—1, MS—2	80—90
S2—1	80—95

Transparency of Selected Selected Polymers (Continued)
Listed by Polymer

Polymer	Transparency (visible light) (ASTM D791) (%)
Cellulose Acetate Butyrate; Molded, Extruded	
ASTM Grade:	
H4	75—92
MH	80—92
S2	85—95
Cellusose Acetate Propionate; Molded, Extruded	
ASTM Grade:	
1	80—92
3	80—92
6	80—92
Chlorinated Polymers	
Chlorinated polyether	Opaque
Chlorinated polyvinyl chloride	Opaque
Polycarbonates	
Polycarbonate	75—85
Polycarbonate (40% glass fiber reinforced)	Translucent
Fluorocarbons; Molded,Extruded	
Polytrifluoro chloroethylene (PTFCE)	80—92

Transparency of Selected Selected Polymers (Continued)
Listed by Polymer

Polymer	Transparency (visible light) (ASTM D791) (%)
Epoxies; Cast, Molded, Reinforced	
Standard epoxies (diglycidyl ethers of bisphenol A)	
Cast rigid	
Cast flexible	90
Molded	85
General purpose glass cloth laminate	Opaque
High strength laminate	Opaque
Filament wound composite	Opaque
Epoxies—Molded, Extruded	
High performance resins	
(cycloaliphatic diepoxides)	
Cast, rigid	
Molded	Opaque
Glass cloth laminate	Opaque
Epoxy novolacs	
Glass cloth laminate	Opaque
Melamines; Molded	
Filler & type	
Unfilled	Good
Cellulose electrical	Opaque

Transparency of Selected Selected Polymers (Continued)
Listed by Polymer

Polymer	Transparency (visible light) (ASTM D791) (%)
Nylons; Molded, Extruded	
6/6 Nylon	
General purpose molding	Translucent
Glass fiber reinforced	Opaque
Glass fiber Molybdenum disulfide filled	Opaque
General purpose extrusion	Opaque
6/10 Nylon	
General purpose	Opaque
Glass fiber (30%) reinforced	Opaque
ABS–Polycarbonate Alloy	Opaque
PVC–Acrylic Alloy	
PVC–acrylic sheet	Opque
PVC–acrylic injection molded	Opaque
Polymides	
Unreinforced	Opaque
Unreinforced 2nd value	Opaque
Glass reinforced	Opaque
Polyesters: Thermosets	
Reinforced polyester moldings	
High strength (glass fibers)	Opaque
Heat and chemical resistsnt (asbestos)	Opaque
Sheet molding compounds, general purpose	Opaque

Transparency of Selected Selected Polymers (Continued)
Listed by Polymer

Polymer	Transparency (visible light) (ASTM D791) (%)
Phenylene Oxides	
SE—100	Opaque
SE—1	Opaque
Glass fiber reinforced	Opaque
Phenylene oxides (Noryl)	
Glass fiber reinforced	Opaque
Polypropylene:	
General purpose	Translucent—opaque
High impact	Translucent—opaque
Asbestos filled	Opaque
Glass reinforced	Opaque
Flame retardant	Opaque
Polyphenylene sulfide:	
Standard	Opaque
40% glass reinforced	Opaque
Polystyrenes; Molded Polystyrenes	
General purpose	Transparent
Medium impact	Opaque
High impact	Opaque
Glass fiber -30% reinforced	Opaque
Styrene acrylonitrile (SAN)	Transparent
Glass fiber (30%) reinforced SAN	Opaque

Transparency of Selected Selected Polymers (Continued)
Listed by Polymer

Polymer	Transparency (visible light) (ASTM D791) (%)
Silicones; Molded, Laminated	
Fibrous (glass) reinforced silicones	Opaque
Granular (silica) reinforced silicones	Opaque
Woven glass fabric/ silicone laminate	Opaque
Ureas; Molded	
Alpha—cellulose filled (ASTM Type 1)	21.8
Cellulose filled (ASTM Type 2)	Opaque
Woodflour filled	Opaque

Source: data from *1973 Materials Selector*, Reinhold Publishing, Stamford, Conn., 1972 and ASM **Jun ref**.

REFRACTIVE INDEX OF SELECTED POLYMERS
Listed by Polymer

Polymer	Refractive index, (ASTM D542) (nD)
Acrylics; Cast, Molded, Extruded	
Cast Resin Sheets, Rods:	
General purpose, type I	1.485—1.500
General purpose, type II	1.485—1.495
Moldings:	
Grades 5, 6, 8 1.489—1.493	
High impact grade	1.49
Thermoset Carbonate	
Allyl diglycol carbonate	1.5
Cellulose Acetate; Molded, Extruded	
ASTM Grade:	
H6—1	1.46—1.50
H4—1	1.46—1.50
H2—1	1.46—1.50
MH—1, MH—2	1.46—1.50
MS—1, MS—2	1.46—1.50
S2—1	1.46—1.50
Cellulose Acetate Butyrate; Molded, Extruded	(D543)
ASTM Grade:	
H4	1.46—1.49
MH	1.46—1.49
S2	1.46—1.49

Refractive Index of Selected Selected Polymers (Continued)
Listed by Polymer

Polymer	Refractive index, (ASTM D542) (nD)
Cellusose Acetate Propionate; Molded, Extruded	
ASTM Grade:	
1	1.46—1.49
3	1.46—1.49
6	1.46—1.49
Polycarbonates	
Polycarbonate	1.586
Fluorocarbons; Molded,Extruded	
Polytrifluoro chloroethylene (PTFCE)	1.43
Polytetrafluoroethylene (PTFE)	1.35
Fluorinated ethylene propylene(FEP)	1.34
Polyvinylidene— fluoride (PVDF)	1.42
Epoxies; Cast, Molded, Reinforced	
Standard epoxies (diglycidyl ethers of bisphenol A)	
Cast rigid	
Cast flexible	1.61
Molded	1.61

Refractive Index of Selected Selected Polymers (Continued)
Listed by Polymer

Polymer	Refractive index, (ASTM D542) (nD)
Polyacetals	
Homopolymer:	
Standard	Opaque
20% glass reinforced	Opaque
22% TFE reinforced	Opaque
Copolymer:	
Standard	Opaque
25% glass reinforced	Opaque
High flow	Opaque
Polyesters: Thermosets	
Cast polyyester	
Rigid	1.53—1.58
Flexible	1.50—1.57
Phenylene oxides (Noryl)	
Standard	1.63
Polyarylsulfone	1.651
Polyethylenes; Molded, Extruded	
Type I—lower density (0.910—0.925)	
Melt index 0.3—3.6	1.51
Melt index 6—26	1.51
Melt index 200	1.51
Type II—medium density (0.926—0.940)	
Melt index 20	1.51

Refractive Index of Selected Selected Polymers (Continued)
Listed by Polymer

Polymer	Refractive index, (ASTM D542) (nD)
Melt index 1.0—1.9	1.51
Type III—higher density (0.941—0.965)	
Melt index 0.2—0.9	1.54
Melt Melt index 0.1—12.0	1.54
Melt index 1.5—15	1.54
Polystyrenes; Molded	
Polystyrenes	
General purpose	1.6
Medium impact	Opaque
High impact	Opaque
Glass fiber -30% reinforced	Opaque
Styrene acrylonitrile (SAN)	1.565—1.569
Glass fiber (30%) reinforced SAN	Opaque
Polyvinyl Chloride And Copolymers; Molded, Extruded	
Vinylidene chloride	1.60—1.63

Source: data compiled by J.S. Park from Charles T. Lynch, *CRC Handbook of Materials Science*, Vol. 3, CRC Press, Boca Raton, Florida, 1975 and *Engineered Materials Handbook*, Vol.2, Engineering Plastics, ASM International, Metals Park, Ohio, 1988.

Chemical Properties

COMPOSITION OF SEA WATER
Elements in Solution (Excluding Dissolved Gasses)

Element	Concentration (parts per million)	Percent by weight
Oxygen	857,000	85.7000
Hydrogen	108,000	10.8000
Chlorine	19,000	1.9000
Sodium	10,500.	1.0500
Magnesium	1,275.	0.1275
Sulfur	885.	0.0885
Calcium	400.	0.0400
Potassium	380.	0.0380
Bromine	65.	0.0065
Carbon	30.	0.0030
Strontium	13.	0.0013
Boron	4.6	0.00046
Silicon	2.	0.0002
Fluorine	1.3	0.00013
Aluminum	1.	0.0001

Practical Handbook of Materials Science, Charles T. Lynch, Ed., CRC Press, Boca Raton, Fla, (1989)

ANIONS IN SEA WATER
34.4 Salinity per Mil, or 3.44% by weight

Ion	Percent by weight
Chloride	1.897
Sulfate	0.265
Bicarbonate	0.014
Bromide	0.0065
Borate	0.0027

Practical Handbook of Materials Science, Charles T. Lynch, Ed., CRC Press, Boca Raton, Fla, (1989)

WATER ABSORPTION OF SELECTED POLYMERS
Listed by Polymer

Polymer	Water absorption (24 hr), % (ASTM D570)
ABS Resins; Molded, Extruded	
Medium impact	0.2—0.4
High impact	0.2—0.45
Very high impact	0.2—0.45
Low temperature impact	0.2—0.45
Heat resistant	0.2—0.4
Acrylics; Cast, Molded, Extruded	
Cast Resin Sheets, Rods:	
General purpose, type I	0.3—0.4
General purpose, type II	0.2—0.4
Moldings:	
Grades 5, 6, 8	0.3—0.4
High impact grade	0.2—0.4
Thermoset Carbonate	
Allyl diglycol carbonate	0.2
Alkyds; Molded	
Putty (encapsulating)	0.10—0.15
Rope (general purpose)	0.05—0.08
Granular (high speed molding)	0.08—0.12
Glass reinforced (heavy duty parts)	0.007—0.10

Water Absorption of Selected Polymers (Continued)
Listed by Polymer

Polymer	Water absorption (24 hr), % (ASTM D570)
Cellulose Acetate; Molded, Extruded	
ASTM Grade:	
H4—1	1.7—2.7
H2—1	1.7—2.7
MH—1, MH—2	1.8—4.0
MS—1, MS—2	2.1—4.0
S2—1	2.3—4.0
Cellulose Acetate Butyrate; Molded, Extruded	
ASTM Grade:	
H4	2
MH	1.3—1.6
S2	0.9—1.3
Cellusose Acetate Propionate; Molded, Extruded	
ASTM Grade:	
1	1.6—2.0
3	1.3—1.8
6	1.6
Chlorinated Polymers	
Chlorinated polyether	0.01
Chlorinated polyvinyl chloride	0.11

Water Absorption of Selected Polymers (Continued)
Listed by Polymer

Polymer	Water absorption (24 hr), % (ASTM D570)
Polycarbonates	
Polycarbonate	0.15
Polycarbonate (40% glass fiber reinforced)	0.08
Diallyl Phthalates; Molded	(122 °F, 48 hr), %
Orlon filled	0.2—0.5
Dacron filled	0.2—0.5
Asbestos filled	0.4—0.7
Glass fiber filled	0.2—0.4
Fluorocarbons; Molded,Extruded	
Polytrifluoro chloroethylene (PTFCE)	0
Polytetrafluoroethylene (PTFE)	0.01
Ceramic reinforced (PTFE)	>0.2
Fluorinated ethylene propylene(FEP)	<0.01
Polyvinylidene— fluoride (PVDF)	0.03—0.06
Epoxies; Cast, Molded, Reinforced	
Standard epoxies (diglycidyl ethers of bisphenol A)	
Cast rigid	0.1—0.2
Cast flexible	0.4—0.1
Molded	0.3—0.8
General purpose glass cloth laminate	0.05—0.07
High strength laminate	0.05
Filament wound composite	0.05—0.07

Water Absorption of Selected Polymers (Continued)
Listed by Polymer

Polymer	Water absorption (24 hr), % (ASTM D570)
Epoxies—Molded, Extruded	
High performance resins	
(cycloaliphatic diepoxides)	
Molded	0.11—0.2
Glass cloth laminate	0.04—0.06
Epoxy novolacs	
Cast, rigid	0.1—0.7
Melamines; Molded	
Filler & type	
Unfilled	0.2—0.5
Cellulose electrical	0.27—0.80
Glass fiber	0.09—0.60
Alpha cellulose and mineral	0.3—0.5
Nylons; Molded, Extruded	
Type 6	
General purpose	1.3—1.9
Glass fiber (30%) reinforced	0.9—1.2
Cast	0.6
Flexible copolymers	0.8—1.4
Type 8	9.5
Type 11	0.4
Type 12	0.25

Water Absorption of Selected Polymers (Continued)
Listed by Polymer

Polymer	Water absorption (24 hr), % (ASTM D570)
Nylons; Molded, Extruded	
6/6 Nylon	
General purpose molding	1.5
Glass fiber reinforced	0.8—0.9
Glass fiber Molybdenum disulfide filled	0.5—0.7
General purpose extrusion	1.5
6/10 Nylon	
General purpose	0.4
Glass fiber (30%) reinforced	0.2
Phenolics; Molded	
Type and filler	
General: woodflour and flock	0.3—0.8
Shock: paper, flock, or pulp	0.4—1.5
High shock: chopped fabric or cord	0.4—1.75
Very high shock: glass fiber	0.1—1.0
Phenolics: Molded	
Arc resistant—mineral	0.5—0.7
Rubber phenolic—woodflour or flock	0.5—2.0
Rubber phenolic—chopped fabric	0.5—2.0
Rubber phenolic—asbestos	0.10—0.50
ABS–Polycarbonate Alloy	0.21
PVC–Acrylic Alloy	
PVC–acrylic sheet	0.06
PVC–acrylic injection molded	0.13

Water Absorption of Selected Polymers (Continued)
Listed by Polymer

Polymer	Water absorption (24 hr), % (ASTM D570)
Polymides	
Unreinforced	0.47
Unreinforced 2nd value	0.24—0.40
Glass reinforced	0.2
Polyacetals	
Homopolymer:	
Standard	0.25
20% glass reinforced	0.25
22% TFE reinforced	0.2
Copolymer:	
Standard	0.22
25% glass reinforced	0.29
High flow	0.22
Polyester; Thermoplastic	
Injection Moldings:	
General purpose grade	0.08
Glass reinforced grades	0.06—0.07
Glass reinforced self extinguishing	0.07
General purpose grade	0.09
Glass reinforced grade	0.07
Asbestos—filled grade	0.1

Water Absorption of Selected Polymers (Continued)
Listed by Polymer

Polymer	Water absorption (24 hr), % (ASTM D570)
Polyesters: Thermosets	
Cast polyyester	
Rigid	0.20—0.60
Flexible	0.12—2.5
Reinforced polyester moldings	
High strength (glass fibers)	0.5—0.75
Heat and chemical resistsnt (asbestos)	0.25—0.50
Sheet molding compounds, general purpose	0.15—0.25
Phenylene Oxides	
SE—100	0.07
SE—1	0.07
Glass fiber reinforced	0.06
Phenylene oxides (Noryl)	
Standard	0.22
Glass fiber reinforced	0.22, 0.18
Polyarylsulfone	0.4
Polypropylene:	
General purpose	<0.01—0.03
High impact	<0.01—0.02
Asbestos filled	0.02—0.04
Glass reinforced	0.02—0.05
Flame retardant	0.02—0.03

Water Absorption of Selected Polymers (Continued)ed)
Listed by Polymer

Polymer	Water absorption (24 hr), % (ASTM D570)
Polyethylenes; Molded, Extruded	
Type I—lower density (0.910—0.925)	
Melt index 0.3—3.6	<0.01
Melt index 6—26	<0.01
Melt index 200	<0.01
Type II—medium density (0.926—0.940)	
Melt index 20	<0.01
Melt index 1.0—1.9	<0.01
Type III—higher density (0.941—0.965)	
Melt index 0.2—0.9	<0.01
Melt Melt index 0.1—12.0	<0.01
Melt index 1.5—15	<0.01
High molecular weight	<0.01
Polystyrenes; Molded	
General purpose	0.30—0.2
Medium impact	0.03—0.09
High impact	0.05—0.22
Glass fiber –30% reinforced	0.07
Styrene acrylonitrile (SAN)	0.20—0.35
Glass fiber (30%) reinforced SAN	0.15

Water Absorption of Selected Polymers (Continued)
Listed by Polymer

Polymer	Water absorption (24 hr), % (ASTM D570)
Polyvinyl Chloride And Copolymers;	
Molded, Extruded	(ASTM D635)
Nonrigid—general	0.2—1.0
Nonrigid—electrical	0.40—0.75
Rigid—normal impact	0.03—0.40
Vinylidene chloride	>0.1
Silicones; Molded, Laminated	
Fibrous (glass) reinforced silicones	0.1—0.15
Granular (silica) reinforced silicones	0.08—0.1
Woven glass fabric/ silicone laminate	0.03—0.05
Ureas; Molded	
Alpha—cellulose filled (ASTM Type l)	0.4—0.8

Source: data compiled by J.S. Park from Charles T. Lynch, *CRC Handbook of Materials Science,* Vol. 3, CRC Press, Boca Raton, Florida, 1975 and *Engineered Materials Handbook,* Vol.2, Engineering Plastics, ASM International, Metals Park, Ohio, 1988.

FLAMMABILITY OF SELECTED POLYMERS
Listed by Polymer

Polymer	Flammability, (ASTM D635) (ipm)
ABS Resins; Molded, Extruded	
Medium impact	1.0—1.6
High impact	1.3—1.5
Very high impact	1.3—1.5
Low temperature impact	1.0—1.5
Heat resistant	1.3—2.0
Acrylics; Cast, Molded, Extruded	(0.125 in.)
Cast Resin Sheets, Rods:	
General purpose, type I	0.5—2.2
General purpose, type II	0.5—1.8
Moldings:	
Grades 5, 6, 8	0.9—1.2
High impact grade	0.8—1.2
Thermoset Carbonate	
Allyl diglycol carbonate	0.35
Alkyds; Molded	
Putty (encapsulating)	Nonburning
Rope (general purpose)	Self extinguishing
Granular (high speed molding)	Self extinguishing
Glass reinforced (heavy duty parts)	Nonburning

Flammability of Selected Polymers (Continued)
Listed by Polymer

Polymer	Flammability, (ASTM D635) (ipm)
Cellulose Acetate; Molded, Extruded	
ASTM Grade:	
H6—1	0.5—2.0
H4—1	0.5—2.0
H2—1	0.5—2.0
MH—1, MH—2	0.5—2.0
MS—1, MS—2	0.5—2.0
S2—1	0.5—2.0
Cellulose Acetate Butyrate; Molded, Extruded	
ASTM Grade:	
H4	0.5—1.5
MH	0.5—1.5
S2	0.5—1.5
Cellusose Acetate Propionate; Molded, Extruded	
ASTM Grade:	
1	0.5—1.5
3	0.5—1.5
6	0.5—1.5
Chlorinated Polymers	
Chlorinated polyether	Self extinguishing
Chlorinated polyvinyl chloride	Nonburning

Flammability of Selected Polymers (Continued)
Listed by Polymer

Polymer	Flammability, (ASTM D635) (ipm)
Polycarbonates	
Polycarbonate	Self extinguishing
Polycarbonate (40% glass fiber reinforced)	Self extinguishing
Diallyl Phthalates; Molded	ignition time (s)
Orlon filled	68 s
Dacron filled	84—90 s
Asbestos filled	70 s
Glass fiber filled	70—400 s
Fluorocarbons; Molded,Extruded	
Polytrifluoro chloroethylene (PTFCE)	Noninflammable
Polytetrafluoroethylene (PTFE)	Nonintlammable
Ceramic reinforced (PTFE)	Noninflammable
Fluorinated ethylene propylene(FEP)	Noninflammable
Polyvinylidene— fluoride (PVDF)	Self extinguishing
Epoxies; Cast, Molded, Reinforced	
Standard epoxies (diglycidyl ethers of bisphenol A)	
Cast rigid	0.3-0.34
Cast flexible	-
Molded	Self extunguishing
General purpose glass cloth laminate	Slow burn to Self extinguishing
High strength laminate	Self extinguishing
Filament wound composite	Self extinguishing

Flammability of Selected Polymers (Continued)
Listed by Polymer

Polymer	Flammability, (ASTM D635) (ipm)
Epoxies—Molded, Extruded	
High performance resins	
(cycloaliphatic diepoxides)	
Cast, rigid	Self extinguishing
Molded	Self extinguishing
Glass cloth laminate	Self extinguishing
Melamines; Molded	
Filler & type	
Unfilled	Self extinguishing
Cellulose electrical	Self extinguishing
Glass fiber	Self extinguishing
Alpha cellulose and mineral	Self extinguishing
Nylons; Molded, Extruded	
Type 6	
General purpose	Self extinguishing
Glass fiber (30%) reinforced	Slow burn
Cast	Self extinguishing
Flexible copolymers	Slow burn, 0.6
Type 8	Self extinguishing
Type 11	Self extinguishing

Flammability of Selected Polymers (Continued)
Listed by Polymer

Polymer	Flammability, (ASTM D635) (ipm)
6/6 Nylon	
General purpose molding	Self extinguishing
Glass fiber reinforced	Slow burn
Glass fiber Molybdenum disulfide filled	Slow burn
General purpose extrusion	Self extinguishing
6/10 Nylon	
General purpose	Self extinguishing
Glass fiber (30%) reinforced	Slow burn
Phenolics; Molded	
Type and filler	
General: woodflour and flock	Self extinguishing
Shock: paper, flock, or pulp	Self extinguishing
High shock: chopped fabric or cord	Self extinguishing
Very high shock: glass fiber	Self extinguishing
Phenolics: Molded	
Arc resistant—mineral	Self extinguishing
Rubber phenolic—woodflour or flock	Self extinguishing
Rubber phenolic—chopped fabric	Self extinguishing
Rubber phenolic—asbestos	Self extinguishing
ABS–Polycarbonate Alloy	0.9

Flammability of Selected Polymers (Continued)
Listed by Polymer

Polymer	Flammability, (ASTM D635) (ipm)
PVC–Acrylic Alloy	
PVC–acrylic sheet	Nonburning
PVC–acrylic injection molded	Nonburning
Polymides	
Unreinforced	IBM Class A
Unreinforced 2nd value	IBM Class A
Glass reinforced	UL SE—0
Polyacetals	
Homopolymer:	
Standard	1.1
20% glass reinforced	0.8
22% TFE reinforced	0.8
Copolymer:	
Standard	1.1
25% glass reinforced	1
High flow	1.1
Polyester; Thermoplastic	
Injection Moldings:	
General purpose grade	Slow burn
Glass reinforced grades	Slow burn
Glass reinforced self extinguishing	Self extinguishing
General purpose grade	Slow burn
Glass reinforced grade	Slow burn

Flammability of Selected Polymers (Continued)
Listed by Polymer

Polymer	Flammability, (ASTM D635) (ipm)
Polyesters: Thermosets	
Cast polyyester	
Rigid	0.87 to self extinguishing
Flexible	Slow burn to self extinguishing
Reinforced polyester moldings	
High strength (glass fibers)	Self extinguishing
Heat and chemical resistant (asbestos)	Self extinguishing
Sheet molding compounds, general purpose	Self extinguishing
Phenylene Oxides	
SE—100	Self extinguishing
SE—1	Self extinguishing
Glass fiber reinforced	Self extinguishing
Phenylene oxides (Noryl)	
Standard	Self extinguishing
Glass fiber reinforced	Self extinguishing
Polyarylsulfone	Self extinguishing
Polypropylene:	
General purpose	0.7—1
High impact	1
Asbestos filled	1
Glass reinforced	1
Flame retardant	Self extinguishing

Flammability of Selected Polymers (Continued)
Listed by Polymer

Polymer	Flammability, (ASTM D635) (ipm)
Polyphenylene sulfide:	
Standard	Non—burning
40% glass reinforced	Non—burning
Polyethylenes; Molded, Extruded	
Type I—lower density (0.910—0.925)	
Melt index 0.3—3.6	1
Melt index 6—26	1
Melt index 200	1
Type II—medium density (0.926—0.940)	
Melt index 20	1
Melt index 1.0—1.9	1
Type III—higher density (0.941—0.965)	
Melt index 0.2—0.9	1
Melt Melt index 0.1—12.0	1
Melt index 1.5—15	1
High molecular weight	1
Polystyrenes; Molded	
Polystyrenes	
General purpose	1.0—1.5
Medium impact	0.5—2.0
High impact	0.5—1.5
Styrene acrylonitrile (SAN)	0.8

Flammability of Selected Polymers (Continued)
Listed by Polymer

Polymer	Flammability, (ASTM D635) (ipm)
Polyvinyl Chloride And Copolymers; Molded, Extruded	
Nonrigid—general	Self extinguishing
Nonrigid—electrical	Self extinguishing
Rigid—normal impact	Self extinguishing
Vinylidene chloride	Self extinguishing
Silicones; Molded, Laminated	
Fibrous (glass) reinforced silicones	Nonburning
Granular (silica) reinforced silicones	Nonburning
Woven glass fabric/ silicone laminate	0.12
Ureas; Molded	
Alpha—cellulose filled (ASTM Type l)	Self extinguishing
Cellulose filled (ASTM Type 2)	Self extinguishing
Woodflour filled	Self extinguishing

Source: data compiled by J.S. Park from Charles T. Lynch, *CRC Handbook of Materials Science,* Vol. 3, CRC Press, Boca Raton, Florida, 1975 and *Engineered Materials Handbook,* Vol.2, Engineering Plastics, ASM International, Metals Park, Ohio, 1988.

Index

LEADING PHRASE

Water

Thermal Expansion and Thermal Conductivity of Selected

Electrical Resistivity of Selected

Density of Selected

Composition Ranges for Selected

Yield Strength of Selected Cast Aluminum

Yield Strength of Selected Cast Aluminum

Yield Strength of Selected Wrought Aluminum

Yield Strength of Selected Wrought Aluminum

Tensile Strength of Selected Aluminum Casting

Tensile Strength of Selected Aluminum Casting

Tensile Strength of Selected Wrought Aluminum

Tensile Strength of Selected Wrought Aluminum

Total Elongation of Selected Cast Aluminum

Total Elongation of Selected Cast Aluminum

Shear Strength of Selected Wrought Aluminum

Shear Strength of Selected Wrought Aluminum

Hardness of Selected Wrought Aluminum

Hardness of Selected Wrought Aluminum

Fatigue Strength of Selected Wrought Aluminum

Fatigue Strength of Selected Wrought Aluminum

Temper Designation System for Aluminum

Composition Ranges for Selected Cast Aluminum

Composition Ranges for Selected Wrought Auminum

Yield Strength of Selected Cast

Yield Strength of Selected Cast

KEY WORD AND TRAILING PHRASE

A

Leading Phrase

Yield Strength of Selected Wrought

Yield Strength of Selected Wrought

Tensile Strength of Selected

Tensile Strength of Selected

Tensile Strength of Selected Wrought

Tensile Strength of Selected Wrought

Total Elongation of Selected Cast

Total Elongation of Selected Cast

Shear Strength of Selected Wrought

Shear Strength of Selected Wrought

Hardness of Selected Wrought

Hardness of Selected Wrought

Fatigue Strength of Selected Wrought

Fatigue Strength of Selected Wrought

Temper Designation System for

Composition Ranges for Selected Cast

Bond

Bond

Composition Ranges for Selected Wrought

Key Word and Trailing Phrase (Continued)

B

Leading Phrase

Carbon
Carbon

The Fourteen

Composition Ranges for Selected
Composition Ranges for Selected Resulfurized
Mechanical Properties of Selected Gray
Yield Strength of Selected
Yield Strength of Selected
Total Elongation of Selected
Total Elongation of Selected
Electrical Resistivity of Selected Alloy
Composition Ranges for Selected
Thermal Expansion and Thermal Conductivity of Selected Alloy
Density of Selected Alloy
Composition Limits of Selected Gray
Tensile Strength of Selected Aluminum
Tensile Strength of Selected Aluminum
Mechanical Properties of Selected Malleable Iron
Dispersion of Glass
Young's Modulus of Selected

Key Word and Trailing Phrase (Continued)

C

Leading Phrase

Poisson's Ratio for Selected

Tensile Strength of Selected

Hardness of Selected

Resistivity of Selected

Transmission Range of Glass

The Elements in

Engineering

Heat Capacity of Selected

Thermal Conductivity of Selected

Thermal Expansion of Selected

Structure of Selected

Density of Selected

Typical Composition of Selected Glass-

Melting Points of

Melting Points of

Refractories,

Thermal Expansion Coefficients for Materials used in Integrated

Thermodynamic

Thermal Expansion

Thermodynamic

Thermodynamic

Key Word and Trailing Phrase (Continued)

Leading Phrase

Typical

Structure,

Engineering

Heat of Fusion For Selected Elements and Inorganic

Melting Points of Selected Elements and Inorganic

Melting Points of Selected Elements and Inorganic

Thermal Conductivity of Special

Thermal Expansion and Thermal

Thermal

Thermal

Thermal

Thermal

Thermal

Thermal

Thermal

Dielectric

Type II Superconducting Compounds:

High Temperature Superconducting Compounds:

Selected Superconductive Compounds And Alloys:

Thermal Conductivity of

Key Word and Trailing Phrase (Continued)

Leading Phrase

Thermal Conductivity of Metals at

Type II Superconducting Compounds: Critical Temperature and

High Temperature Superconducting Compounds: Critical Temperature and

The Seven

Bond Strengths in

Bond Strengths in

Mechanical Properties of Selected

Average Mechanical Properties of Treated

Composition Limits of Selected

Elements in the

Modulus of

Key Word and Trailing Phrase (Continued)

D

E

Leading Phrase

Modulus of

Melting Points of Selected

Melting Points of Selected

Heat of Fusion For Selected

The

The Periodic Table of The

The Metallic

Available Stable Isotopes of the

Properties of Selected

Melting Points of Selected

Densities of Selected

Crystal Structure of the

Atomic and Ionic Radii of the

Atomic Radii of the

Ionic Radii of the

Selected Properties of Superconductive

Thermal Conductivity for Thin Films of Superconductive

Surface Tension of Liquid

Phase Change Thermodynamic Properties for Selected

Thermodynamic Coefficients for Selected

Vapor Pressure of the

Bond Angle Values Between

Bond Length Values Between

Bond Length Values Between

Bond Angle Values Between

Vapor Pressure of the

Key Word and Trailing Phrase (Continued)

Leading Phrase

Vapor Pressure of the
Specific Heat of Selected
Specific Heat of Selected

The
The
The

Total
Total
Total

Elements for
Values of The
Thermal
Thermal
Thermal
Thermal
Thermal

Dissipation

Modulus of Elasticity in
Heat of

Key Word and Trailing Phrase (Continued)

F

Leading Phrase

Coefficient of Static
Values of The Error
Heat of

Dispersion of
Transmission Range of
Typical Composition of Selected
Mechanical Properties of Selected
Composition Limits of Selected

Specific
Specific

Specific

T_c Data for

Refractive

Key Word and Trailing Phrase (Continued)

Leading Phrase

Melting Points of Selected Elements and
Heat of Fusion For Selected Elements and
Melting Points of Selected Elements and
Heat of Formation of Selected
Thermal Conductivity of Cryogenic
Thermal Expansion Coefficients for Materials used in
Atomic and

Mechanical Properties of Selected Malleable
Thermal Expansion and Thermal Conductivity of Selected Alloy Cast
Density of Selected Alloy Cast
Composition Limits of Selected Gray Cast
Composition Limits of Selected Ductile
Composition Ranges for Selected Malleable
Mechanical Properties of Selected Gray Cast
Mechanical Properties of Selected Ductile
Average Mechanical Properties of Treated Ductile
Electrical Resistivity of Selected Alloy Cast
Available Stable

Bonding, Thermodynamic, and

The Fourteen Bravais
Bond
Bond
Carbon Bond
Carbon Bond

Key Word and Trailing Phrase (Continued)

K

L

Leading Phrase

Surface Tension of

Electrical,
Mechanical Properties of Selected
Composition Ranges for Selected

Average

The
Diffusion in Selected
Heats of Sublimation (at 25°C) of Selected
Thermal Conductivity of
Thermal Conductivity of
Diffusivity Values of
Young's

Bond Strengths in Diatomic

Key Word and Trailing Phrase (Continued)

Key Word and Trailing Phrase (Continued)

Leading Phrase

Modulus of Elasticity in Flexure of Selected

Flexural Strength of Selected

Coefficient of Static Friction for Selected

Abrasion Resistance of Selected

Volume Resistivity of Selected

Dielectric Strength of Selected

Dielectric Constant of Selected

Dissipation Factor for Selected

Arc Resistance of Selected

Transparency of Selected

Refractive Index of Selected

Water Absorption of Selected

Flammability of Selected

The Elements in

Specific Heat of Selected

Thermal Conductivity of Selected

Thermal Expansion of Selected

Specific Gravity of Selected

Vapor

Vapor

Vapor

Atomic and Ionic

Atomic

Ionic

Transmission

Poisson's

Key Word and Trailing Phrase (Continued)

R

Leading Phrase

Abrasion
Arc
Electrical

Volume
Composition Ranges for Selected

Refractories, Ceramics, and
Composition of
Anions in
The Elements in
Diffusion in

Tool Steel

Available
Coefficient of
Tool
Thermal Expansion of Selected Tool
Density of Selected Tool
Composition Limits of Selected Tool
Composition Ranges for Selected Carbon

Key Word and Trailing Phrase (Continued)

S

Key Word and Trailing Phrase (Continued)

Leading Phrase

Crystal

Heats of
Type II

High Temperature

Selected
Critical Temperature Data for Type II
T_c Data for High Temperature
Selected Properties of
Thermal Conductivity for Thin Films of

Modulus of Elasticity in
Surface

Key Word and Trailing Phrase (Continued)

T

Leading Phrase

Phase Change

Phase Change

Bonding,

Thermal Conductivity for

Mechanical Properties of Selected

Thermal Expansion of Selected

Density of Selected

Composition Limits of Selected

Key Word and Trailing Phrase (Continued)

Leading Phrase

Average Mechanical Properties of
Critical Temperature Data for

Composition of Sea
Anions in Sea
Yield Strength of Selected
Yield Strength of Selected
Tensile Strength of Selected
Tensile Strength of Selected
Shear Strength of Selected
Shear Strength of Selected
Hardness of Selected
Hardness of Selected
Fatigue Strength of Selected
Fatigue Strength of Selected
Composition Ranges for Selected

Key Word and Trailing Phrase (Continued)

Leading Phrase

Elongation at

Compressive

Key Word and Trailing Phrase (Continued)